ELECTRON CYCLOTRON RESONANCE ION SOURCES

Related Titles from AIP Conference Proceedings

737 Advanced Accelerator Concepts: Eleventh Advanced Accelerator Concepts Workshop
Edited by Vitaly Yakimenko, December 2004, CD-ROM included, 0-7354-0220-5

737 Advanced Accelerator Concepts: Eleventh Advanced Accelerator Concepts Workshop
Edited by Vitaly Yakimenko, December 2004, CD-ROM included, 0-7354-0220-5

693 Beam-Halo Dynamics, Diagnostics, and Collimation: 29th ICFA Advanced Beam Dynamics Workshop on Beam-Halo Dynamics, Diagnostics, and Collimation and the Beam-Beam'03 Workshop
Edited by Jie Wei, Wolfram Fischer, and Pamela Manning, December 2003, CD-ROM included, 0-7354-0166-7

691 High Energy Density and High Power RF: 6th Workshop on High Energy Density and High Power RF
Edited by Steven H. Gold and Gregory S. Nusinovich, November 2003, CD-ROM included, 0-7354-0164-0

680 Application of Accelerators in Research and Industry: 17th International Conference on the Application of Accelerators in Research and Industry
Edited by Jerome L. Duggan and I. L. Morgan, August 2003, CD-ROM included, 0-7354-0149-7

650 Beams 2002: 14th International Conference on High-Power Particle Beams
Edited by Thomas A. Mehlhorn and Mary Ann Sweeney, December 2002, 0-7354-0107-1

648 Beam Instrumentation Workshop 2002: Tenth Workshop
Edited by Gary Smith and Thomas Russo, December 2002, CD-ROM included, 0-7354-0103-9

647 Advanced Accelerator Concepts: Tenth Workshop
Edited by Christopher E. Clayton and Patrick Muggli, December 2002, CD-ROM included, 0-7354-0102-0

642 High Intensity and High Brightness Hadron Beams: 20th ICFA Advanced Beam Dynamics Workshop on High Intensity and High Brightness Hadron Beams; ICFA-HB2002
Edited by Weiren Chou, Yoshiharu Mori, David Neuffer, and Jean-François Ostiguy, November 2002, CD-ROM included, 0-7354-0097-0

639 Production and Neutralization of Negative Ions and Beams: Ninth International Symposium on the Production and Neutralization of Negative Ions and Beams
Edited by Martin P. Stockli, November 2002, 0-7354-0094-6

625 High Energy Density and High Power RF: 5th Workshop on High Energy Density and High Power RF
Edited by B. E. Carlsten, August 2002, 0-7354-0078-4

600 Cyclotrons and Their Applications 2001: Sixteenth International Conference
Edited by Felix Marti, December 2001, 0-7354-0044

592 High Quality Beams: Joint US-CERN-JAPAN-RUSSIA Accelerator School
Edited by S. I. Kurokawa, S. Y. Lee, J. Miles, E. A. Perevedentsev, November 2001, 0-7354-0034-2

To learn more about these titles, or the AIP Conference Proceedings Series, please visit the webpage http://proceedings.aip.org/proceedings

ELECTRON CYCLOTRON RESONANCE ION SOURCES

16th International Workshop on ECR Ion Sources
ECRIS '04

Berkeley, California   26 – 30 September 2004

EDITOR
Matthaeus Leitner
Lawrence Berkeley National Laboratory
Berkeley, California

SPONSORING ORGANIZATIONS
Lawrence Berkeley National Laboratory, Berkeley, CA
Communications and Power Industries, Palo Alto, CA
Isoflex Isotopes, San Francisco, CA
GMW Associates, Ion Beam Equipment, San Carlos, CA

Melville, New York, 2005
AIP CONFERENCE PROCEEDINGS ■ VOLUME 749

Editor:

Matthaeus Leitner
Lawrence Berkeley National Laboratory
1 Cyclotron Road
Building 47R0112
Berkeley, CA 94720-8201
U.S.A.

E-mail: mleitner@lbl.gov

Authorization to photocopy items for internal or personal use, beyond the free copying permitted under the 1978 U.S. Copyright Law (see statement below), is granted by the American Institute of Physics for users registered with the Copyright Clearance Center (CCC) Transactional Reporting Service, provided that the base fee of $22.50 per copy is paid directly to CCC, 222 Rosewood Drive, Danvers, MA 01923. For those organizations that have been granted a photocopy license by CCC, a separate system of payment has been arranged. The fee code for users of the Transactional Reporting Service is: 0-7354-0234-5/05/$22.50.

© 2005 American Institute of Physics

Individual readers of this volume and nonprofit libraries, acting for them, are permitted to make fair use of the material in it, such as copying an article for use in teaching or research. Permission is granted to quote from this volume in scientific work with the customary acknowledgment of the source. To reprint a figure, table, or other excerpt requires the consent of one of the original authors and notification to AIP. Republication or systematic or multiple reproduction of any material in this volume is permitted only under license from AIP. Address inquiries to Office of Rights and Permissions, Suite 1NO1, 2 Huntington Quadrangle, Melville, NY 11747-4502; phone: 516-576-2268; fax: 516-576-2450; e-mail: rights@aip.org.

L.C. Catalog Card No. 2005921466
ISBN 0-7354-0234-5
ISSN 0094-243X

Printed in the United States of America

CONTENTS

Preface ix
Committees xi
Chronology and Sponsors xiii

I. NEW DEVELOPMENTS AND STATUS REPORTS

First Results of the Superconducting ECR Ion Source VENUS with 28 GHz 3
    D. Leitner, C. M. Lyneis, S. R. Abbott, R. D. Dwinell, D. Collins,
    and M. Leitner

Recent Development of IMP ECR Ion Sources 10
    H. W. Zhao, Z. M. Zhang, L. T. Sun, Y. Cao, W. He, X. Z. Zhang, X. H. Guo, L. Ma, P. Yuan,
    M. T. Song, W. L. Zhan, and B. W. Wei

Production of Highly Charged Heavy Ions by Means of a Hybrid Source in DC Mode and in Afterglow Mode 15
    S. Gammino, G. Ciavola, L. Torrisi, L. Andò, L. Celona, M. Presti, S. Manciagli, A. Picciotto,
    A. M. Mezzasalma, J. Krása, L. Láska, M. Pfeifer, J. Wolowski, E. Woryna, P. Parys, G. D. Shirkov, and
    D. Hitz

First High Temperature Superconducting ECRIS 19
    D. Kanjilal, G. O. Rodrigues, P. Kumar, C. P. Safvan, U. K. Rao, A. Mandal, A. Roy, C. Bieth, S. Kantas,
    and P. Sortais

Production of Intense Beam of Medium Charge State Heavy Ions from RIKEN ECRISs 23
    T. Nakagawa, Y. Higurashi, M. Kidera, T. Aihara, M. Kase, and Y. Yano

Recent ECRIS Related Research and Development Work at JYFL 27
    H. Koivisto, P. Suominen, O. Tarvainen, J. Ärje, E. Lammentausta, P. Lappalainen, T. Kalvas,
    T. Ropponen, and P. Frondelius

ECRIS Operation with Multiple Frequencies 31
    R. Vondrasek, R. Scott, R. Pardo, H. Koivisto, O. Tarvainen, P. Suominen, and D. H. Edgell

Status Report of ECRIS at KVI 35
    J. P. M. Beijers, I. Formanoy, H. R. Kremers, J. Mulder, J. Sijbring,
    and S. Brandenburg

II. BEAM TRANSPORT

High Current Beam Transport with Phoenix 28 GHz: Experiment and Simulation 41
    T. Thuillier, J. C. Curdy, T. Lamy, A. Lachaize, A. Ponton, P. Sole, P. Sortais, and J. L. Vieux-Rochaz

Use of Simulations Based on Experimental Data 47
    P. Spädtke, K. Tinschert, R. Lang, and R. Iannucci

The WARP Code: Modeling High Intensity Ion Beams 55
    D. P. Grote, A. Friedman, J.-L. Vay, and I. Haber

III. DIAGNOSTICS AND TRANSPORT

Plasma Potential Measurements with a New Instrument 61
    O. Tarvainen, P. Suominen, T. Ropponen, H. Koivisto, R. C. Vondrasek, and R. H. Scott

Developments and Plasma Studies at the ATOMKI-ECRIS 67
    S. Biri, A. Valek, E. Takács, B. Radics, J. Pálinkás, J. Karácsony, L. Kenéz, A. Kitagawa, and
    M. Muramatsu

Effect of the Plasma Electrode Position on the Beam Intensity and Emittance of the RIKEN 18 GHz ECRIS 71
    Y. Higurashi, T. Nakagawa, M. Kidera, T. Aihara, M. Kase, and Y. Yano

Radioactive Beams from ^{252}CF Fission Using a Gas Catcher and an ECR Charge Breeder at Atlas 75
    G. Savard, R. C. Pardo, E. F. Moore, A. A. Hecht, and S. Baker

IV. DIAGNOSTICS AND APPLICATIONS

High Resolution He-like Argon and Sulfur Spectra from the PSI ECRIT 81
    M. Trassinelli, S. Biri, S. Boucard, D. S. Covita, D. Gotta, B. Leoni, A. Hirtl, P. Indelicato, E.-O. Le Bigot, J. M. F. dos Santos, L. M. Simons, L. Stingelin, J. F. C. A. Veloso, A. Wasser, and J. Zmeskal

Novel Technique for Trace Element Analysis Using the ECRIS and Heavy Ion Linear Accelerator (ECRIS-AMS) 85
    M. Kidera, T. Nakagawa, K. Takahashi, S. Enomoto, R. Hirunuma, K. Igarashi, M. Fujimaki, E. Ikezawa, O. Kamigaito, M. Kase, and Y. Yano

Photoionization of Multiply Charged Ions at the Advanced Light Source 88
    A. S. Schlachter, A. L. D. Kilcoyne, A. Aguilar, M. F. Gharaibeh, E. D. Emmons, S. W. J. Scully, R. A. Phaneuf, A. Müller, S. Schippers, I. Alvarez, C. Cisneros, G. Hinojosa, and B. M. McLaughlin

Formation of Electron Distribution Function in ECR Discharge Sustained by Strong Microwave Emission in an Open Trap 92
    V. L. Erukhimov and V. E. Semenov

V. TECHNIQUES AND GYROTRONS

Analysis of the SERSE Ion Output by Using Klystron-Based or TWT-Based Microwave Generators 99
    L. Celona, S. Gammino, G. Ciavola, F. Consoli, and A. Galatà

A New Method for Enhancing the Performances of Conventional B-Geometry ECR Ion Sources 103
    G. D. Alton, Y. Kawai, Y. Liu, and H. Bilheux

Ghost Signals in Allison Emittance Scanners 108
    M. P. Stockli, M. Leitner, D. P. Moehs, R. Keller, and R. F. Welton

Multicharged Ion Generation in Plasma Confined in a Cusp Magnetic Trap at Quasigasdynamic Regime 112
    V. Skalyga and V. Zorin

Multiple Ionization of Metal Ions by ECR Heating of Electrons in Vacuum Arc Plasmas 116
    A. V. Vodopyanov, S. V. Golubev, D. A. Mansfeld, A. G. Nikolaev, E. M. Oks, S. V. Razin, and K. P. Savkin

VI. NEW SOURCES

An All-Permanent Magnet ECR Ion Source for the ORNL MIRF Upgrade Project 123
    D. Hitz, M. Delaunay, A. Girard, L. Guillemet, J. M. Mathonnet, J. Chartier, and F. W. Meyer

GTS-LHC: A New Source for the LHC Ion Injector Chain 127
    C. E. Hill, D. Küchler, R. Scrivens, D. Hitz, L. Guillemet, R. Leroy, and J. Y. Pacquet

Design of SuSI—Superconducting Source for Ions at NSCL/MSU — I. The Magnet System 131
    P. A. Zavodszky, B. Arend, D. Cole, J. DeKamp, G. Machicoane, F. Marti, P. Miller, J. Moskalik, J. Ottarson, J. Vincent, and A. Zeller

VII. RADIOACTIVE ION BEAMS

Radioactive Ion Beam Production at GANIL: Status and Prospectives 137
    R. Leroy, C. Barue, C. Canet, M. Dubois, M. Dupuis, F. Durantel, W. Farabolini, J.-L. Flambard, G. Gaubert, S. Gibouin, C. Huet-Equilbec, Y. Huguet, P. Jardin, N. Lecesne, P. Leherissier, F. Lemagnen, J. Y. Pacquet, F. Pellemoine, M. G. Saint Laurent, O. Tuske, and A. C. C. Villari

Recent Results with the 2.45 GHz ECRIS at TRIUMF-ISAC 143
    P. Bricault, K. Jayamanna, D. H. L. Yuan, M. Olivo, and P. Schmor

Charge State Breeding with an ECRIS for ISAC at TRIUMF ... 147
    F. Ames, K. Jayamanna, D. H. L. Yuan, M. Olivo, R. Baartman, P. Bricault, M. McDonald, P. Schmor, and T. Lamy

Radioactive Beams Using the AECR-U and the 88-Inch Cyclotron ... 151
    M. A. McMahan, D. Leitner, J. Powell, and C. Silver

VIII. POSTERS

Design Study of a Hybrid ECRIS ... 157
    D. Hitz, A. Girard, D. Guillaume, L. Guillemet, P. Seyfert, J. M. Poncet, and L. T. Sun

Characterisation and Performance of the CERN ECR4 Ion Source ... 161
    C. Andresen, J. Chamings, V. Coco, C. E. Hill, D. Küchler, A. Lombardi, E. Sargsyan, and R. Scrivens

A 2.45 GHz Singly Charged ECR Ion Source for RIB Production ... 165
    H. Y. Zhao, X. Z. Zhang, H. W. Zhao, Z. M. Zhang, L. T. Sun, H. Wang, and B. H. Ma

The Latest Results of the All Permanent ECR Ion Source LAPECR2 ... 168
    L. T. Sun, H. W. Zhao, Z. M. Zhang, W. He, H. Wang, B. W. Ma, and X. W. Ma

New Design of the ECRIS Plasma Chamber Using a Modified Multipole Structure ... 171
    P. Suominen, O. Tarvainen, P. Frondelius, V. Nieminen, and H. Koivisto

Design and Calculations for the New ECRIS at KVI ... 175
    H. R. Kremers, J. P. M. Beijers, S. Brandenburg, I. Formanoy, J. Mulder, J. Sijbring, H. Koivisto, and K. Rantilla

Permanent Magnet Microwave Source for Generation of EUV Light ... 179
    S. K. Hahto, K.-N. Leung, J. Reijonen, Q. Ji, D. Schneider, R. Bruch, S. Kondagari, and H. Merabet

Tests of New NIRS Compact ECR Ion Source for Carbon Therapy ... 183
    M. Muramatsu, A. Kitagawa, Y. Sakamoto, S. Sato, Y. Sato, Hirotsugu Ogawa, S. Yamada, Hiroyuki Ogawa, Y. Yoshida, and A. G. Drentje

Traveling Wave vs. Cavity RF Injection for the ORNL HRIBF Volume-Type ECR Ion Source ... 187
    Y. Liu, F. M. Meyer, and J. M. Cole

Performances of Volume versus Surface ECR Ion Sources ... 191
    Y. Liu, G. D. Alton, H. Z. Bilheux, and F. M. Meyer

Testing of the "Flat-B" 6-GHz ECR Ion Source Equipped with a RF Polarizer (*abstract only*) ... 195
    H. Z. Bilheux, Y. Liu, J. M. Cole, and G. D. Alton

The Texas A&M ECR Ion Sources: A Status Report ... 196
    D. P. May, G. J. Derrig, F. P. Abegglen, and H. Peeler

Design of an All-Permanent-Magnet, Volume-Type, 10 GHz ECR Ion Source with Field-Forming Iron Yoke ... 199
    M. Stalder, C. T. Steigies, and R. F. Wimmer-Schweingruber

Improved Spindle Cusp Magnetic Field for ECRIS ... 202
    M. H. Rashid, C. Mallik, and R. K. Bhandari

Preliminary Bremsstrahlung Measurements on VENUS at 18 and 28 GHz (*abstract only*) ... 206
    C. M. Lyneis and D. Leitner

The 28 GHz, 10 kW, CW Gyrotron Generator for the VENUS ECR Ion Source at LBNL ... 207
    M. Marks, S. Evans, H. Jory, D. Holstein, R. Rizzo, P. Beck, B. Cisto, D. Leitner, C. M. Lyneis, D. Collins, and R. D. Dwinell

Influence of Wall-Current-Compensation and Secondary-Electron- Emission on the Plasma Parameters and on the Performance of Electron Cyclotron Resonance Ion Sources ... 211
    L. Schachter, S. Dobrescu, and K. E. Stiebing

2Q-LEBT Prototype for the RIA Facility ... 215
    N. E. Vinogradov, V. N. Aseev, M. R. L. Kern, P. N. Ostroumov, R. C. Pardo, R. Scott, and R. C. Vondrasek

Formation of Ion Beam from High Density Plasma of ECR Discharge ... 219
    I. Izotov, S. Razin, A. Sidorov, V. Skalyga, V. Zorin, R. Geller, T. Lamy, P. Sortais, and T. Thuillier

Plasma Potential Measurements for a "Volume"-Type ECR Ion Source ... 223
    Y. Kawai, Y. Liu, G. D. Alton, H. Z. Bilheux, and J. M. Cole

Theoretical and Experimental Studies of the Extracted MCI Beam from an ECR Ion Source ... 227
    L. T. Sun, Y. Cao, H. W. Zhao, X. H. Guo, Z. M. Zhang, Y. C. Feng, J. Y. Li, L. Ma, J. Li, H. Y. Zhao, W. He, X. X. Li, D. Hitz, and A. Girard

An ESS System for ECRIS Emittance Research ... 231
    Y. Cao, L. T. Sun, W. He, L. Ma, Z. M. Zhang, H. Y. Zhao, H. W. Zhao, X. Z. Zhang, X. H. Guo, B. H. Ma, J. Li, H. Wang, J. Y. Li, X. X. Li, Y. C. Feng, and W. Lu

The Development of New Space Charge Compensation Methods for Multi-components Ion Beam Extracted from ECR Ion Source at IMP ... 234
    L. Ma, H. W. Zhao, Y. Cao, H. Y. Zhao, T. M. Song, W. He, and Z. M. Zhang

Recent Development of IMP LECR3 Ion Source ... 238
    Z. M. Zhang, H. W. Zhao, J. Y. Li, L. T. Sun, Y. C. Feng, H. Wang, B. H. Ma, X. Z. Zhang, X. H. Guo, X. X. Li, Y. Cao, and H. Y. Zhao

Experiments on Beam Extraction from the CAPRICE ECRIS (*abstract only*) ... 241
    K. Tinschert, P. Spädtke, J. Bossler, R. Iannucci, and R. Lang

Beam Simulation Studies of the LEBT for RIA Driver Linac ... 242
    Q. Zhao, X. Wu, V. Andreev, A. Balabin, M. Doleans, D. Gorelov, T. L. Grimm, W. Hartung, D. Leitner, C. M. Lyneis, F. Marti, S. O. Schriber, and R. C. York

Efficient Plasma Ion Source Modeling with Adaptive Mesh Refinement (*abstract only*) ... 246
    J. S. Kim, J. L. Vay, A. Friedman, and D. P. Grote

Development of the 3D Parallel Particle-In-Cell Code IMPACT to Simulate the Ion Beam Transport System of VENUS (*abstract only*) ... 247
    J. Qiang, D. Leitner, D. S. Todd, and R. D. Ryne

A Microwave Driven Ion Source for Continuous-Flow AMS (*abstract only*) ... 248
    J. Wills, R. J. Schneider, K. F. von Reden, J. M. Hayes, M. L. Roberts, and A. Benthien

ECRIS'04 Registered Attendees ... 249
Conference Photographs ... 257
Author Index ... 261

PREFACE

The 16th International Workshop on ECR Ion Sources, ECRIS'04, took place at the Lawrence Berkeley National Laboratory (LBNL) in Berkeley, California, USA, from September 26 through September 30, 2004. Seventy three participants from 14 countries attended the workhop. The program covered main topics related to the physics, design and operation of ECR ion sources, including: status reports and new developments of ECR ions sources, radioactive ion beam production and charge breeders, next generation sources, high intensity ion beam production, plasma physics and plasma diagnostics, ion beam extraction and ion beam transport, and applications of ECR ion sources.

Framed by the impressive backdrop of the San Francisco bay area the intensive four day program also included the workshop barbeque with live music at the 88-Inch Cyclotron at LBNL, the afternoon excursion to Alcatraz Island and the tour of major LBNL facilities. More than 26 years after the first ECR Ion Source Workshop, the ECR ion source field is still very active and many new major ion source developments were reported. ECR ion sources continue to push towards higher magnetic fields and higher microwave frequencies in order to increase the highest extractable charge states or to increase the current of medium charged ions. Large accelerator proposals around the world demand intense ion beams from ECR sources and it is no surprise that a significant number of papers were related to the ion beam transport from ECR sources. Further, driven by inventive diagnostics techniques, the understanding of basic ECR plasma behavior continues to grow.

The topics and success stories covered by the paper contributions of the ECRIS'04 participants are too manifold to be listed in this preface, and the reader is encouraged to enjoy browsing through the papers in this edition. The high scientific quality of the presentations during ECRIS'04 is reflected in the many contributions of these proceedings. Hopefully, these proceedings will find a valuable place on your bookshelf, where you can return to find more detailed information on specific ion sources or where you can reflect on the current state of ECR ion source technology.

The ECRIS'04 workshop became a success due to the efforts of the participants and contributors, which are gratefully acknowledged. In the name of all ECRIS'04 participants the editor would like to express thanks to the ECRIS International Advisory Committee. Its chair, Claude Lyneis, organized the 6th ECRIS workshop at the 88-Inch Cyclotron at LBNL almost 20 years ago, at the time when the first ECR ion source became online in Berkeley. At this 16th ECRIS workshop the first results of the next-generation superconducting ECR ion source VENUS operating at 28 GHz were presented by the Berkeley group.

Much of the success of the ECRIS'04 workshop is a result of the outstanding organizational work done by Daniela Leitner, the workshop chair. Her efforts will be remembered by all of the ECRIS participants with gratitude. Special appreciation for organizing the ECRIS'04 workshop goes to Kathleen Brower, Patricia Butler, Todd Damon, Michelle Galloway, Joy Kono, Peggy McMahan, Jim Morel, and in particular to the workshop secretary Patti Kobayashi. In addition, the editor would like to thank Todd Damon and Michelle Galloway for help in formatting the manuscripts.

The next ECRIS workshop will be held at the Institute of Modern Physics, Chinese Academy of Sciences, in Lanzhou, China, in 2006. In the meantime, we hope that these proceedings remind you of pleasant and interesting days spent in Berkeley, California!

Matthaeus Leitner
Editor ECRIS'04
Lawrence Berkeley National Laboratory, November 2004

Local Organizing Committee:

Daniela Leitner, Chair	DLeitner@lbl.gov
Kathleen Brower	KHBrower@lbl.gov
Todd Damon	DSTodd@lbl.gov
Michelle Galloway	MLGalloway@lbl.gov
Patti Kobayashi, Secretary	PTKobayashi@lbl.gov
Joy Kono	JNKono@lbl.gov
Matthaeus Leitner	MLeitner@lbl.gov
Claude Lyneis	CMLyneis@lbl.gov
Peggy McMahan	P_McMahan@lbl.gov
Jim Morel	JRMorel@lbl.gov

International Advisory Committee:

Claude Lyneis, Chair
Santo Gammino
Dennis Hitz
Hannu Koivisto
Daniela Leitner
Don May
Takahide Nakagawa
Hong Wei Zhao

ECRIS'04 Proceedings Editor:

Matthaeus Leitner
MLeitner@lbl.gov

Workshop Secretariat:

Patti Kobayashi
1 Cyclotron Road
MS88R0192
Berkeley, CA 94720
Fax: 01-510-486-7983
Phone: 01-510-486-7849
PTKobayashi@lbl.gov

Chronology

1.	1978	6 November	Karlsruhe, Germany
2.	1979	12 November	Louvain la Neuve, Belgium
3.	1980	8 December	Darmstadt, Germany
4.	1982	14-15 January	Grenoble, France
5.	1983	21-22 April	Louvain la Neuve, Belgium
6.	1985	17-18 January	Berkeley, California, USA
7.	1986	22-23 May	Jülich, Germany
8.	1987	16-18 November	East Lansing, Michigan, USA
9.	1988	15-16 September	Grenoble, France
10.	1990	1-2 November	Knoxville, Tennessee, USA
11.	1993	6-7 May	Groningen, Netherlands
12.	1995	25-27 April	Tokyo, Japan
13.	1997	26-28 February	College Station, Texas, USA
14.	1999	3-6 May	Geneva, Switzerland
15.	2002	12-14 June	Jyväskylä, Finland
16.	2004	26-30 September	Berkeley, California, USA

Sponsors

CPI Communications and Power Industries
(http://www.cpii.com)

ISOFLEX Isotopes
(http://www.isoflex.com)

GMW Associates, Ion Beam Equipment
(http://www.gmw.com)

I NEW DEVELOPMENTS AND STATUS REPORTS

First Results of the Superconducting ECR Ion Source Venus with 28 GHz

D. Leitner, C.M. Lyneis, S.R. Abbott, R.D. Dwinell, D. Collins, M. Leitner

Lawrence Berkeley National Laboratory, 1 Cyclotron Rd. MS88R0192, Berkeley, CA 94708

Abstract. VENUS (Versatile ECR ion source for NUclear Science) is a next generation superconducting ECR ion source, designed to produce high current, high charge state ions for the 88-Inch Cyclotron at the Lawrence Berkeley National Laboratory. VENUS also serves as the prototype ion source for the RIA (Rare Isotope Accelerator) front end. The magnetic confinement configuration consists of three superconducting axial coils and six superconducting radial coils in a sextupole configuration. The nominal design fields of the axial magnets are 4T at injection and 3T at extraction; the nominal radial design field strength at the plasma chamber wall is 2T, making VENUS the world most powerful ECR plasma confinement structure. From the beginning, VENUS has been designed for optimum operation at 28 GHz with high power (10 kW).

In 2003 the VENUS ECR ion source was commissioned at 18 GHz, while preparations for 28 GHz operation were being conducted. During this commissioning phase with 18 GHz, tests with various gases and metals have been performed with up to 2000 W RF power. At the initial commissioning tests at 18 GHz, 1100 eμA of O^{6+}, 160 eμA of Xe^{20+}, 160 eμA of Bi25+ and 100 eμA of Bi^{30+} and 11 eμA of Bi^{41+} were produced.

In May 2004 the 28 GHz microwave power has been coupled into the VENUS ECR ion source. At initial operation more than 320 eμA of Xe^{20+} (twice the amount extracted at 18 GHz), 240 eμA of Bi^{24+} and Bi^{25+}, and 245 eμA of Bi^{29+} were extracted. The paper briefly describes the design of the VENUS source, the 28 GHz microwave system and its beam analyzing system. First results at 28 GHz including emittance measurements are presented.

I. INTRODUCTION

The goal of the VENUS ECR ion source project as the RIA R&D injector is the production of 200eμA of U^{30+}, a high current medium charge state beam. On the other hand, as an injector ion source for the 88-Inch Cyclotron the design objective is the production of 5eμA of U^{48+}, a low current very high charge state beam. To achieve those ambitious goals, the VENUS ECR ion source has been designed for optimum operation at 28 GHz.

The Venus ECR ion source project was started in 1997 with the development of the superconducting structure and cryostat which was completed in 2001. At the last ECR ions source workshop in June 2002 at the University of Jyvaskyla, the first plasma ignition using 18 GHz microwave was reported. The source was commissioned with 18 GHz in 2003 and in 2004 28 GHz was coupled for the first time. Table 1 summarizes the major milestones of the project.

TABLE 1. Major Milestones of the VENUS Project

Date	Milestone
09/1997	Prototype Magnet completed
09/2001	Final Magnet Tests: 4T Injection, 3T Extraction, 2.4 T Sextupole achieved
06/2002	First Plasma at 18 GHz
09/2003	160 eμA of Bi^{24+}, 160 eμA Xe^{20+}
09/03-11/03	Cryostat Modification for 28 GHz operation
01/04-04/04	Gyrotron system assembly at CPI
05/26/04	First 28 GHz Plasma
06/04	320 eμA Xe^{20+}
08/04	245 eμA Bi^{29+}, 15 eμA Bi^{41+}

II. THE VENUS ECR ION SOURCE

The following sections describe briefly the design of the VENUS ECR ion source, the superconducting magnets, the cryostat, the microwave system, and the beam analyzing system. A detail description of the various components can be found in the paper referenced in each section.

The mechanical design

Fig. 1 shows the mechanical layout of the VENUS ECR ion source. The vacuum system design uses only UHV compatible components and metal seals. It is optimized for good plasma chamber pumping. Therefore the turn around time after the source has been vented to air is only about two to three hours before the source reaches high performance again. Two off-axis wave guides (18 GHz and 28 GHz), two high temperature ovens, 2 gas feeds, and a water cooled biased disk are inserted from the injection tank. All surfaces exposed to the plasma are made from aluminum. The mechanical design is described in more detail in [1, 2].

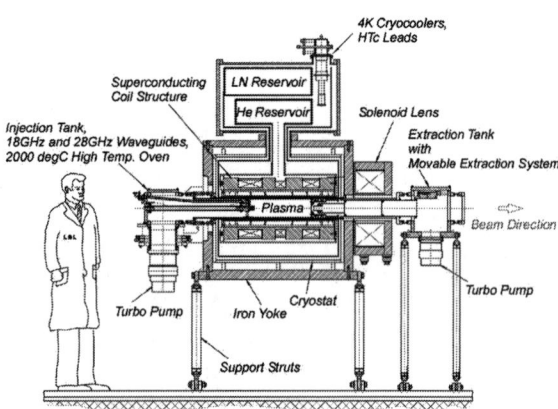

FIGURE 1. Mechanical layout of the VENUS ion source and cryogenic systems

The Superconducting Magnets

The design and development of the superconducting magnets are described in [2, 3]. The sextupole coils are wound around a pole with iron in the center, which enhances the peak field about 10%. The main challenge for the superconducting magnet design comes from the strong forces that the sextupole coils experience strong in the axial field of the solenoids. These forces, if not sufficiently counteracted, cause azimuthal movements of the sextupole coils and lead to quenches. VENUS is the first superconducting ECR ion source that uses a new clamping scheme utilizing liquid metal filled bladders to prevent any movement of the energized coils [3]. During commissioning of the superconducting magnets, the sextupole reached 110% of its design field after a few training quenches (2.4T) with the solenoids operating at design field (4T at injection and 3T at extraction). Another important step for the magnet commissioning was the development of a PLC (Programmable Logic Controller) based external regulation loop for the superconducting magnet power supplies. It allows ramping of the magnets in a reasonable time and stabilizes the magnets at the requested currents without fast oscillations, which can cause quenches.

The Cryogenic System

The cryogenic system for VENUS operates at 4.2 K with three cryocoolers each providing up to 45 W of cooling power at 50 K and 1.5 W at 4 K in a closed loop mode without further helium transfers [4, 5]. In addition, the cryostat has provisions for a fourth cryocooler. The main modification during the 18 GHz commissioning phase was the development of a novel heat exchanger for the cryocoolers, which efficiently couples the cryocoolers to the LHe reservoir [4] and minimizes the temperature gradient between the cryocooler heads and the helium reservoir. The present system provides up to 2 W of cooling power to remove heat generated by bremstrahlung, which is produced by the plasma electrons and deposited in the cryostat. However, the preliminary 28 GHz tests showed that improved x-ray shielding will be necessary to run VENUS at the full capacity of the 10 kW 28 GHz gyrotron power supply. The heat leak related to bremstrahlung is discussed in more detail in section V.

The Low Energy Beam Transport System

The low energy ion beam transport system consists of a movable accel-decel extraction system (operating at up to 30 kV extraction voltage), and a large gap, 90 degree double focusing analyzing magnet [5, 6]. The beam transport system was designed for high current, high charge state extraction. Therefore, to minimize beam blow up due to space charge, the extracted ion beam is directly matched into the analyzing magnet. After the mass analyzing section, a two-axis emittance scanner has been installed. Emittance measurements results are described in section IV.

The Microwave System

The VENUS ECR ion source plasma can be heated with 2 kW of 18 GHz power and/or up to 10 kW of 28 GHz power. A "traditional" microwave set-up is used for the 18 GHz microwave power. The 18 GHz system consists of a 18 GHz solid state oscillator, an 18 GHz klystron amplifier, a quartz HV break and a quartz vacuum window. The 28 GHz power is provided by a VIA-301 Heatwave™ gyrotron system that is able to deliver 100 watts to 10 kW continuous wave (CW) RF output at 28 GHz [7]. The gyrotron may be operated locally via its front panel or remotely via either RS-232 and/or Ethernet connections. The microwave components for 10 kW, 28 GHz operation are significantly different from those systems using lower frequency, lower power klystron amplifiers. The 28 GHz system propagates the microwave in an over-moded circular wave guide system in the TE_{01} mode. This mode has low attenuation but requires specialized bends, mode filters, and other microwave components to prevent the propagation of unwanted modes. The schematic of the microwave layout is shown in Fig. 2.

III. COMMISSIONING RESULTS AT 18 AND 28 GHZ

The VENUS source was initially tested with various gases at 18 GHz and 28 GHz in order to be able to compare VENUS to other high performance sources. But more extensive measurements have been performed using bismuth for the Rare Isotope Accelerator (RIA) ion beam development program. Bismuth was chosen since its mass is close to uranium, which means that the extraction and ion beam transport characteristics are very similar. However Bi is easier to use, since it is less reactive than uranium, not radioactive, and evaporates at modest temperatures. Furthermore, it has only one isotope and provides a clean spectrum for systematic emittance measurements.

The 18 GHz commissioning was carried out in 2002 and 2003 while preparations for the 28 GHz operation were progressing. During the 18 GHz commissioning period, a number of improvements were made to the cryostat system, the 18 GHz microwave system, and the magnet power supply control system [8, 9]. Following these improvements, VENUS is now operational at the full capacity of the 2 kW, 18 GHz klystron. The operation experience has been excellent in terms of stability, repro-

FIGURE 2. Schematic layout of the VENUS 28 GHz microwave system.

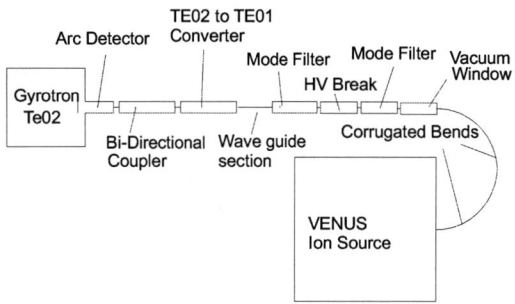

ducibility, and reliability. However, since VENUS has a large plasma volume of about 9 liters, the maximum microwave power density available for VENUS is only .22 kW/liter at 2 kW. At this power density, VENUS cannot reach its performance peak at 18 GHz. In comparison, the power density used in the AECR-U at peak performance is 1.7 kW/liter in double frequency mode and about 1 kW/liter in single frequency mode [10].

With the installation of the 10 kW 28 GHz gyrotron in May 2004, the maximum power level available is 10 kW, which would provide a power density of 1.1 kW/liter. 4.5 kW is the maximum power injected so far in the early test. Fig. 3 shows the analyzed current dependency to the microwave

FIGURE 3. Dependence of the extracted current for several ions to the coupled 28 GHz microwave power

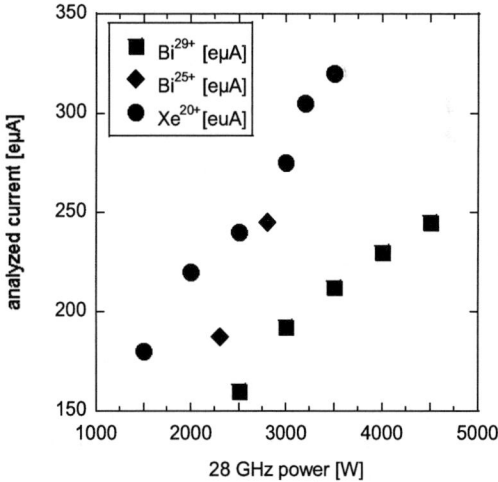

power coupled into the plasma for a few sample ion beams. The approximately linear increase in current for all the ions with rf power shows that 4.5 kW is well below the saturation point.

Table 2 states the initial performance of VENUS at 18 GHz and preliminary results at 28 GHz for oxygen, xenon, and bismuth. For comparison, the published data from other high performance ion sources are included.

TABLE 2. Preliminary commissioning results of VENUS at 18 GHz and 28 GHz in comparison with three other high performance ECR ion source, the double frequency heated AECR-U [10] and the 18 GHz ECR ion source GTS [11] and SERSE 28 GHz [12]

		VENUS	VENUS	AECR-U	GTS	SERSE
f(GHz)		18	28	10+14	18	28
^{16}O	6$^+$	1100	1200	840*	1950	
	7$^+$	324	>360	360*		
Xe	20$^+$	164	320		310	380
	27$^+$	84	120	30	168	
Bi	24$^+$		243			
	25$^+$	160	243	70		
	27$^+$	150		75		
	28$^+$	128	240	60		
	29$^+$	115	245	55		
	30$^+$	102	225	57		
	31$^+$	86	203	48		
	32$^+$	60	165	41		
	33$^+$	43		32		
	34$^+$	34		25		
	36$^+$	26		16		
	37$^+$	23		11.9		
	38$^+$	20		9.4		
	41$^+$	11	15	4.4		
	43$^+$	5.4	11.5	3.0		
	44$^+$	4.5	7.7	2.2		
	46$^+$		3.6	1.2		
	47$^+$		2.4	0.90		
	48$^+$		1.4	0.60		
	49$^+$		1.0	0.25		
	50$^+$		0.5	0.15		

* 3 frequency heating (8.6, 10, 14 GHz)

Figure 4 displays three charge state distribution (CSD) spectra as the source tune is shifted from low (4a), to medium (4b) and high charge state production (4c). The low to medium charge states are relevant for RIA, the high charge states with an M/Q lower than 5 are of interest to the 88-Inch Cyclotron. The CSD- peak was shifted from 24+ to 37+ between spectrum 5a and 5c. In the latter spectrum, the lower bismuth charge states disappear. These wide shifts in the CSD distribution are possible since VENUS has a strong plasma confinement, which allows reaching several different charge state distribution equilibria.

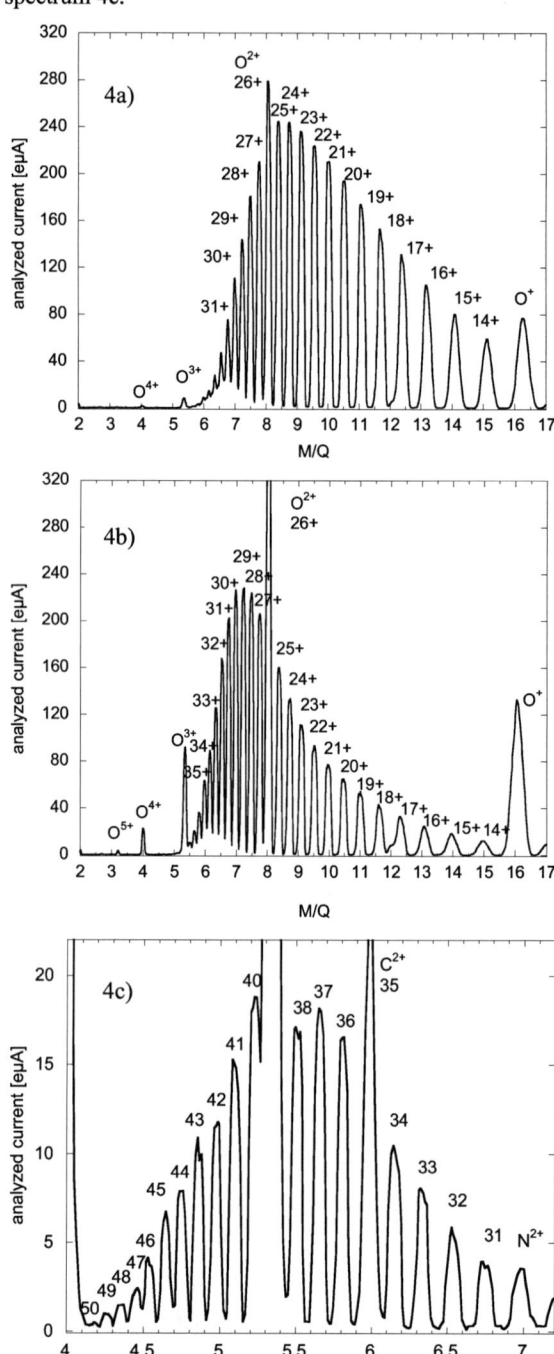

FIGURE 4. Analyzed Bi current for an ion source tune at 28 GHz optimized for low (4a), medium (4b), and high (4c) charge states. Note the different current scales in the spectrum 4c.

The ratio of support (mixing) gas ions to bismuth ions can be used to shift the charge state distribution. To illustrate this fact, the spectrum 4c is plotted again in Fig. 5. By comparing the spectra Fig. 4a), Fig. 4b) and Fig. 5), it can be seen that the oxygen support gas spectrum emerges from the bismuth spectrum as the Bi charge state distribution is shifted to higher charge states. If the source is tuned for the low charge Bi states, the high charge states of oxygen completely disappear from the charge state distribution (see Fig. 4a and Fig. 4b). As the source is tuned for Bi^{41+}, the support (mixing gas) dominates the spectrum, and the oxygen spectrum peaks again on the He like ion O^{6+} (see Fig.5). The same rf power and very similar confinement fields were used to obtain 5b) and 6). However, the bismuth and the oxygen flux were reduced lowering the plasma chamber pressure about 12%. In addition, the bias voltage was lowered from 100V to 36 V. This 'gas mixing' effect is well known and used in ECR ion sources as well as EBIS/EBIT sources [13].

FIGURE 5. Oxygen charge state distribution from the Bi spectrum of Fig. 4c

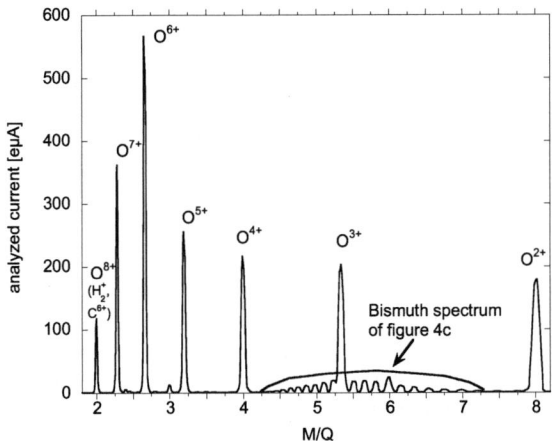

IV. EMITTANCE MEASUREMENTS

Two main contributions to the ion beam emittance have to be considered for an ECR ion source extraction system: (1) the ion beam transverse temperature, and (2) the induced beam rotation (angular momentum) due to the decreasing axial magnetic field in the extraction region. Considering that the ions in an ECR plasma are relatively cold with temperatures in the order of 1eV or less, the emittance contribution due to the magnetic field becomes the dominant factor for most modern ECR ion sources [6]. Assuming an uniform plasma density distribution across the plasma outlet hole, the emittance due to beam rotation induced by the decreasing magnetic field in the vicinity of the extractor can be described by Busch's theorem (assuming $\varepsilon^{100\%} = 5 \cdot \varepsilon^{rms}$, a waterbag distribution)

$$\varepsilon_{MAG}^{xx'-rms-norm} = 0.032 \, r^2 B_0 \frac{1}{M/Q} \quad (1)$$

where ε is the normalized x-x' rms emittance in $\pi \cdot mm \cdot mrad$, r is the plasma outlet hole radius in mm,

B_0 is the axial magnetic field strength at the extractor in T, and M/Q is the dimensionless ratio of ion mass in amu to ion charge state [6]. Following this dependence, the emittance should decrease with ion mass and increase with charge state for a charge state distribution. However, the experimental results don't show this behavior.

Preliminary emittance measurements were performed for bismuth ion from Bi^{23+} to Bi^{41+} using 18 GHz heating and for Bi^{29+} and Bi^{31+} using 28 GHz. The results are plotted in Fig. 6. For the ion beam transport the main difference between the two heating frequencies is the extraction mirror fields, which were 1.2 T for 18 GHz and 2.1 T for 28 GHz. The minimum theoretical emittance values using those field values and the VENUS extraction hole radius of 4mm are also plotted in Fig. 6. It can be clearly seen that the measured emittance does not follow the predicted dependence. On the contrary, the higher the charge state the lower the measured emittance value. Similar results have been previously measured on different ECR ion sources [14, 15]. These results are consistent with a possible model that the highly charged ions are trapped closer to the axis and therefore would be extracted from a virtual extraction hole that is smaller than the real extraction hole decreasing the measured effect of the magnetic field on the emittance [14, 15].

FIGURE 6. Dependence of the emittance value from the ion charge state for bismuth for the 18 GHz heated plasma and the 28 GHz heated plasma. In addition, the predicted emittance dependence from the magnetic field is also shown.

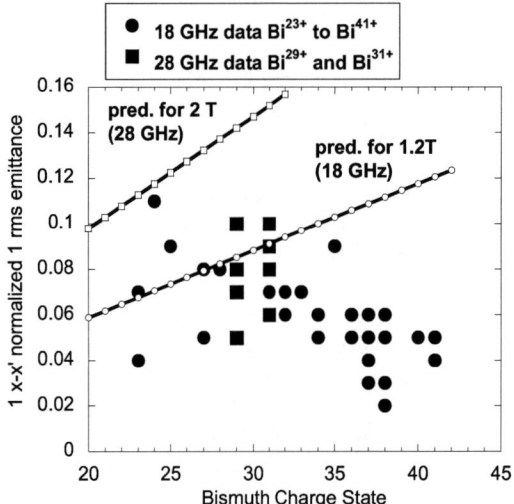

V. PRELIMINARY BREMSTRAHLUNG MEASUREMENTS ON VENUS

Bremstrahlung produced by the hot plasma electrons colliding with the plasma walls are particularly troublesome for superconducting ECR ion sources. The high energy bremstrahlung that go through the radial plasma and cryostat walls cause an additional cryogenic heat load [16] and localized heating in the superconducting coils that may lead to quenches [12]. Generally, higher frequency sources produce higher x-ray fluxes although the precise scaling has not been measured. Model calculations of electron cyclotron resonance-heated plasmas predict that the mean energy of the hot electrons increases approximately linearly with frequency [17].

FIGURE 7. Axial bremstrahlung measured for 2 kW of 28 GHz microwave power.

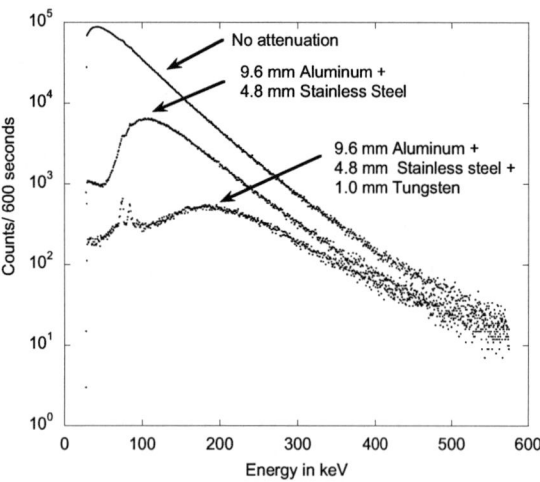

The VENUS cryostat has several calibrated carbon glass resistors, which are located in the cryostat and on the cold heads of the cryocoolers. The temperatures can be measured to an accuracy of about 5 mK and the response of the system to an additional heat load has been calibrated by using a small heater located at liquid helium temperature. The results at 18 GHz showed that the heat loading was sensitive to the tuning of the source and was on the order of 150 mW per kW of microwave power [9]. Measurements at 28 GHz, however, show that the bremstrahlung heating rate is higher and on the order of 1 W per kW of microwave power.

The bremstrahlung spectra at 28 GHz were measured using a 12.5 cm thick tungsten alloy collimator and germanium detector located on the straight-through port of the analyzing magnet. While it would have been preferable to measure the radial bremstrahlung flux, the thick iron yoke surrounding VENUS makes this difficult. Figure 7 shows the axial bremstrahlung spectrum with no attenuation, with sheets of aluminum and stainless steel to simulate the VENUS plasma chamber and cryostat wall and with an additional 1 mm thick sheet of tungsten. The relative energy in each spectrum was calculated and the addition of 1 mm of tungsten reduced the transmitted energy by a factor of 4.5.

VI. RADIAL ELECTRON LOSSES

There are two indications that the radial electron losses occur predominately in localized areas on the plasma wall. First, when the polyester high voltage insulation for the plasma chamber was recently removed, it was found to be discolored due to x-ray damage at the location along the plasma flutes where there is a local minimum in the magnetic field. Second, a hole was melted in the water-cooled aluminum chamber during an uncontrolled power excursion by the 28 GHz gyrotron. During the power excursion the superconducting magnets quenched, which was probably caused by the intense bremstrahlung heating.

The location of the hole was measured prior to the successful repair of the chamber and it coincided with the location of one of the two discolored areas of the insulation. The hole in the plasma chamber occurred along one of the plasma flute although it was rotated 3 degrees from the flute centerline. The axial position was 6 cm displaced toward injection from the midpoint between the injection and extraction coils whose centers are 50 cm apart. An analysis of the magnetic field strength as a function of axial position shows that the hole occurred where the field has a local minimum. At this location, the large gradient in the solenoid field produces a radial component that partially cancels the radial field produced by the sextupole. The VENUS plasma chamber has a hexagonal symmetry with V-shaped grooves that increase the radius by about 4 mm at the plasma flutes [5]. As a result the wall field is greatest at the flute centerline and decreases as one rotates azimuthally. At the hole location, magnet flux lines guide the hot trapped electrons along a trajectory that brings them tangent to the ECR heating surface. This may provide enhanced heating since the electrons will stay in resonance longer than when the flux line is normal to the ECR surface and the trapped electrons pass through the resonance rapidly. At a 6-degree rotation at the edge of the V, the flux lines no longer intersect the ECR surface and are not heated. It appears that the hot electrons are preferentially lost along a narrow line in the axial direction. The x-ray discolorations on the insulation were roughly 1.5 cm wide azimuthally and 5 cm long axially, which is consistent with a line loss. They occurred at the location of the hole and at a 120-degree rotation. The field has threefold symmetry, but there were only two discolored spots. This probably indicates that the plasma wall and the sextupole have a slight misalignment.

FUTURE PLANS

During the next year we are planning to continue the commissioning at 28 GHz power levels of up to 10 kW. First test for the production of uranium ion beams are also planned to verify the performance measured for Bi also for U.

A major focus will be the design and construction of an improved plasma chamber that is able to absorb the high power x-ray radiation emitted from the ECR plasma and reduce the bremstrahlung loading in the cryostat and in the superconducting coils. This should also protect against bremstrahlung-induced quenches. The present aluminum chamber is only 4 mm thick at the plasma flutes. Two mm of a high density, high z material such as tungsten, tantalum or gold will reduce the heating by a factor of 10 or more, which would reduce the 28 GHz the heating rate to about 0.1 W per kW of microwave power or 1.0 W at 10 kW. Since without shielding VENUS has operated at 4.5 kW and about 5 W of bremstrahlung heating, the new shielding should also provide sufficient protection against induced quenches.

REFERENCES

[1] M. A. Leitner, D. Leitner, S. R. Abbott, C. E. Taylor, and C. M. Lyneis, *Proceedings of the 15th International Workshop on ECR Ion Sources, ECRIS'02*, (University of Jyväskylä, University of Jyväskylä, Finland, 2002).

[2] M.A. Leitner, C.M. Lyneis, D.C. Wutte, C.E. Taylor, and S. R. Abbott, Physica Scripta **T92**, 171-173 (2001).

[3] C. E. Taylor, S. Caspi, M. Leitner, S. Lundgren, C. Lyneis, D. Wutte, S.T. Wang, and J. Y. Chen, IEEE Transactions on Applied Superconductivity **10**, 224 (2000).

[4] C.E. Taylor, S.R. Abbott, D. Leitner, M. Leitner, and C. M. Lyneis, Advances in Cryogenic Engineering **CEC-49**, 1818 (2004).

[5] M. A. Leitner, D. Leitner, S. R. Abbott, and C. M. Lyneis, *Proceedings of the 15th International Workshop on ECR Ion Sources, ECRIS'02*, (University of Jyväskylä, University of Jyväskylä, Finland, 2002).

[6] M. A. Leitner, D. C. Wutte, and C. M. Lyneis, *Proceedings of the Particle Accelerator Conference (PAC'01)*, (The American Physical Society, IEEE, Chicago, 2001).

[7] S. Evans, H. Jory, D. Holstein, R. Rizzo, P. Beck, M. Marks, D. Leitner, C.M. Lyneis, D. Collins, and R. D. Dwinell, these *proceedings*.

[8] D. Leitner, S.R. Abbott, R.D. Dwinell, M. Leitner, C. Taylor, and C. M. Lyneis, *Proceedings of the Partcle Accelerator Conference PAC'03*, (American Physics Society, IEEE, Portland, Oregon, 2003).

[9] D. L. C. M. Lyneis, S. R. Abbott, R. D. Dwinell, M. Leitner, C. S. Silver, and C. Taylor, RSI **75**, 1389 (2004).

[10] Z. Q. Xie, Rev. Sci. Instrum **69**, 625 (1998).

[11] D. Hitz, A. Girard, K. Serebrennikov, G. Melin, D. Cormier, J. M. Mathonnet, L. Sun, J. P. Briand, and M. Benhachoum, RSI **75**, 1403 (2004).

[12] S. Gammino, G. Ciavola, L. Celona, D. Hitz, A. Girard, and G. Melin, Rev. Sci. Instrum **72**, 4090 (2001).

[13] R. Geller, *Electron Cyclotron Resonance Ion Source and ECR Plasmas*, 394-397, (Institute for Physics Publishing, Bristol, 1996).

[14] D. Wutte, M. Leitner, and C. M. Lyneis, Physica Scripta **T92**, 247 (2001).

[15] P. Sortais, L. Maunoury, A. C. Villari, R. Leroy, J. Mandin, J. Y. Pacquet, and E. Robert, *Proceedings of the Proceedings of the 13th International Workshop on ECR Ion Sources*, (Texas A&M University, College Station, Texas A&M University, 1997).

[16] Y. Jongen, private communication, , 1984).

[17] A. Girard, P. Pernot, G. Melin, and C. Lécot, Phys. Rev. E **62**, 1189 (2000).

Recent Development of IMP ECR Ion Sources

H. W. Zhao, Z. M. Zhang, L.T. Sun, Y.Cao, W. He, X. Z. Zhang, X. H. Guo,
L. Ma, P.Yuan, M. T. Song, W. L. Zhan, B. W. Wei

Institute of Modern Physics (IMP), Chinese Academy of Sciences, Lanzhou, 730000, China

Great efforts have been made to develop highly charged ECR ion sources for application of heavy ion accelerator and atomic physics research at IMP in the past few years. The latest development of ECR ion sources at IMP is briefly reviewed. Intense beams with high and intermediate charge states have been produced from IMP LECR3 by optimization of the ion source conditions including rf frequency extended up to 18GHz. 1.1 emA of Ar^{8+} and 325 eμA of Ar^{11+} were produced. Dependence of beam emittance on those key parameters of ECR ion source, beam extraction and space charge compensation were experimentally studied at LECR3. Furthermore, an advanced superconducting ECR ion source named SECRAL is being constructed. SECRAL is designed to operate at rf frequency 18-28GHz with axial mirror magnetic fields 3.6-4.0 Tesla at injection, 2.2 Tesla at extraction and sextupole field 2.0 Tesla at the wall. The superconducting magnet with sextupole and three solenoids was tested in a test-cryostat and 95% of designed fields were reached. Construction status and planed schedule of SECRAL are presented. .

INTRODUCTION

HIRFL (Heavy Ion Research Facility in Lanzhou) is a cyclotron and storage ring complex which consists of two cyclotrons and a heavy ion cooling storage ring (HIRFL-CSR)[1]. HIRFL is dedicated for heavy ion nuclear physics research and highly charged atomic physics research. Heavy ion beams with intensity 5×10^{11}—8×10^{12} pps for Ca, Ni, Zn, Ge, Xe, Pb, are requested from the cyclotrons for radioactive ion beam physics, super-heavy nuclei and super-heavy element research. To satisfy the requirements, 50-100 eμA stable beams for $^{58}Ni^{17+}$, $^{70}Zn^{17+}$, $^{74}Ge^{20+}$, $^{86}Kr^{23+}$, $^{129}Xe^{34+}$ are expected from ion source. For this reason, great efforts have been made to develop intense highly charged ECR ion sources at IMP in the past few years including successful operation of LECR2 (Lanzhou ECR ion source No.2), intense highly charged ion beam production by LECR3 (Lanzhou ECR ion source No.3), systematic studies for ECRIS beam emittance, beam extraction and space charge compensation, and construction of an advanced superconducting ECR ion source named as SECRAL (Superconducting ECR ion source with Advanced design in Lanzhou). On the other hand, researches for highly charged atomic physics and surface physics based on low energy highly charged ion beams from ECR ion source are very active at IMP. LECR3 has been put into operation for a setup of atomic physics research since 2002 and each year 3000-4000 hours of heavy ion beams have been delivered including Ar^{5-18+}, Xe^{15-33+} and Pb^{20-40+}. Furthermore, a 300 kV high voltage platform dedicated for atomic and surface physics research is being built. An all permanent magnet ECR source named as LAPECR2[2] (Lanzhou All Permanent magnet ECR ion source No.2) for highly charged ion beam production will be put at this high voltage platform. LAPECR2 assembling has been completed and the first plasma is expected by the end of 2004.

The emphasis of this article is focused on the most recent results from LECR3 for intense beam production and beam quality studies, and in particular, SECRAL construction status and preliminary test of the SECRAL magnet are reported.

RECENT STUDIES AND RESULTS FROM LECR3

A. Test of LECR3 at rf 18 GHz[2-3]

LECR3 was tested very quickly at rf 18 GHz to enhance the ion source performance. The beam intensity of Ar^{11+} can be raised up to 325 eμA at 18 GHz 1.2-1.5 kW rf power with 20-25 kV extraction voltage and φ 9 mm extraction aperture. During the test at 18 GHz, the axial magnetic field peak in the injection side was finally optimized to 1.7 Tesla which is the maximum value of the coil safety permitted.

In order to produce intense intermediate charge state ion beams, φ 13 mm extraction aperture of a plasma electrode was tested. The ion source LECR3 was optimized for Ar^{8+} at 18 GHz and 22 kV extraction voltage. The ion source can produce 1.1mA of Ar^{8+} with 1.0-1.2 kW rf power, while only 0.5 mA of Ar^{8+} can be produced at similar condition with φ 10 mm extraction aperture. Emittance measurements demonstrate that Ar^{8+} beam emittance with φ 13 mm aperture could be 70% higher than that with φ 10 mm extraction aperture.

B. Studies of beam quality [2,4].

To study beam quality, dependences of beam emittance and beam image on the key parameters of ECR ion source were experimentally measured at LECR3. Surprisingly, beam emittance could be increased by more than a factor 2 for different axial magnetic field distributions and different rf power while keeping the other conditions constant. Beam image and beam phase pattern could be changed dramatically by different plasma conditions as shown in Fig.1, which are in consistence with the measurements by D.Wutte[5] and D.Hitz[6]. Detailed measurements can be found in Ref[2,4].

C. Preliminary studies for beam extraction and space charge compensation [7].

An accel-decel extraction system with aluminum electrodes was tested at LECR3 to study

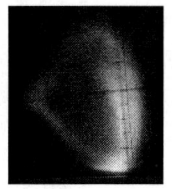 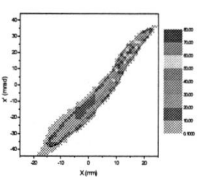

Fig.1 Beam image and emittance phase pattern.

beam extraction and space charge compensation[8]. All the three electrodes (the plasma electrode, the electrode at negative voltage and the electrode at ground voltage) are made of aluminum material to make use of much more secondary electron emissions. The ion source was optimized for O^{6+} at rf 14.5 GHz 500 W and 16 kV voltage on the plasma electrode. The ion source was not at a good condition and only 235 eμA stable O^{6+} beam was produced when both the negative electrode and the ground electrode were at ground potential. When the negative electrode is applied a negative voltage and ramping to higher voltage step by step, the beam intensity of O^{6+} keeps rise with increase of the negative voltage, correspondingly, current of the solenoid lens should be reduced step by step with the negative voltage, while other conditions of the ion source are kept constant. The beam intensity of O^{6+} goes up to 310 eμA when the negative voltage is ramped to 8 kV, and then saturates, even begins to decrease as the negative voltage continues to increase. During the test, when the negative voltage is more than 4 kV, the beam current becomes unstable and fluctuation could be more than 5%. Anyway, 30% beam enhancement can be clearly achieved. Whether this phenomenon is mainly caused by accel-decel lens effect or space charge compensation by the electrons in the extraction region is still an open question. Further experiments and theoretical simulations are needed.

Inspired by Dudnikov's idea that electron-negative gas such as SF_6 and BF_3 can be used to compensate space charge effect in very intense singly charged ion beam transport line, an appropriate flow rate of SF_6 gas was injected into the LECR3 beam transport line through a chamber located between the ion source and the analyzing magnet. 10%-30% enhancement of the beam currents for different charge states of Argon and Oxygen can be seen. During the experiment, the total beam was about 2 mA which was not intensive as expected, and the extraction voltage was 10 kV. This effect could be explained by compensation of space charge through the electron-negative gas. But almost no such effect occurs when the extraction voltage is more than 15 kV. Some further experiments need to be done so that clear conclusion could be drawn.

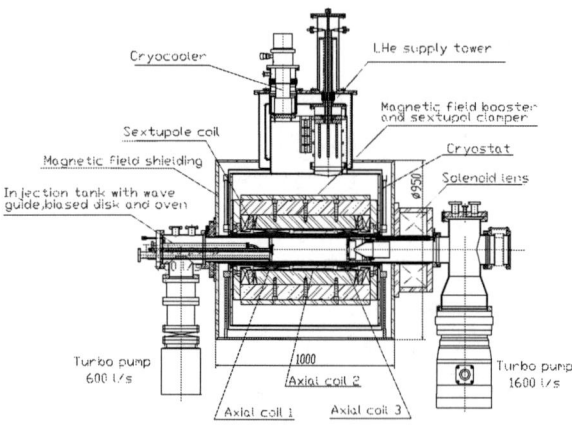

Fig.2 Schematic view of SECRAL ion source

SECRAL CONSTRUCTION STATUS

A. SECRAL source design

SECRAL aims for developing a very compact superconducting ECR ion source with a completely new structure and high performances for highly charged ion beam production. The ion source will work at 18 GHz at initial test operation, and finally will be extended to 28 GHz. SECRAL will be operated at 30-40 kV extraction voltage so that very intense highly charged ion beams could be efficiently extracted and transported.

The design of SECRAL is illustrated in Figure 2. The superconducting magnet assembly consists of three axial solenoid coils and six sextupole coils with a cold iron structure as field booster and clamp. At full excitation, this magnet assembly will produce peak mirror fields on axis 3.6-4.0 Tesla at injection, 2.2 Tesla at extraction and a radial sextupole field of 2.0 Tesla at plasma chamber wall. What is different from the traditional design, such as LBNL VENUS[9] and LNS SERSE[10], is that the three axial solenoid coils are located inside of the sextupole bore in order to reduce the interaction forces between the sextupole coils and the solenoid coils. The superconducting coil configuration is shown in Fig.3. The three axial solenoid coils polarized by the same direction current are supported by one piece stainless steel bobbin. Each coil of the sextupole assembly is wound around one piece of iron pole as a sextupole field booster. Another six iron segments are fixed around the sextupole coils as magnet clamp and field booster. Four aluminum shrinking rings surround the sextupole assembly for tight clamping.

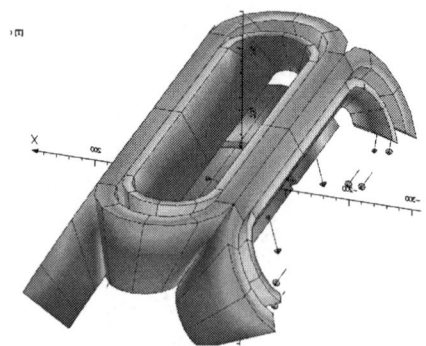

Fig.3 SECRAL superconducting coil configuration. The three solenoids are located inside the sextupole bore.

New coil configuration of the SECRAL superconducting magnet not only reduces the interaction forces between the sextupole and the three solenoids, but also enables the ion source much more compact, only 1 m in length and 1 m in diameter, which is the most compact superconducting ECR ion source. The compact structure is very beneficial to rf coupling and beam extraction. The cold iron structure of the sextupole magnet with iron segments as field booster and magnet clamping can reduce stray magnetic field coming out of the superconducting coils to a very low level (less than 50 GS). This results in a very low interaction force between the magnet assembly and the iron yoke, which makes support of the magnet inside the cryostat much easier and also good to operation of the cryo-cooler.

The superconducting magnet assembly with those superconducting coils and the iron parts are all immersed into 4.2 K liquid helium for cooling down. The cryogenics is designed to operate at low liquid helium consumption. This is realized by use of one stage cryo-cooler providing pre-cooling at 50 K and radiation shields around the liquid helium container, and also by use of high critical temperature (HTc) current leads that conduct electric current and minimize the heat leakage. The current leads are mounted within the cryostat supply tower and are thermally linked to the cryo-cooler at the hot end. The total heat load on the liquid helium reservoir is approximately 1 W at 4.2 K. The consumption of liquid helium is expected to be less than 1.5 l/h. The cryostat is surrounded by a warm iron shielding yoke to further reduce stray magnetic field.

The plasma chamber in diameter ϕ 125 mm is made of a double wall aluminum tube with water cooling channel in between. The microwave guides, metallic evaporation ovens and biased disk system are inserted into the ion source through an injection vacuum tank by the way off-axis. A 700 l/s and 1600 l/s turbo molecular pump is located at injection and extraction side respectively to pump the plasma chamber.

The low energy (30-40 kV extraction voltage) transmission line should be able to transport more than 15 mA total current of intense highly charged heavy ion beams. The beam transport line consists of an accel-decel extraction system, a solenoid lens and 110 degree analyzing magnet. To reach high transmission efficiency and high resolution, a 110 degree analyzing magnet with a large gap 120 mm and 600 mm bending radius is particularly designed. Considering large beam emittance and beam blow up due to intense space charge effect, the beam transport line is designed as short as possible and the solenoid lens is directly attached on the extraction side flange of the source body so as to focus the beam immediately after extraction, furthermore, ϕ 150 mm beam pipe between the ion source and the analyzing magnet is used.

Fig.4 SECRAL superconducting magnet assembly.

B. SECRAL construction status

The SECRAL superconducting magnet assembly is being constructed at ACCEL Instruments Inc, Germany. The magnet assembly including the superconducting axial solenoid coils and the sextupole has been completed and the tests are being conducted. The magnet assembly with the wiring and quench protection diodes is shown in Fig.4 which is the most crucial part. Fabrication of the cryostat is almost completed. The other components of the ion source and its beam line including the plasma chamber, the injection tank, the extraction system, the solenoid lens, the analyzing magnet and so on have been fabricated in Lanzhou and ready for assembling. Fig.5 shows the analyzing magnet and the beam line.

C. SECRAL superconducting maget test in a test-cryostat

The first test of the superconducting magnet in

a test-cryostat was performed in March 2004. The three solenoid coils fully reached the designed currents without energizing of the sextupole. The sextupole

Fig.5 SECRAL analyzing magnet.

only reached 70% of the designed maximum current after a few quenches when the three solenoid coils were not energized. Failure of the sextupole test is due to malfunction of the four stainless steel shrinking rings, which are caused by non-appropriate shrinking tolerances. The four stainless steel shrinking rings are finally replaced by aluminum one and some other procedures were also taken to strengthen clamping of the magnet, and then the second test was done.

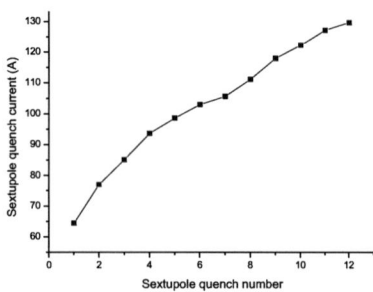

Fig.6 Sextupole quench numbers and quench current. 95% of designed current was reached.

The second test of the superconducting magnet in a test-cryostat was conducted in August 2004. The sextupole was separately ramped to 95% of designed current after 13 quenches as shown in Fig.6. It does not make much sense only to ramp the sextupole separately. So the magnet was tested by ramping the sextupole firstly, or by ramping the three solenoids firstly, or ramping the sextupole and the three solenoids in parallel. Generally speaking, the tests demonstrate that the SECRAL magnet has reached 95% of designed fields when the sextupole and the axial solenoids are energized simultaneously. This is a milestone of SECRAL project. The magnet will be accommodated into the real cryostat and a formal magnet test with real cryostat is expected in November 2004.

D. SECRAL planed schedule

It is expected that formal test of the magnet with the real cryostat will be done in November 2004, and preliminary acceptance test of the SECRAL magnet for reaching the designed specifications will be scheduled by January 2005 at ACCEL. The final acceptance test of the magnet will be conducted at Lanzhou by March 2005. The ion source assembling will be in April 2005 and the first plasma from SECRAL at 18GHz is planed in May 2005.

ACKNOWLEDGEMENTS

The work was supported by Knowledge Innovation Program of Chinese Academy Sciences under contract No. KJCX1-09 and National Natural Foundation for Distinguished Young Scientist under contract No. 10225523.

REFERENCES

1. J.W. Xia, Nucl . Instr. and Meth. A488 (2002)11.

2. L.T.Sun, et.al. This proceedings.

3. Z.M.Zhang, et.al. This proceeding.

4. Y.Cao , et.al. This proceedings.

5. D.Wutte, et.al. Rev. Sci. Inst. **75** (2004).

6. D.Hitz, et.al. Rev. Sci. Inst. **75** (2004).

7. L.Ma, et.al. This proceedings.

8. P. Spaedtke, private discussion.

9. C.M.Lyneis, et.al. Rev. Sci. Inst. **75** (2004).

10. S. Gammino, et.al. Rev. Sci. Inst. **72**, 1090 (2001).

Production of Highly Charged Heavy Ions by means of a Hybrid Source in Dc mode and in Afterglow Mode

S. Gammino[1], G. Ciavola[1], L. Torrisi[1], L. Andò[1], L. Celona[1], M.Presti[1],
S. Manciagli[1], A. Picciotto[2], A. M. Mezzasalma[2], J. Krása[3], L.Láska[3], M. Pfeifer[3],
J. Wolowski[4], E. Woryna[4], P. Parys[4], G.D. Shirkov[5], D. Hitz[6]

[1]*INFN-LNS, Via S. Sofia 44, 95123 Catania, Italy*
[2]*Dip. Fisica della Materia-T.F.A., Univ. di Messina, Ctr. Papardo-Sperone 31, Messina, Italy*
[3]*Institute of Physics-ASCR, Na Slovance 2, 182 21 Prague 8, Czech Republic*
[4]*Institute of Plasma Physics and Laser Microfusion, Hery Street, Warsaw, Poland*
[5]*Joint Institute for Nuclear Research-Laboratory of Particle Physics, Dubna, Russia*
[6]*CEA-DRFMC-SBT, 17 rue des Martyrs, 38054 Grenoble, France*

Abstract. The ECLISSE experiment has been carried out by coupling a Laser Ion Source (based on a Nd:YAG laser (0.9 J / 9 ns, laser power densities <10^{11} W/cm^2) to the SERSE superconducting ECR ion source. Cw beams of highly charged ions from metal samples without the use of ovens or sputtering technics were obtained in a variety of experimental conditions. The maximum charge states obtained from the ECRIS were 38$^+$ for Ta and 41$^+$ for Au. The peak current was obtained for 25$^+$ and 29$^+$ respectively and it was in the order of some tens of μA. In this work the analysis of some preliminary results obtained in afterglow mode will be also presented. We employed microwave pulse (length 4 msec) and laser pulse (length 9 nsec) with the same frequency (30 Hz) and variable relative phase. For appropriate phase values, a current enhancement of about one order of magnitude was observed.

HIGH DENSITY SHORT PULSE PLASMAS FROM LASER ION SOURCES: COUPLING WITH ECRIS

A novel method for the production of intense beams of highly charged ions was proposed in 1998 at INFN-LNS. The concept of the hybrid ion source is presented in Fig. 1: the first stage consists of a Laser Ion Source (LIS) which gives intense currents of electrons and of multiply charged ions (q/m = 1/10 or lower), then the ECR ion source acts as a charge state multiplier [1]. The ECLISSE project (ECR ion source Coupled to a Laser Ion Source for charge State Enhancement) aims to obtain intense beam in dc and pulsed mode, mainly from refractory elements. The SERSE source is particularly suited for the ECLISSE method because of its dense plasma, which guarantees a good trapping of LIS-generated ions [2]. The LIS ions have a Boltzmann energy distribution, which is more shifted towards high energy when their charge state increases. At lower laser energy intensity the ion energy can be reduced below 200 eV per charge state [3]. This feature is particularly relevant as the coupling efficiency between LIS-generated beams and the ECR plasma is rapidly decreasing with the increase of the energy of the incoming beam. The ion beam absorption by a plasma of length L is given by the formula $N_i = N_0 (1 - exp(-L\,?l\,))$, being l the mean free path of ion in the plasma, N_i and N_0 the number of ions which are coupled and injected, respectively. Ion beam absorption in the plasma decreases with increasing ion energy per charge (E/q=$?_i$). For the values of plasma density obtained with SERSE, a high coupling efficiency is obtained only for $?_i$ = 200 eV [1,2]. The coupling efficiency is close to zero for $?_i$ =1 keV. Moreover the ion beam energy content should be not larger than the plasma energy content [2]. Preliminary studies were carried out to determine:

a) The efficiency of the coupling process of ions from the LIS beam to the ECRIS plasma.
b) The energy distribution and charge state distribution (CSD) produced by the LIS at different laser power density.
c) The etching rates and the amount of ions and neutrals extracted from the target.

d) The magnetic field effect on LIS output.
e) The effect of biasing the metal target.

In spite of the low intensity of focused laser pulses, the ion yield from the LIS was high enough to feed the ECRIS plasma. In ref. [4] a detailed study of the neutrals behavior and of the ionization fraction is reported. These parameters were relevant to determine the effectiveness of the hybrid source.

The effect of the magnetic field was also studied [5] and it was clearly demonstrated that it helps as it focuses the LIS-generated beam and maintains it close to the axis. The expected dependence of the performance of the hybrid ion source on the bias voltage was confirmed by the experiments [1].

the metal target (20 mm diameter) was placed on an insulated rotating rod. A displacement system has been prepared in order to irradiate the sample on a fresh surface over an annular shape. The Nd:YAg laser has been aligned along the normal to the target surface, by means of a He-Ne laser. A focusing lens (4 m focal distance) was placed in air at 10 cm from the window placed on the 0° flange of the analysis magnet. The bottleneck of the laser beam optics was clearly the extraction electrode. During the last experiment a 6 mm collimator was mounted on the electrode and erosion by the laser halo was evident.

The target was connected to a power supply placed on the high voltage insulated box, so that it can be biased with respect to the source at a voltage variable between 0 and –3 kV. Only ions within a narrow window of ion energy distribution can be caught by the plasma, then the presence of a retarding potential permits to use the most intense part of the ion energy distribution window and to increase their stopping power in the plasma, i.e. by decelerating ions, we may increase the number of ions that are caught by the plasma. Anyway a large amount of ions are not captured by the plasma, which phenomenon is reflected negatively on the source performance as it makes the chamber pressure higher.

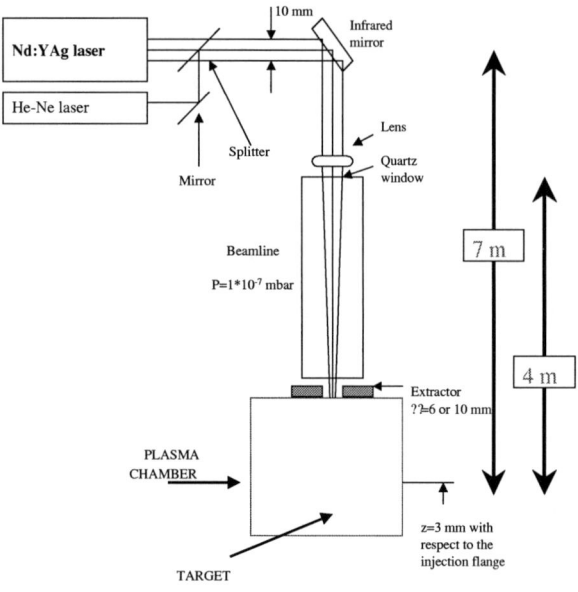

FIGURE 1. The experimental setup optical scheme (the drawing is not in scale).

FIGURE 2. The injection flange and the target, featuring an annular mark due to the laser ablation.

THE EXPERIMENTAL SETUP

Because of the lack of radial accessibility to the plasma chamber of SERSE the easier solution is to inject the laser beam from the extraction hole. The laser beam was injected into the beamline from the 0° port of the 90° analysis magnet, on-axis with respect to the extracted beam [1].

The SERSE plasma chamber contained a Ta or Au target placed on a rotating rod in the injection side of the chamber. The Nd:YAg (0.9J/9 ns, 1.06 μm, 1?30 Hz repetition rate) laser beam hits the target, with a beam spot dimension variable from 1 to 8 mm. Laser power density was in the range $1 \cdot 10^9$ to $5 \cdot 10^{10}$ W/cm^2. In Fig. 2 the SERSE injection flange is shown;

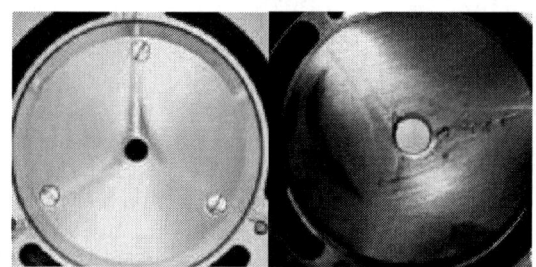

FIGURE 3. The extraction electrode featuring the traces of deposited tantalum and gold.

A clear demonstration came from the photos in Fig. 2 and 3, taken at the end of the experiment; they features some target erosion and the presence of deposited material on the extractor. In general, we estimated that the ionisation efficiency was about one order of magnitude lower than for the ionisation of gases or oven-evaporated materials, which means conversely that large margins for the improvement of the hybrid ion source are still present.

EXPERIMENTAL RESULTS

Tests with the ECLISSE facility have been carried out with tantalum and gold target and they have confirmed that cw or quasi-cw beams can be created with good reliability and reproducibility.

The employed repetition rate was 30 Hz or 1 Hz. Preliminary measurements with Argon plasma, not optimised for high charge states have given 40 µA of Ta^{25+} and Ta^{26+}, 11µA of Ta^{31+}, 1 µA of Ta^{33+}, with a bias voltage of 240 V and a chamber pressure in the order of $3 \cdot 10^{-7}$ mbar; even 150 nA of Ta^{38+} could be obtained. The rf power was about 1500 W and the laser repetition rate was 30 Hz. Laser power was 450 mJ only, because at higher rate there was an excessive target outgassing, and high voltage sparks occurred for 20 kV extraction voltage. With higher laser energy, more than 50 µA of Ta^{25+} have been obtained [1] after a long tuning. A more favourable coupling can be obtained by pulsing the laser at 1 Hz (Fig. 4). In this case up to 600 mJ could be used and the beam current exceeded 25 µA for all species from Ta^{23+} to Ta^{28+}; currents of 8.5 and 4 µA were obtained for Ta^{32+} and Ta^{33+}, by increasing the rf power to 1.7 kW.

In this framework we observed for the first time the "two bumps phenomenon", i.e. the CSD had not a typical bell-like shape, but two superimposed CSD appeared (Fig. 5). One possible explanation can consist of the sum of two CSD, one due to the ions caught by the ECR plasma and ionised to the higher charge states permitted by the conflict with the charge exchange process; the other CSD being due to the atoms and clusters emitted from the Tantalum target, which are ionised at a lower charge state. The two bumps phenomenon could be reproduced many times, even if different peak currents were obtained.

A typical Au spectrum is presented in Fig. 6. It must be underlined the CSD is shifted of two charge states with respect to a similar CSD produced by oven evaporation at the same microwave power rate. In some cases the highest peak was 32^+ or 33^+. More than 3 µA of Au^{36+} were obtained as well as 16 µA of Au^{29+} and 12 µA of Au^{32+}. By decreasing the laser energy, more than 16 µA of Au^{30+} were produced and the current of high charge states was quite high, up to Au^{41+} in spite of the high pressure ($1.8 \cdot 10^{-7}$ mbar).

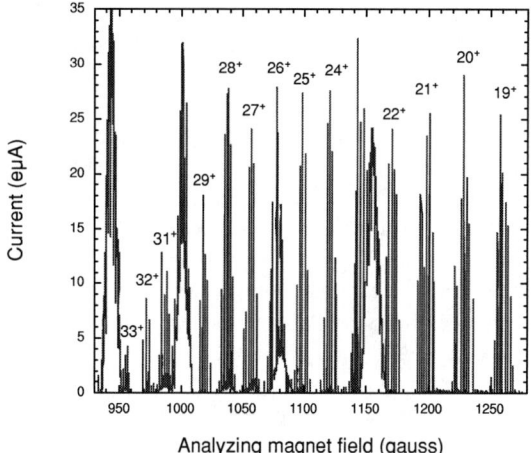

FIGURE 4. Tantalum CSD at 1 Hz (Argon-plasma, 10 mm aperture), featuring a not usual flat distribution.

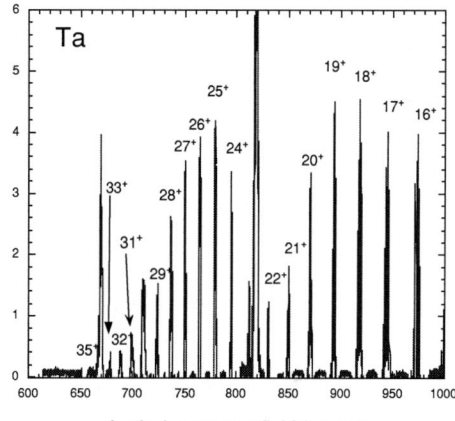

FIGURE 5. A typical 'two bump' CSD.

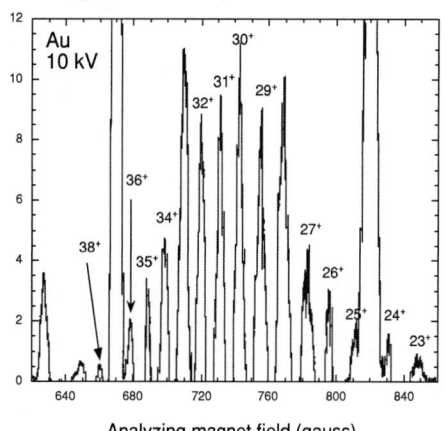

FIGURE 6. Gold CSD at 10 kV and 1.7 kW RF power.

AFTERGLOW MODE

Preliminary experiments using the pulsed regime were carried out only for the laser repetition rate of 30 Hz; different laser energy, different width and time delay of RF pulses were tested. In this experiment neither gas mixing nor biased disk were added for sake of simplicity; in addition, a 6 mm collimator and 10 kV extraction voltage were used, that limited the total current from the source. The maximum current of Ta^{28+} in this condition was only 2 ?A and about 10 ?A of Ta^{24+} were obtained. Rf power was 1.5 kW and the laser energy was 600 mJ. The rf pulse lasted 4 ms and the 9 ns laser pulse relative phase could be changed continuously. When we changed the relative phase we found a maximum of 14 µA. In Fig. 7 the series of peaks for phases variable by 60° are presented. It is to be remarked that the ion beam pulse width for the 0° case could be adapted from 0.5 to almost 2 ms by changing the phase.

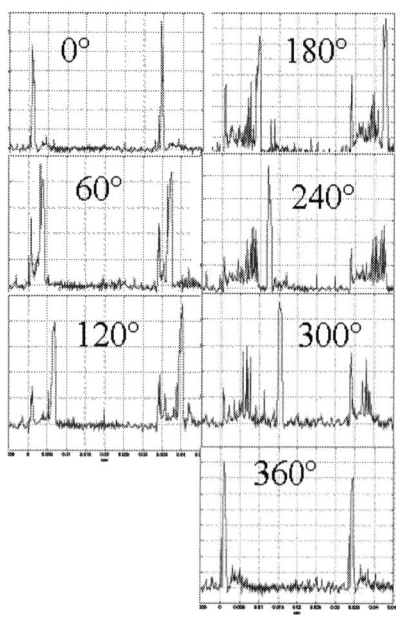

FIGURE 7. Afterglow peaks for variable relative phase between rf pulse and laser pulse (full scale is 45ms for x-axis and 15µa for y-axis; 60° corresponds to 5.5 ms).

COMMENTS AND PERSPECTIVES

The proof-of-principle test have shown that a cw or quasi – cw regime is obtained for the highest charge states, that is relevant for cyclotron – based facilities. The main results are here summarized:

1) the coupling between the LIS ions and the ECRIS plasma is effective;
2) the coupling is more effective in the presence of Argon ECR plasma rather than oxygen plasma; in the latter case, the maximum currents were about a factor three lower;
3) given a certain charge state, the ECLISSE method may permit a current increase (up to 50 µA); improvement of CSD with respect to the conventional method (oven, sputtering, sample insertion) is limited by the high pressure in the plasma chamber.

These results are certainly satisfactory, but they could be even better, provided that the target is better cooled (thus limiting the outgassing) and provided that a tool to measure the main plasma parameters be available. The next experiment, yet approved by the LNS Advisory Committee, is aimed to improve the experimental conditions in terms of vacuum by a different target positioning, of materials, of laser synchronization with RF pulse. The 2nd harmonic of the laser frequency will be applied, that option not being available during previous tests. The measurements with increasing laser repetition rate, from 1 Hz to 30 Hz, will be carried out for different values of biased disk on the target, in dc and afterglow mode.

Over the long term period the availability of 3rd generation ECRIS with higher plasma energy content will make easier the constraints for the ECLISSE method and 2 mA peak current for $q>25^+$ could be produced [6].

ACKNOWLEDGEMENTS

This work was supported by the Fifth Committee of INFN (ECLISSE and PLAIA experiments) and partially supported by the grant IAA 1010405 from the Grant Agency of the Academy of Sciences of the Czech Republic. Studies performed at the IPPLM in Warsaw are partially supported by the Polish State Committee for Scientific Research within the KBN grant No. 5 P03B 108 20.

The support of F. Consoli, A. Galatà, S. Marletta, F. Chines, C. Percolla is acknowledged.

REFERENCES

1. S. Gammino et al., *Jour. of Appl. Phys* **96** (5) 2961 (2004)
2. S. Gammino et al., INFN/TC 02-06 report, (2002)
3. S. Gammino et al., *Rev. Sci. Instr.* **73** (2) 650 (2002)
4. L. Torrisi et al, *Rev. Sci. Instr.* **72** (1) 68 (2001)
5. J. Wolowski, *Laser Part and Beams* **20** (1) 113 (2002)
6. G. Shirkov, private communication

First High Temperature Superconducting ECRIS

D.Kanjilal[1], G.O.Rodrigues[1], P.Kumar[1], C.P.Safvan[1], U.K.Rao[1], A.Mandal[1], A.Roy[1], C.Bieth[2], S.Kantas[2] and P.Sortais[3]

[1]*Nuclear Science Centre, Aruna Asaf Ali Marg, New Delhi – 110067*

[2]*Pantechnik, 14000 CAEN, France,* [3]*Laboratoire de Physique Subatomique et de Cosmologie, Grenoble, France*

Abstract. The first High Temperature Superconducting Electron Cyclotron Resonance Ion Source (HTS-ECRIS) called PKDELIS has been developed as a collaborative project. The source has been designed for suitable use on a high voltage platform with minimum requirements of electrical power and water cooling. The design is based on the required A/q of ~ 7 for the High Current Injector (HCI) of the Superconducting Linear Accelerator (SC-LINAC) at Nuclear Science Centre and to provide relatively higher beam currents of multiply charged ions. High Temperature Superconducting coils (Bi-2223) have been chosen to reduce the power and cooling requirements for obtaining large axial magnetic fields corresponding to a frequency of 18 GHz. The HTS coils are operated in a superconducting mode in a temperature range of about 20 to 22 K using Gifford-McMahon type cryo-refrigerators. A 36 element hexapole was designed using NdFeB to obtain higher fields at the chamber wall. The source is tested thoroughly by producing beams of carbon, oxygen, neon, argon, xenon, tantalum and lead at various charge states having analysed current up to 2 mA. The detailed design aspects and test results are presented.

1 INTRODUCTION

The first high temperature superconducting electron cyclotron resonance ion source (HTS-ECRIS) called PKDELIS has been designed and developed to operate at 14 to 18 GHz and suitable for use on a 300 kilovolt (kV) platform with minimum requirements of electrical power and cooling water. Other sources like SHIVA at the University of Tsukuba and RAMSES in RIKEN [1,2] which also utilise Gifford-McMahon type refrigerators for cooling the coils at much lower temperatures viz., below 5 K used conventional superconducting coils. The PKDELIS being operational at higher superconducting temperature (> 20 K) requires a simpler cryostat [3]. This source is suitable for operation on a high voltage deck and the initial test results have been reported earlier [4,5].

2 DESIGN SPECIFICATIONS

The design of this high performance positive ion source is based on the required mass to charge ratio of ~7 for the high current injector (HCI) of NSC. PKDELIS is suitable to provide high current multiply charged positive ion beams for injection into the superconducting linear accelerator (SC-LINAC) after pre-accelerating the beams to matching energy using radio frequency quadrupoles and low velocity resonators.

FIGURE 1. View of the HTS ECR ion source PKDELIS

2.1 HTS coils and axial field

Since no cryogen can be transferred across the high potential of 300 kV, high temperature superconducting coils of Bi-2223 have been chosen to reduce the power and cooling requirements for producing the large axial magnetic fields corresponding to a frequency of 18 GHz. The HTS coils are operated in a temperature range of about 20 to 22 K by using Gifford-McMahon cryo-refrigerators. The calculation of the axial field profile by the POISSON group of codes [6] using solenoid coils made of BSSCO-2223 HTS wires at an operational current density of 90 A mm^{-2} is shown in figure 2.

2.2 Radial magnet configuration

The hexapole design is based on the design of the Halbach [7] configuration and is made of permanent magnets comprising of Nd, Fe, and B. Surface treated magnets are used for high temperature and high humidity applications. The 3D calculations of hexapole fields using minimum possible values of B_r have been carried out for both 24 sectors and 36 sectors and the results are shown in figure 3 and 4. The design is aimed for maxium field using 36 wedge shaped magnets.

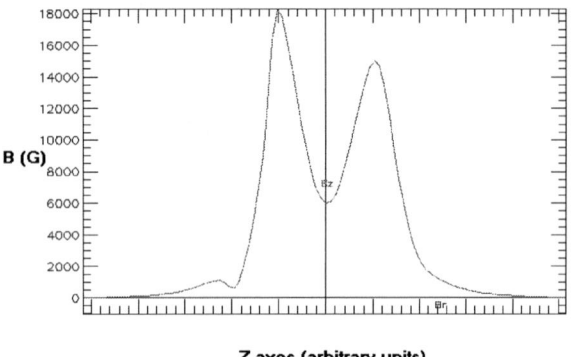

FIGURE 2. Axial field of the HTS coils

2.3 Extraction system

The ion optical calculations of the extraction system in the presence of the strong axial field produced by the HTS coils has been worked out using the IGUN code [8].The total extraction system comprises of the plasma electrode, puller electrode, focus electrode and a last electrode with the same potential as the puller electrode. The puller electrode is polarised to a negative potential of -20 kV in order to obtain better optics. For example, trajectories of oxygen beams in various charge states, for an extraction voltage of 35 kV and puller voltage of -20 kV for a total source current of 10 mA inside the puller electrode is shown in figure 5. The calculated geometrical emittances are shown in figure 6 for various plasma electrode diameters and magnetic fields.

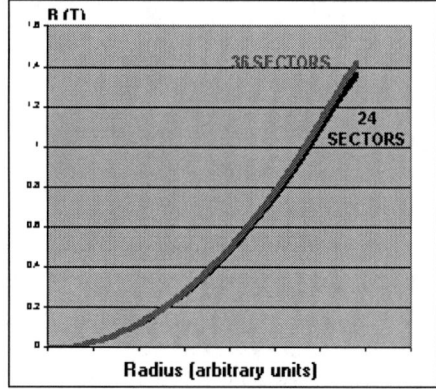

FIGURE 3. Radial field of Hexapole

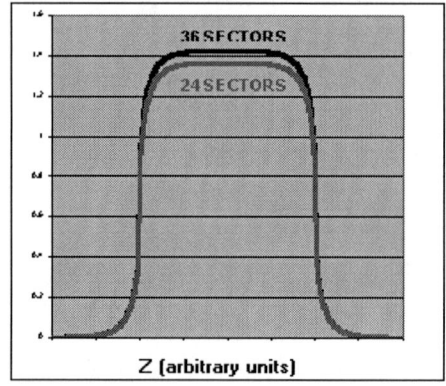

FIGURE 4. Longitudinal variation of the radial field

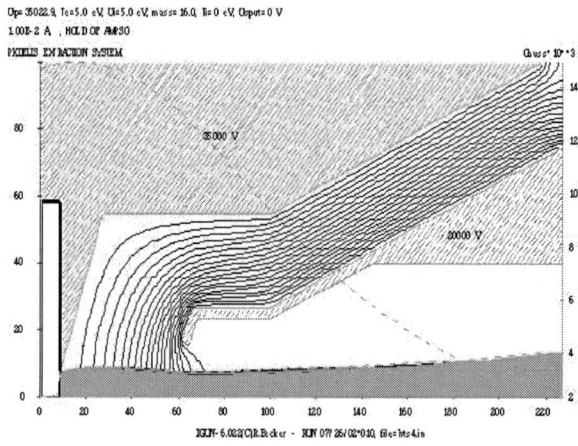

FIGURE 5. Optics of the extraction system

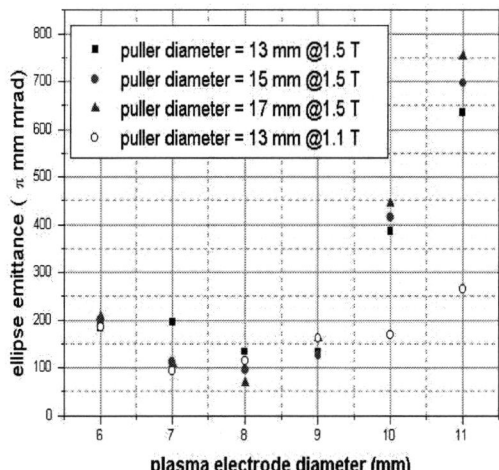

FIGURE 6. Calculated geometrical emittances

2.4 Source configuration

The plasma chamber is made of high purity aluminium so that the oxide formation on the aluminium surface would possibly be a source of secondary electrons inside the source. High quality materials have been chosen for the iron yoke in order to achieve relatively high fields. The cryostat which houses the coils are manufactured entirely from 300 series stainless steel and is in two main sections. A very useful design feature is that access can be gained to the cryo-coolers without dismantling the iron yoke. Each coil is made up of 10 pancake layers and wound as double pancakes and separated by epoxy-glass insulation. The coils are wound onto a high conductivity copper former for support and to provide a path for heat to be conducted away from the coils to the cold head of the cryo-cooler.

3 Test Results of HTS Coils and Hexapole

Fabrication and testing of the HTS coils were carried out at Space CryoMagnetics Ltd, UK. Each of the coils were tested stand-alone for reliability of continous operation. The cool down process for the coils takes about 12 hours to reach the operating temperature. For example, when the individual coils were excited to 150 A, the equilibrium temperatures of the coil former for the injection and extraction coils were at 22 K and 20 K repectively. The test results for both the coils were found to be comparable. Figure 7 shows the axial field measurements at maximum currents of 181 A and 145 A on the injection and extraction coils respectively. The measurements show very well the agreement with the field simulations using POISSON and OPERA 3D [9]. The field mapping of the hexapole was done at the factory of Pantechnik. The measured radial field on the chamber wall was measured to be 1.35 T.

4 Test Results of Various Beams

The beam tests were performed using the test bench of Pantechnik. 14.5 GHz and 18 GHz RF generators were used for the tests. A medium resolution dipole magnet having a total aperture of 50 mm pertaining to the chamber, and a multi-electrode extraction system were used for transporting and analysing the beams. For metal ions, a special micro-oven and sputtering system was used. Figure 8 shows the charge state distribution for ^{129}Xe optimised on ^{129}Xe^{14+}. Table 1. shows some of the optimised beams which have been extracted successfully.

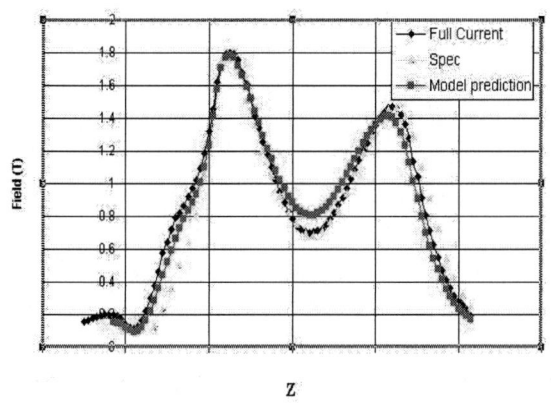

FIGURE 7 Axial field of HTS coils

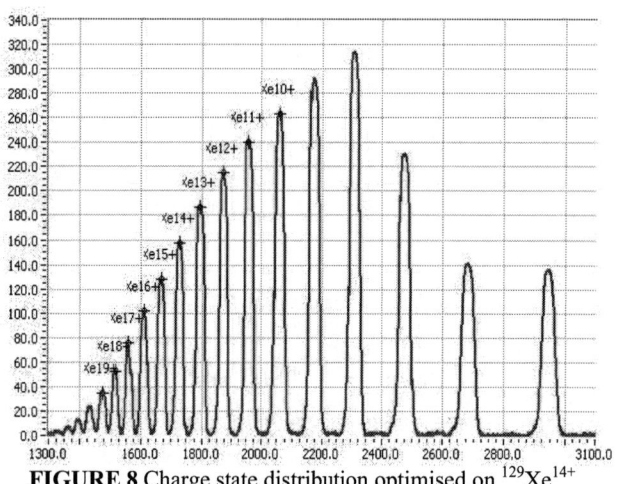

FIGURE 8 Charge state distribution optimised on ^{129}Xe^{14+}

TABLE 1. Extracted currents for various beams

Ion species	Rf power (Watts)	Beam current (eµA)
C^{2+}	597	2000
O^{2+}	193	2037
Ne^{2+}	391	2044
Ne^{3+}	391	1533
Ar^{4+}	496	1023
Ar^{7+}	488	617
Ar^{8+}	521	732
Xe^{14+}	614	157
Xe^{21+}	652	28
Ta^{20+}	426	65
Ta^{25+}	476	27
Au^{21+}	898	28
Pb^{29+}	738	12

5 Future Plans and Upgradation

The future plan is to try the frequency mixing with 28 GHz to see the possible improvements. Improvements of the axial field especially at the injection side will be explored. Possibilities of a newly designed extraction system for improved beam transport with space charge compensation will be looked into. Space charge neutralisation studies are expected to give further information for improved beam transport. The diffusion processes of the electrons and ions will be studied for enhancing the improvements of the outputs of highly charged ions. Due to the large emittances from ECR sources and in particular for this source, as shown in figure 6., a large acceptance, combined function, analysing magnet has been designed to accept a substantial fraction of the total beam intensity. The magnet has been designed to include higher order field components to minimise higher order aberrations. To achieve vertical focussing the entrance and exit pole faces have been designed with particular shim angles with negative radii of curvatures to generate increasing sextupole field components. The poles have been specially shaped to introduce decreasing sextupole field components in the horizontal plane.

6 Conclusion

The performance of the first high temperature superconducting ECR ion source PKDELIS is excellent. The use of BSSCO-2223 HTS wires for the axial field is very suitable for minimising the total power and cooling requirements of the source to be operated on a high voltage platform.

7 References

1. T.Kurita et al., *Rev.Sci. Instrum.* **71**, 909 (2000)
2. T. Nakagawa, T.Kurita, M. Kidera, M. Imanaka, Y. Higurashi, M. Tsukuba, S.M.Lee, M.Kase and Y.Yano, *Rev. Sci. Instrum,* **73**, 513 (2002)
3. D.Kanjilal, G.Rodrigues, C.Bieth, S.Kantas, P.Sortais, C.P.Safvan, P.Kumar, U.K.Rao, A.Mandal, A.Roy, Proceedings of Indian Particle Accelerator Conf. ,Indore, India, February 3-6, 2003, pp.144.
4. D.Kanjilal, G.Rodrigues, P.Kumar, C.P.Safvan, U.K.Rao, A.Mandal, A.Roy, C.Bieth, S.Kantas, P.Sortais, Proceedings of the Asian Particle Accelerator Conf., Korea, March 22-26, 2004 (in press).
5. C.Bieth, S.Kantas, P.Sortais, D.Kanjilal, G.Rodrigues, S.Milward, S. Harrison, R.McMahon, *Production of highly charged ions with an ECRIS using high temperature super-conducting coils,* 12th International Conference on the Physics of Highly Charged Ions, Lithuania, September 2004
6. *POISSON/SUPERFISH group of codes*, Los Alamos National Laboratory, Los Alamos
7. K. Halbach, *Nuclear Instruments and Methods,* **169**, 1 (1980).
8. R. Becker, *Rev. Sci. Intrum.,* **63**, 2756(1992)
9. *OPERA 3D,* Vector Fields , U.K

Production of intense beam of medium charge state heavy ions from RIKEN ECRISs

T. Nakagawa, Y. Higurashi, M. Kidera, T. Aihara*, M. Kase and Y. Yano

RIKEN, Hirosawa 2-1, Wako, Saitama 351-0198, Japan

SHI, Accelerator Service Ltd, Kita-shinagawa 5-9-11, Shinjuku-ku, Tokyo 141-0001, Japan

Abstract. Beam intensities of medium charge-state Ar, Kr and Xe ions have been measured under the various conditions. Beam intensities were strongly dependent on the plasma electrode position and B_{min}. By optimizing these parameters, 2 mA of Ar^{8+}, 0.6 mA of Kr^{13+}, and 0.3 mA of $Xe^{18,20+}$ were obtained. We observed that the RF power of 100~150W was enough to produce 1mA of Ar^{8+} form RIKEN 18 GHz ECRIS by optimizing these parameters.

INTRODUCTION

Production of intense beams of multi-charged ions from ion source, such as Ar^{8+}, Kr^{13+}, Xe^{20+} and U^{35+}, is strongly demanded for RIKEN radioisotope beam (RIB) factory project [1]. For this reason, we have constructed high performance ECRISs (RIKEN 18 GHz ECRIS [2] and the liquid-He-free SC-ECRIS[3]). To increase the beam intensity, many laboratories modified the structure of the ion source and increase the microwave frequency and its power. It is natural to think that the boundary condition of the plasma (plasma chamber geometry, property of its surface), the magnetic field strength and its configuration play essential role to increase the beam intensity. In recent our experimental studies, we recognized that the plasma electrode position and magnetic field configuration are key parameters to increase the medium charge state of heavy ions.

In this paper, we report how to increase the beam intensity by optimizing these parameters and its mechanisms.

RIKEN 18 GHZ ECRIS

A detailed description of the RIKEN 18 GHz ECRIS and its present performance are described in Ref. 2. In order to increase beam intensities, a negatively biased disc made of stainless steel was installed in the plasma chamber and an aluminum cylinder was used to cover the inner wall of the plasma chamber. The diameter of the plasma electrode hole was 10 mm. The distance between the extraction electrode and the plasma electrode was ~15 mm. The hole of the extraction electrode was 12 mm. The detailed description of beam transport system was reported in Ref. 4

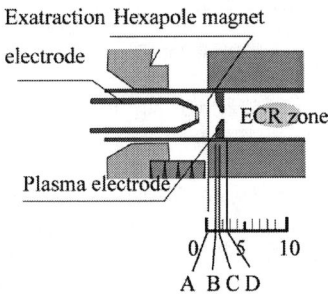

Fig. 1. Cross sectional view of movable plasma electrode and beam extraction side of the RIKEN 18 GHz ECRIS.

Experimental results and discussions

Plasma electrode position

In a previous paper [5], the plasma electrode position for maximizing the beam intensities of the $Ar^{8+,9+}$ was found to be at electrode position C (see Fig. 1)). Figure 2 shows the beam intensity of Ar^{8+} as a function of plasma electrode position. The other parameters (B_{ext}, gas pressure, biased disc position, and negative bias voltage of the disc) were tuned to maximize the beam intensity. It is clearly seen that the best plasma electrode position exist in case of RIKEN 18 GHz ECRIS. Figure 3 shows the optimum plasma electrode position for various charge state heavy ions. More detailed experimental results and discussion are described in ref. 6

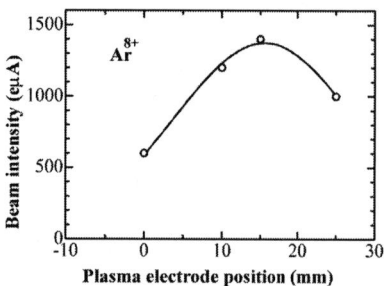

Fig. 2. Beam intensity of Ar^{8+} as a function of plasma electrode position.

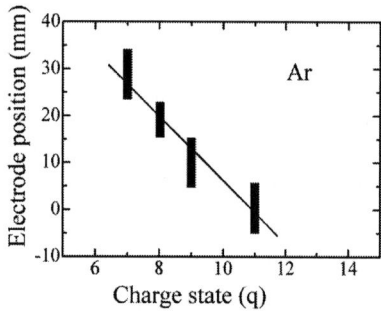

Fig. 3. Optimum plasma electrode position for various charge state of heavy ions.

The electrode position affects the density of the plasma electrode hole; the plasma density should be higher near ECR zone. It means that we can obtain higher beam intensity near ECR zone. However, the position also affects the beam trajectory and emittance as described in Ref. 7. In the present stage, we observed that the beam intensities of lower charge state heavy ions increased and the intensities of higher charge state decreased when moving the plasma electrode toward the ECR zone. Although it is useful method to increase the beam intensity of the medium charge state of heavy ions in the practical point of view, we need further investigation to clarify its mechanisms and to optimize it.

RF power dependence

Up to now, the RF power of the RIKEN ECRISs has been defined as the value subtracted from the injected RF power to reflected one. Actually, the several components of the microwave transmission (rectangular wave guide, flexible waveguide, E-H tuner etc) should absorb microwave. It means that the absorbed RF power in the plasma chamber is strongly dependent on the distance between the RF power supply and ion source, and the number of components of microwave transmission line. In order to evaluate the performance of the ion source, it is better to use the absorbed RF power in the plasma chamber instead of the RF power described above. For this reason, we determined the absorbed RF power in the plasma chamber with measurement of increase of the cooling water temperature of the plasma chamber. Detailed experimental method and results are described in Ref.6. Figure 4 shows the beam intensities of Ar as a function of absorbed RF power in the plasma chamber. In this experiment, the main parameters (gas pressure, magnetic field configuration, biased disc position and its negative voltage) of the ion source except RF power were tuned to maximize the beam intensity. Remarkably, the 1mA of Ar^{8+} can be obtained at the absorbed RF power of 100~150W. If we minimize the wave guide length and number of microwave transmission components, we only need RF power of lower than 500W to obtain intense beam of medium charge state heavy ions. Such low RF power can be reached by the TWT microwave power supply.

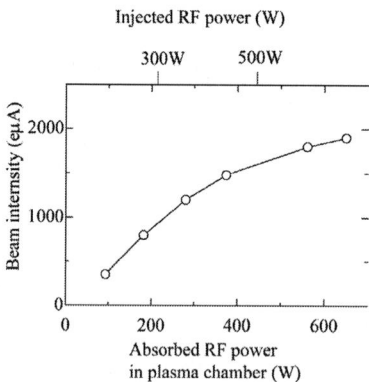

Fig.4. Beam intensity of Ar^{8+} as a function of absorbed RF power in the plasma chamber.

. Effect of magnetic field configuration

In the previous works, we observed that the existence optimum value of B_{min} for maximizing the beam intensity of heavy ions. These values were almost constant and independent on the charge state. In this experiment, we observed that the optimum gas pressure to maximize the beam intensity increased with increasing the B_{min}.

It is obvious that the gradient of the magnetic field at the resonance zone decreases with increasing the B_{min}. It we assume that the stochastic heating plays essential role to transfer the energy from microwave to electrons in plasma, the energy transfer at resonance zone increases with decreasing the gradient.

Consequently, the electron temperature becomes higher. Since the gas pressure strongly affects the ion confinement time and temperature of the electrons as described in Ref.8, we can control the temperature and confinement time with changing the gas pressure. In our case, the temperature may be too high and confinement time to too long to produce medium charge state heavy ions at lower gas pressure. This may be the reason why we have to increase the gas pressure when increasing the B_{min} (decreasing the gradient of magnetic field strength).

In order to investigate this mechanism, we measured the density of electrons, the temperature of electrons and ion confinement time by laser ablation method when changing the B_{min}.

This experiment was performed in SHIVA. An Nd:YAG laser system was used at a wavelength of 1.06μm, a pulse width of 15 ns (Q switch mode of operation). Laser flux was focused onto the aluminum target by a focusing lens. The ablated fluxes of neutral particles go directly into the plasma.

The repetition rate of the laser pulse was 1 Hz. The power of the laser was 120 mJ. The extraction voltage of SHIVA was kept at 15 kV for all experiment. The experiment has been performed with pure oxygen gas.

Simple rate equations assuming a sequential charge stripping of ions in the plasma are used to analyze these data as described in Ref.9
Recently, Grenoble group reported that the ion confinement time is proportional to the charge state of the ion.[10] We can represent the confinement time as follows: $\tau_i = (i/i_{max})\tau_c$

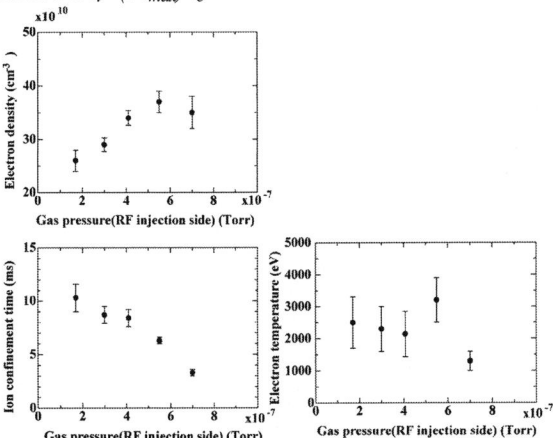

Figure 5. resulting parameters as a function of gas pressure

where i_{max} is the number of charge for fully-striped ion, i.e. atomic number, and τ_c is the ion confinement time of charge state i_{max} ion. For obtaining three plasma parameters (n_e, T_e and τ_c), we used a least square method. Figure 5 shows the resulting plasma parameters as a function of gas pressure. As predicted by the model calculation [8], the ion confinement time gradually decreases with increasing the gas pressure. Figure 6 shows the resulting plasma parameters as a function of B_{min}. Both of T_e and n_e increased monotonically with increasing B_{min}, while τ_c decreases. The value of $n_e\tau_c$ becomes maximum around $B_{min} \sim$ 0.39T, which is correspond to the optimum B_{min} ~0.39T.

Figure 7 plots the effective B_{ext} and B_{min} versus charge state for O^{q+}, Ar^{q+}, Kr^{q+} and Xe^{q+}, when the plasma electrode was placed at position C of RIKEN 18 GHz ECRIS. For the plasma confinement

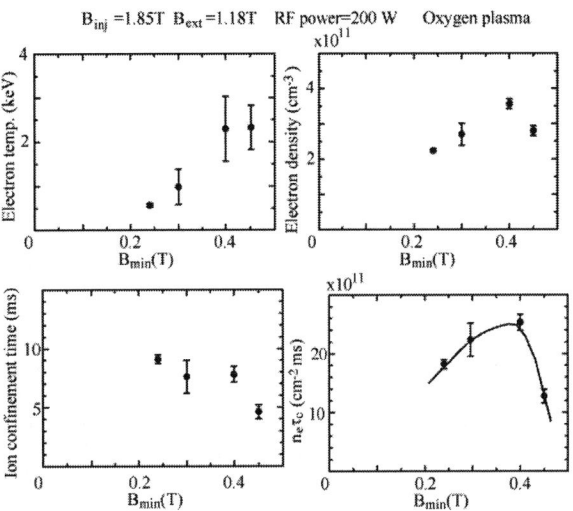

Fig. 6. Plasma parameters as a function of B_{min}

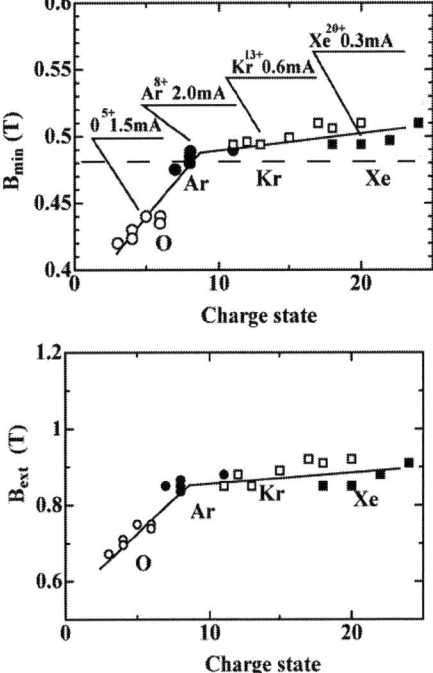

Fig.7. B_{min} and effective B_{ext} for production of multi-charged O, Ar, Kr and Xe ions.

the effective B_{ext} should not be the maximum magnetic field at the beam extraction side, but the magnetic field at the plasma electrode position. The dashed line indicates the optimum value for B_{min} for maximizing the beam intensity of O, Ar, Kr and Xe ions from the RIKEN liquid He-free super conducting ECR ion source RAMSES[9]. Unfortunately, RIKEN 18 GHz ECRIS has only two set of solenoid coils. It means that we cannot change B_{min}, B_{ext} and B_{inj} independently. For Ar^{8+}, B_{min} is almost same as the optimum value for B_{min} obtained with RAMSES. However, to maximize beam intensities of $Kr^{13\sim17+}$ and $Xe^{13\sim24+}$, B_{min} had to be set slightly higher than the optimum value for B_{min} under this experimental condition. For O^{5+}, the B_{min} is significantly lower than the optimum value of B_{min}. When B_{min} is set to the optimum value of B_{min} (~0.48 T) for production of O, Kr and Xe ions, more beam intensity results for the same RF power.

Beam intensity form RIKEN 18 GHz ECRIS

Figure 8 displays beam intensities extracted from RIKEN 18 GHz ECRIS. In order to maximize the beam intensity, the magnetic field strength, gas pressure, biased disc position and its negative biased voltage were optimized. Typically maximum beam intensities were obtained whenever the biased disc was positioned near $(B_{inj})_{max}$.

2mA of Ar^{8+}, 0.6mA of Kr^{13+}, 0.3mA of $Xe^{18+,20+}$ were extracted at the extraction voltage of 12~17kV and RF power of 700W.

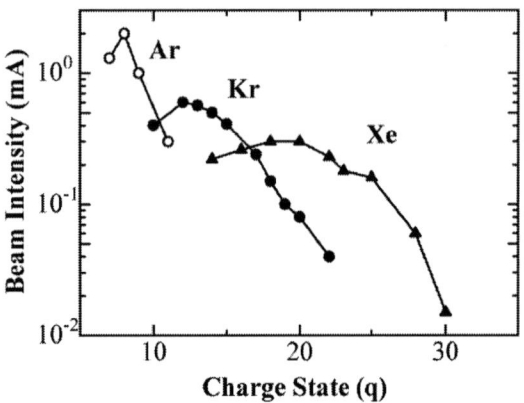

Fig.8. Beam intensity of multi-charged Ar, Kr and Xe ions.

When comparing the performance of the RIKEN 18 GHz ECRIS with other source such as SERSE and GTS, we note that the maximum magnetic field strength of axial direction (~ 1.4 T) is weaker than those(~2.5T). The plasma chamber volume of RIKEN 18 GHz ECRIs (~ 1 L) is quite small compared to the SERSE (~6 L). The intensities of very highly charged heavy ions, such as Xe^{30+}, extracted form SERSE (38.5 eμA) and GTS (28 eμA)are much higher than from the RIKEN 18 GHZ ECRIS (15eμA). This difference may be due to increase in ion confinement time. The beam intensity of medium charge state heavy ions from RIKEN 18 GHz ECRIS(e.g. ~300 eμA of Xe^{20+}) is higher than those from SERSE (~160 eμA)and GTS(~220 eμA). From these experimental results, it can be concluded that the small plasma chamber size and low magnetic field strength in the RIKEN 18 GHz ECRISECRIS is suitable for production of medium charge-state ions at moderate RF power (lower than 1 kW) .

Conclusions

The effects of plasma electrode position on the beam intensities were measured. Medium charge-state were strongly enhanced by moving the plasma electrode toward the ECR zone. Consequently, 2mA of Ar^{8+}, 0.6mA of Kr^{13+}, 0.3mA of Xe^{20+} were extracted at the extraction voltage of 12~17kV. To obtain 1mA of Ar^{8+}, we only need 100~150W of absorbed RF power in the plasma chamber. The electron density, temperature and ion confinement time were strongly dependent on the B_{min}. The value of $n_e\tau_c$ becomes maximum at the B_{min}~0.39T in case of the SHIVA, which is correspond to the optimum value for B_{min}.

REFERENCES

1. Y. Yano, Proceedings of the 15th International Conference on Cyclotron and Their Application (IOP, Bristol, 1999)p696.
2. T. Nakagawa and Y. Yano, Rev. Sci. Instrum. 71(2000)637 and references therein.
3. T. Nakagawa et al, Rev. Sci. Instrum. 73(2002)513 and references therein.
4. T. Nakagawa et al, NIM B in press
5. Y. Higurashi et al, Nucl. Instrum. Methods A510 (2003) 206.
6. Y. Higurashi et al, these proceedings
7. P. Suominen et al, Rev. Sci. Instrum. 75(2004)1517
8. A. Girarad et al, Rev. Sci. Instrum. 75(2004)1381
9. T. Nakagawa et al, Rev. Sci. Instrum 75(2004)1394
10. A. Girarad et al, Pys. Rev. E62(2000)1182

Recent ECRIS Related Research And Development Work At JYFL

H. Koivisto, P. Suominen, O. Tarvainen, J. Ärje, E. Lammentausta, P. Lappalainen, T. Kalvas, T. Ropponen and P. Frondelius

Department of Physics, University of Jyväskylä, FIN-40014 University of Jyväskylä, Finland

Abstract. The main focus of the JYFL (University of Jyväskylä, Department of Physics) ion source group has recently been on the development of a new plasma chamber and measurements of the plasma potential with a device developed at JYFL. The new plasma chamber is based on an idea described at ICIS'03 in Dubna. The work is mainly presented elsewhere in these proceedings (P. Suominen et al.). The plasma potential measurements are based on the use of a decelerating voltage. With the aid of the device, information about the plasma potential and the temperature of the ions can be obtained. This work is also described elsewhere in these proceedings (O. Tarvainen). The radial feeding of the microwave power into the plasma chamber has been studied. According to the first results approximately the same intensity can be reached if the microwave power is launched into the plasma chamber axially or through the radial port. The phenomena will be investigated in more detail. Development work for the production of metal ion beams has also been carried out. An overview of the work of the JYFL ion source group will be presented.

INTRODUCTION

As a University laboratory the JYFL Accelerator Laboratory has two tasks: education and research. The main object of the JYFL ion source group is to serve the research program of the laboratory. Therefore, new ion beams have to be developed and the ion source performance and the beam quality have to be improved. Due to very interesting challenges related to the ion sources new students have been attracted to the group. This makes it possible to carry out efficient development work for the needs of the nuclear physics research. It also enables the group to carry out ion source related research work and to provide subjects for the Master and Doctoral theses. In this article we will present the work of the JYFL ion source group. One interesting part of our work is the plasma potential measurements [1]. We have found that new information can be obtained related to the temperature of ions, beam emittance and gas mixing effect. A new plasma chamber has been designed, which is based upon the idea of MMPS (Modified MultiPole Structure) [2, 3]. The plasma-EM wave coupling experiments have also been started and initial results will be presented. Some new ion beams have been developed – like Zr and Sr. A contract with European Space Agency (ESA) has been signed, which produces some demands also on the development work of ion source group.

CARBON CONTAMINATION VS. PLASMA POTENTIAL

A new device for the plasma potential measurements has been developed at JYFL [1]. The device has been used with the JYFL 6.4 GHz ECRIS, the JYFL 14 GHz ECRIS and with the 14 GHz ECRIS at Argonne National Laboratory. With the aid of the device we have confirmed the fact that there is a strong dependence between the performance of the ion source for high charge states and the plasma potential of the ECRIS plasma. This can be studied for example with the aid of the MIVOC method. More detailed description of different plasma potential measurements is presented elsewhere in these proceedings (O. Tarvainen et al.)

The MIVOC compounds contain usually carbon (like $Fe(C_5H_5)_2$), which drifts to the wall of the plasma chamber. The consumption rate of the MIVOC material can be more than 1 mg/h if the maximum intensity for medium charge states like Fe^{12+} is required. According to our experiments [4] about 95 % of carbon deposits onto the wall of the plasma chamber. Consequently, after a run of 3 weeks the carbon contamination of the plasma chamber walls can be almost 1 gram. The contamination is not evenly distributed. We have measured [5] that the ions tend to follow the electron flux, i.e. a lot of carbon exist on the magnetic poles of the multipole and on the magnetic loss cones in the injection and the extraction ends (triangle pattern). We have also noticed that the part of the contamination is caused by the second-generation neutral carbon atoms. This contamination occurs only in the injection side of the plasma chamber [6].

How can the carbon contamination be used to study the dependence between the ion source performance and the plasma potential? The plasma chamber of the our 14 GHz ECRIS is made of aluminum. The surface is oxidized due to the oxygen plasma and it behaves as a good source of secondary electrons. For the aluminum oxide the yield of the emission is 2 – 9. In the case of carbon this yield is 1, being about 0.5 for soot. Consequently, the secondary electron emission decreases dramatically due to the carbon contamination if the plasma chamber is made of aluminum. This decreases the intensity of highly charged ions.

The plasma potential of the JYFL 14 GHz ECRIS has been measured to be about 20 V. After a MIVOC run of almost 3 weeks the plasma potential was around 50 V and the corresponding intensity of O^{7+} ion beam was only 30 - 50 µA (normal intensity is around 150 µA). During the run the maximum intensity for $^{46}Ti^{10+}$ ion beam was required for the nuclear physics experiment. The beam was produced from the natural compound and maximum intensity of about 60 nA on target was reached. The experiment caused a strong carbon contamination, which has to be removed in order to restore the performance of the source to the normal level. The contamination can be removed by three different methods: 1) by the oxygen plasma, 2) by the alcohol or by 3) the mechanical cleaning. We have measured that the oxygen plasma can remove about 50 µg/h of carbon. Consequently, this cleaning method is insufficient in the case of heavy contamination. Using the alcohol for the cleaning, the plasma potential decreases to the value of around 30 V and the O^{7+} intensity increases to the value of about 80 – 100 µA. According to our experience the value of about 200 µA can be achieved after the mechanical cleaning of the plasma chamber.

The afore-mentioned experiments confirmed that feeding extra electrons into the ECRIS plasma, which improves the intensity of highly charged ion beams, reduces also the plasma potential. Although, the secondary electron emission is a very efficient method to increase the electron density it is also very sensitive to the contamination. Using an aluminum liner during the MIVOC runs can solve the contamination problem. However, the liner requires extra venting and it also decreases the performance of the ion source. This gives a motivation to seek new methods for the feeding of extra electrons into the ECR plasma.

NEW PLASMA CHAMBER FOR THE JYFL 6.4 GHZ ECRIS

An iron induced modified multipole structure (MMPS) for the ECR ion source was introduced at ICIS'03 in Dubna [2]. In this structure the multipole field is increased only at the magnetic pole with the aid of high permeability material like iron.

FIGURE 1. An old multipole structure of the JYFL 6.4 GHz ECRIS and the idea of MMPS-structure. High permeability material is used to locally increase the multipole field.

After the ion source conference, numerous 3D-simulations were performed for more realistic results [6]. Those simulations confirmed that a remarkable improvement indeed could be achieved on the magnetic poles, which encouraged us to design a new plasma chamber for the JYFL 6.4 GHz ECRIS. The multipole field of about 0.9 T can be achieved on the magnetic pole with the aid of the iron (≈ 0.5 T without the iron). The effect of the iron is local being about 4 mm wide. The chamber is now being manufactured and will be ready for the experiments by the end of 2004. More detailed description about this work can be found elsewhere in these proceeding (P. Suominen et al.).

PLASMA COUPLING

Axial versus radial microwave feed was studied with the JYFL 6.4 GHz ECRIS. Here the axial or radial feed means that the power is launched into the plasma chamber through the injection side or through the radial port of the plasma chamber, respectively. In other words the waveguide is parallel or perpendicular with the long axis of the plasma chamber. With the aid of the power divider the power ratio can be altered continuously. The power was measured about 1 meter before the plasma chamber. The intensity of the ion beam was measured as a function of the power ratio $P_{axial}/P_{tot.}$, where $P_{tot} = P_{axial} + P_{rad}$.

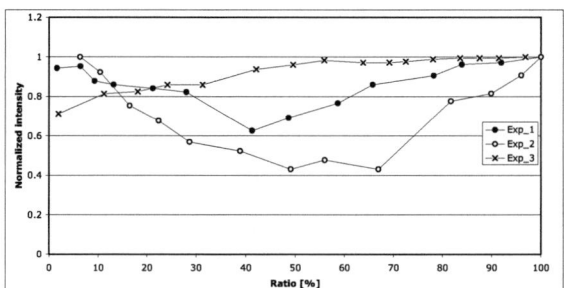

FIGURE 2. The normalized intensity of O^{6+} ion beam as a function of the microwave power ratio.

Figure 2 shows the results of three experiments performed with oxygen. The normalized intensity is used because of the different microwave power in different experiments. No gas mixing was used. In Exp_1 the ion source was tuned for O^{6+} ion beam at the microwave power of 280 W. The intensity was 103 μA when the power was fed axially into the plasma chamber. After that some of the power was fed radially into the plasma chamber by changing the setting of the power divider. At the beginning, the intensity of the ion beam decreased when axially fed power decreased and the radially fed power increased. Surprisingly, the intensity started to increase as the power ratio was further reduced. Finally, when all the power was fed through the radial port, practically the same intensity (103 μA) was reached as in the case of the axial feeding. As a next step the microwave power was decreased to 200 W. Similar behavior was obtained (Exp_2). However, the next experiment (Exp_3) gave a different behavior. In this test, the intensity of O^{6+} was maximized being 90 μA at 280 W when approximately 50 % of the power was fed axially. After that the intensity was measured as a function of the μw-power ratio. As figure 2 shows the intensity was almost constant when 50 – 100 % of the power was fed axially into the plasma chamber. As a next step the microwave power was increased to 430 W. Other parameters of the ECRIS were kept constant (earlier the O^{6+} intensity was maximized for the power ratio of 50 %). The microwave window used for the radial waveguide limited the power launched through the radial port. Figure 3 shows (Exp_4) that the intensity of O^{6+} ion beam increased when the part of the microwave power was fed radially into the plasma chamber. The best performance was reached at the power ratio of around 60 % (128 μA). As a next step all the power was fed axially and the tuning of the source was optimized for the production of O^{6+} ion beam. The microwave power ratio was altered in order to see if the beneficial effect still occurs. As figure shows (Exp_5) the effect was now smaller but again the highest intensity was obtained when part of the power was fed radially. Similar result was achieved for O^{7+} ion beam. As a consequence of the tests a small (5-10 %) improvement is obtained when part of the microwave power is fed radially into the plasma chamber. This effect occurs at the high microwave power. Our future plan is to test different configuration of axial feeding in order to see if different modes of standing waves can be excited in order to maximize the coupling.

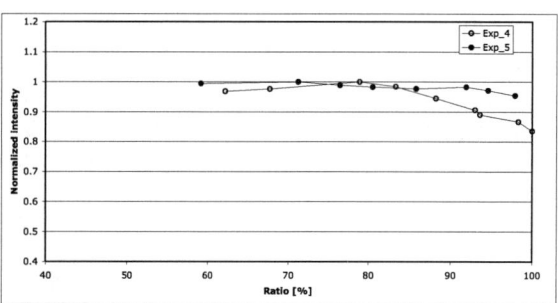

FIGURE 3. The normalized intensity of O^{6+} ion beam as a function of the power ratio fed into the plasma chamber.

BEAM DEVELOPMENT

Several new ion beams were developed for the needs of the accelerator based nuclear and material physics (Sr by oven and Zr by MIVOC using $Zr[C_5H_5]_2[CH_3]_2$) and for the low energy material physics (Mn by oven and Pb from PbO by oven). However, the most requested ion beams (^{46}Ti and Mo) are still unavailable. Very intensive Ti ion beams have been produced with the aid of the MIVOC method using the commercial compound. However, the enriched compound is not available and so far we have not succeeded in synthesis. For enriched titanium, we will construct a new oven, which exceeds the performance of our present miniature oven. In order to keep the radiation losses small and the heating efficiency high the crucible will be heated directly by the tantalum wire. The production of the Mo ion beams was tested with oven using MoO_3. The beam was very unstable and difficult to handle. The sputtering technique seems to be the most promising for Mo ion beams although the intensity is not high enough and the consumption rate is high. As a consequence more development work is needed to solve the problems. An interesting behavior was found concerning the sputtering technique: the consumption rate of the sputtered material can be estimated during the run. This can be done because the ion current and the sputtering voltage of the sputtered sample are known. In addition, the average charge state from the spectrum can be calculated. By using these values and the data for sputtering yields as a function of the energy the mass of sputtered material can be calculated. Proceeding this way we found that the difference between the measured and calculated consumption rate is usually less than 15 %.

DEVELOPMENT WORK FOR RADIATION TESTS

The European Space Agency (ESA) uses the Accelerator Laboratory for the studies of the radiation effects in semiconductor devices. Different beam cocktails are used for efficient testing. The beam intensity needed for the tests is normally of the order of 10^3-10^4 particles/s. However, for the tuning of the cyclotron much higher beam intensity is needed due to the noise coupled to the Faraday cup. The amplitude of the noise is normally of the order of 0.1 nA (see Fig. 4) or higher. Consequently, the beam intensity has to be at least 0.1 nA after the K130 cyclotron in order to make the tuning of the cyclotron possible. This value can be difficult to reach if very high charge states or ions from the rare isotopes are needed. In order to overcome this problem the ion source performance and/or the beam diagnostics has to be improved.

To improve the beam diagnostics we have developed a low current amplifier based on the integrating circuit. Figure 4 shows the comparison between the old and improved beam diagnostic. The dashed line shows the noise measured from the Faraday cup using our old system (Keithley 485 Picoammeter). The solid line corresponds to the beam intensity of 15 pA measured from the same Faraday cup using the integrator. As figure shows the improvement factor of higher than 10 was achieved.

A TWTA was also tested with the JYFL 14 GHz ECRIS in order to confirm the effect of the two-frequency heating. The test was carried out in collaboration with the ANL ion source group. A remarkable improvement of highly charged ion beams was obtained. For example the intensity of Xe^{32+} increased by the factor of 7 (from 0.5 µA to 3.7 µA). As a consequence of the tests a TWTA has been bought to improve the facility for the radiation tests. Together the TWTA and improved beam diagnostic system will significantly improve the operation of JYFL K130 cyclotron

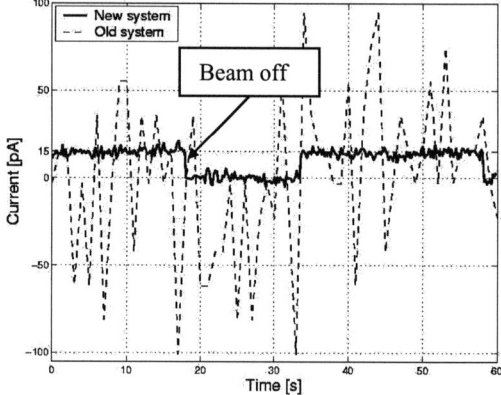

FIGURE 4. The dashed line shows the noise measured from the Faraday cup located (using Keithley) after the cyclotron. Solid line corresponds to the current of 15 pA measured with the integrator from the afore-mentioned point. In both experiments similar conductor arrangement was used.

ACKNOWLEDGMENTS

This work has been supported by the Academy of Finland under the Finnish Centre of Excellence Programme 2000-2005 (Project No. 44875, Nuclear and Condensed Matter Programme at JYFL).

REFERENCES

1. O. Tarvainen, P. Suominen and H. Koivisto, will be published in Rev. Sci. Instr., 75(10), (2004).
2. H. Koivisto, P. Suominen, O. Tarvainen, D. Hitz, Accepted for publication in Rev. Sci. Instr.
3. P. Suominen, O. Tarvainen and H. Koivisto, Accepted for publication in Nucl. Instr. and Meth. In Phys. Res. B..
4. O. Tarvainen, P. Suominen and H. Koivisto, Nucl. Instr. and Meth. in Phys. Res. B 217 (2004), p. 136.
5. O. Tarvainen, P. Suominen and H. Koivisto, Rev. Sci. Instr., 75(5), (2004), p. 1523.
6. P. Suominen, O. Tarvainen and H. Koivisto, Accepted for publication in Nucl. Instr. and Meth. In Phys. Res. B.

ECRIS Operation With Multiple Frequencies

R. Vondrasek*, R. Scott*, R. Pardo*, H. Koivisto[ψ], O. Tarvainen[ψ],
P. Suominen[ψ], and D. H. Edgell[τ]

*Argonne National Laboratory, Argonne, USA
[ψ]Department of Physics, Accelerator laboratory, University of Jyväskylä, FIN-40351, Jyväskylä, Finland
[τ]Laboratory for Laser Energetics, University of Rochester, Rochester, NY 14623-1299

Abstract. The usefulness of two-frequency heating for the production of high-charge state high-intensity beams from an ECRIS has been well established. Factors of 2→5 increase in beam currents have been observed accompanied by a shift to higher charge states. The ECRIS at Argonne National Laboratory has been routinely operated utilizing a 14 GHz klystron and a tunable 11-13 GHz traveling wave tube amplifier (TWTA) and the operating characteristics of the source are well known. However, the characteristics of the multi-frequency heated plasma are less well known. Investigations regarding the changes in the source production have been taking place at Argonne National Laboratory. Parameters such as the charge state distribution (CSD), production times and plasma potential have been measured for a multi-frequency heated plasma with emphasis being given to the effect of the frequency gap between the two RF waves. It has been found that the production times decrease in multi-frequency mode with a corresponding increase in the CSD and the overall beam output. At the same time, the plasma potential appears to not change significantly. It has also been found that a larger frequency gap (14.0 and 10.84 GHz), while producing higher charge-state ions, produces less overall beam of the material of interest but reaches equilibrium more quickly when compared to a smaller gap (14.0 and 12.31 GHz). Possible mechanisms for the observed behavior will be discussed.

INTRODUCTION

The desire of experimentalists to probe reactions which have ever smaller reaction cross-sections has challenged the electron cyclotron resonance ion source (ECRIS) designers to build and operate sources with enhanced production of highly charged ions (HCI). Various techniques have been employed to improve the operating characteristics of an ECRIS such as wall coatings, biased disks, magnetic confinement and the injection of multiple RF frequencies. The Argonne National Laboratory's ATLAS accelerator utilizes two ECR ion sources [1] employing many of the above listed techniques to improve overall performance. Many groups have worked to characterize the changes in the plasma parameters of an ECRIS when these techniques are employed, and the ECR group at ANL has initiated a research program to characterize the changes in the source parameters when multi-frequency heating is employed.

A detailed description of the physics and construction of ECR ion sources is given elsewhere [2], while the operation of ECRIS in multi-frequency mode was first studied by [3]. A simple statement is that multi-frequency heating improves source performance by providing an additional resonance surface for electron heating. It was also determined that the two ECR surfaces should be sufficiently separated in order to produce the maximum amount of HCI [4]. Using computer codes and the measurement of the charge state distribution (CSD), the production times of the various charge states and the settings of the various tunable source parameters – RF power, magnetic field and neutral pressure, the characteristics of the ECRIS plasma in multi-frequency mode can be deduced [5].

MEASUREMENT TECHNIQUE

The measurements took place on the ATLAS ECR2 ion source. The primary frequency was provided by a 14 GHz 2.5 kW klystron with a secondary frequency from a 0.5 kW 11-13 GHz traveling wave tube amplifier (TWTA). For the plasma potential measurements the source was tuned for maximum output of O^{7+}, while for the production time measurements the tunable source parameters – magnetic field, gas feed rate, biased disk – were kept constant. The RF power levels and frequencies were the only parameters varied between each production time measurement series.

A previously employed technique was utilized to measure the production times of the ionic charge states [6]. A high voltage pulse (500 μsec width at -1.5 kV) applied to a sputter sample positioned at the radial wall of the plasma chamber produced a flux of neutral particles which entered the plasma and was ionized in a step-wise fashion. The faraday cup signal was viewed on a digital oscilloscope, and the traces were recorded to disk. Figure 1 illustrates the behavior of Pb^{Q+} ions in an oxygen plasma with 500 W of injected RF. The steady-state CSD is determined by applying a DC voltage of –0.5 kV to the sample. It has been previously shown that the steady state and pulsed CSDs are very similar [6].

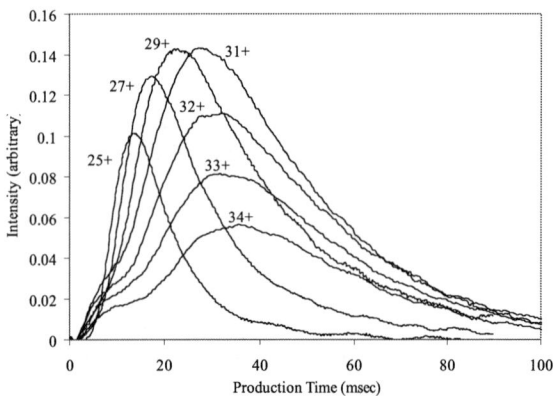

FIGURE 1. Time evolution for the charge states of lead in an oxygen plasma at 500 W using 14 GHz after a 500 μsec pulse of neutrals into the plasma.

The plasma potential was measured utilizing a stopping grid mounted within the beamline after the analyzing magnet. The power supply used to bias this grid is floated at source potential, and the voltage is varied while recording the decay of the beam current until it reaches a baseline value. The plasma potential can then be determined by analyzing the decay curve. A more detailed description of the technique is found elsewhere in these proceedings [See Tarvainen et al. in these proceedings].

EXPERIMENTAL RESULTS

Single Versus Multi-Frequency Heating

The source output in multi-frequency mode as compared to single frequency for lead is shown in Figure 2. For the single frequency measurement the source was operated with 14 GHz at 500 W, while for the multi-frequency measurement an additional 250 W was injected at 10.84 GHz. In addition to the intensities, the production times of the various charge states are also displayed. As can be seen in the figure, the intensities in all of the higher charge states increase and the production times decrease when multi-frequency heating is used. Previous experience has shown that the increase in beam intensity achieved with multi-frequency heating cannot be duplicated with an equivalent amount of RF at a single frequency [4]. The total extracted beam summed over all of the lead charge states increased from 0.76 pμA to 2.6 pμA and the peak of the CSD shifted from 29+ to 31+.

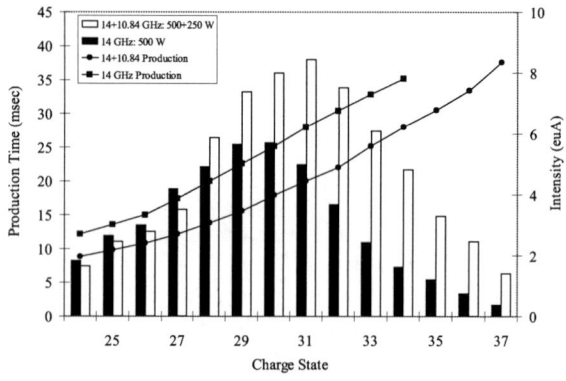

FIGURE 2. Intensities and production times for lead in single and multi-frequency modes.

A more direct comparison of single versus multi-frequency heating was made when the source was operated with 14 GHz at 500 W and measurements were made. The 14 GHz power was then lowered to 250 W and the second frequency was energized at 10.84 GHz and 250 W. This resulted in the same net RF power being launched into the source. As can be seen in Figure 3, the production times decrease in the multi-frequency case even though there is less power going into any single resonance zone. In addition, the

CSD shifts to lower charge states but still maintains an equivalent total summed beam current.

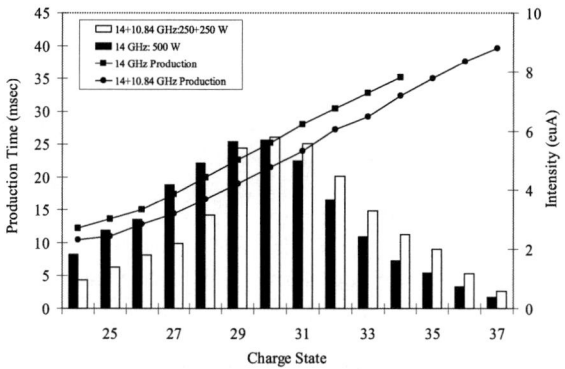

FIGURE 3. Intensities and production times for lead ions with single (14 GHz) versus multi-frequency (14 and 10.84 GHz) heating at the same total RF level of 500 W.

Effect Of The Frequency Gap

It has been observed that the HCI intensities are dependent upon the magnitude of the gap between the two frequencies, with a larger gap improving source performance. In order to gain more understanding with regards to this behavior, the production times and beam intensities were measured as a function of the frequency gap.

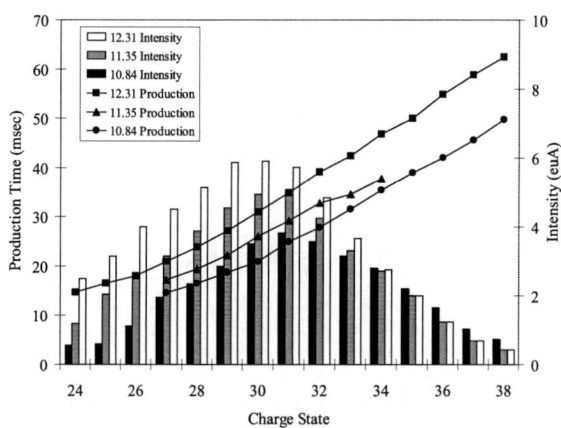

FIGURE 4. Production times and intensities for lead charge states with 14 GHz and 10.84, 11.35 or 12.31 GHz from the TWTA. The total RF power was kept constant at 500 W and no other source parameters were adjusted.

The source was operated with 14 GHz at 300 W with an additional 200 W from the TWTA at varying frequencies which were chosen for maximum beam output, stable operation, and low reflected power. The other tunable source parameters were not varied during these measurements. The resulting TWTA frequencies were 10.84, 11.35 and 12.31 GHz. The intensity and production time results are shown in Figure 4.

Several observations can be made with regards to the figure. The first is that the peak of the CSD shifts from 30+ to 31+ when the frequency gap is enlarged. In addition, the production times are faster with a larger gap. Contrasting this is the observation that the total beam summed over all of the lead charge states is lower when a larger frequency gap is used. The total extracted lead beam decreases from 2.60 pµA with 12.31 GHz as the secondary frequency to 0.87 pµA for 10.84 GHz due to the reduced contributions of the lower charge states; while at the same time the extraction current, a measure of the total amount of beam coming out of the source, increases with the larger gap.

Another variation in source behavior as the frequency gap is narrowed is the appearance of a precursor pulse in the production time trace. The precursor pulse is a flux of beam normally seen in the higher charge state traces appearing 4-9 msec after the voltage pulse has been applied to the sputter sample. The proposed mechanism for the precursor pulse is a disruption of the plasma potential due to the sudden influx of neutrals originating from the sample. These neutrals are ionized in the plasma resulting in a sudden increase in cold electrons. Since these electrons are not well confined there is an increase in the electron loss to the wall. The plasma potential increases to compensate for this new loss term and the HCI, which had previously been trapped by the potential dip, escape until equilibrium can be restored. This pulse predominantly occurs for the higher charge states due to their increased sensitivity to changes in the plasma potential.

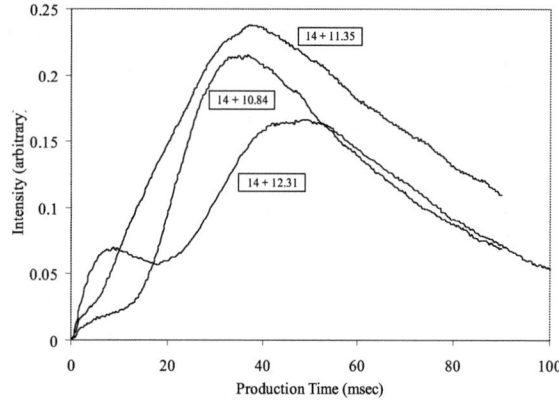

FIGURE 5. Pulse shapes for Pb^{34+} with 14 GHz and 10.84, 11.35 or 12.31 GHz. The total RF power was kept constant at 500 W. The precursor peak is visible at ~8 msec.

It was observed that the appearance of the precursor pulse was delayed for a larger frequency gap. For 29+ the precursor was only observed for the smallest frequency gap where as for 34+ the pulse appeared for all frequency gaps but its magnitude was highest for the smallest gap as shown in Figure 5. This indicates that more cold electrons are escaping the plasma and affecting the plasma potential when the frequency gap is smaller.

Plasma Potential

The plasma potential was measured for the case of single versus multi-frequency heating with O^{7+}. The results show that the plasma potential does not decrease when two frequencies are launched into the plasma with the same total RF power as in single frequency mode. With a total of 500 W launched in the 14 and 10.84 GHz bands the potential is 4.6 V. An identical potential of 4.6 V is achieved with 500 W of 14 GHz alone. The difference between the two modes is the beam production. In multi-frequency mode the O^{7+} current is 68 eμA while it is only 31 eμA in single frequency mode.

DISCUSSION

While specific plasma parameters such as ion confinement time, electron density and temperature have not yet been determined using the data acquired, some general observations can be made.

From the observed increases in beam intensities and decreases in production times, it is reasonable to surmise that the electron density and population increase when multi-frequency heating is used. The peak of the CSD shifting three charge states higher with multi-frequency heating indicates an increase in the confinement time which is associated with a deepening of the proposed plasma potential dip [2].

Assuming that an electron undergoes stochastic heating when it passes through an ECR zone, then the presence of a second resonance increases the heating [2] and results in less electrons being lost. These electrons then undergo additional heating at the ECR zones gaining more perpendicular energy until they are well confined within the inner resonant surface. This leads to an increase in the electron density within the inner resonance. This may explain why when the total extracted current (all Q/M) decreases, as in the case of two frequency heating at the same total RF power level, the high charge states of lead continue to increase in intensity accompanied by a decrease in production times. The decrease in the total extracted current indicates a reduced electron population while both of the latter traits indicate an increase in electron density as well as confinement time.

It is possible that the second lower frequency resonance can be viewed as a separate potential structure located within the potential structure of the higher resonance. The superposition of the potentials alters the overall potential profile and the local densities of electrons and ions. If we assume that the proposed central potential dip of the higher resonance is localized within the core of the plasma then the positively charged sheath of the inner resonance will locate itself within this dip. This would diminish the region that would have conditions suitable for the central dip. A larger frequency gap would more severely affect the overall potential structure and therefore the confinement of the HCI. The plasma potential measurements indicate that the magnitude of the potential does not change when multi-frequency heating is employed, yet a much higher extracted beam current is realized for the higher charge states. This would suggest that the gross potential structure is not the feature responsible for the improved beam production but rather a finer detail of the structure, namely the central potential dip.

ACKNOWLEDGMENTS

This work was supported by the U.S. Department of Energy, Nuclear Physics Division, Contract Number W-31-109-ENG-38.

REFERENCES

1. Schlapp M. et al., *Proceedings of the 13th International Workshop on ECR Ion Sources*, College Station, TX, p.22 (1997)
2. Geller R., *Electron Cyclotron Resonance Ion Sources and ECR Plasmas*, Bristol: Institute of Physics, 1996
3. Xie, Z. Q., and Lyneis, C. M., *Rev. Sci. Instrum.* **66**, 4218-4221 (1995)
4. Vondrasek, R. C. et al., *Proceedings of the 15th International Workshop on ECR Ion Sources*, University of Jyvaskyla, Jyvaskyla, Finland, p.63 (2002)
5. Shirkov, G. et al., *Proceedings of the 15th International Workshop on ECR Ion Sources*, University of Jyvaskyla, Jyvaskyla, Finland, p.80 (2002)
6. Vondrasek, R. C. et al., *Rev. Sci. Instrum.* **73**, 548-551 (2002)

Status Report of ECRIS at KVI

J.P.M. Beijers, I. Formanoy, H.R. Kremers, J. Mulder,

J. Sijbring, S. Brandenburg

Kernfysisch Versneller Instituut (KVI), Zernikelaan 25, 9747 AA Groningen
The Netherlands

Abstract. Recent ECRIS development at KVI is motivated by a new experimental program, called TRIμP, that aims to study fundamental symmetries and interactions by very precise measurements of decaying short-lived radioactive nuclides trapped in ion or atom traps. ECRIS produced heavy-ion beams are accelerated by the AGOR cyclotron and used as driver beams to produce the short-lived radionuclei of interest. This new experimental program puts new demands on the injector ECRIS, i.e. a variety of metal-ion beams and a significant intensity upgrade up to two-orders of magnitude. This report briefly summarizes what has been achieved up to now and what will be done in the near future to meet the challenges.

INTRODUCTION

At KVI two CAPRICE-type ECRIS's are routinely in use, one dedicated for atomic physics research and the other used as heavy-ion injector for the AGOR cyclotron. Both ion sources are basically identical, equipped with room-temperature solenoids and NdFeB permanent hexapoles. The 14.5 GHz radiofrequency power is injected via a coaxial waveguide. The stainless steel plasma chambers have a thin double-wall in order to water-cool the magnetic hexapoles. A more detailed description of the present KVI ECRIS's can be found in Ref. [1]. Before the TRIμP program the ECRIS effort was mainly focussed on the production of gaseous ion beams of moderate intensities. Gas mixing is used to enhance the production of the higher charge state ions, but up to now there was no need to employ additional standard techniques such as a biased-disk or wall-coating.

However, since 2002 a new, long-range research program called TRIμP has started at KVI, which requires a significant change of operation and intensity upgrade of the injector ECRIS. The TRIμP program aims to study physics beyond the Standard Model by performing very accurate and precise measurements on decaying short-lived nuclei trapped in ion or atom traps. Radioactive nuclei of interest, e.g. ^{21}Na or ^{213}Ra, are produced and separated with the ISOL technique using heavy-ion beams in the energy-range between 6 and 50 MeV/amu as driver beams. Following the ion separator are decelerating and trapping stations where the final measurements will be made. Ref. [2] gives a more detailed description of the new TRIμP facility.

The TRIμP program has a major impact on ECRIS operation at KVI: i) The beam menu of AGOR will change from a light-ion dominated to a heavy-ion dominated menu and thus to a heavier demand on ECRIS operation and availability. ii) The TRIμP program requires various metal-ion beams as driver beams in order to produce specific radionuclides of interest, e.g. a Mg beam for ^{21}Na or a Pb beam for ^{213}Ra. And iii) to produce the short-lived radionuclides at a sufficient rate a significant increase of extracted beam currents is mandatory. We aim for an intensity increase of up to two orders of magnitude compared to the performance of the present CAPRICE-type ECRIS. In order to meet these demands we are modifying the present ECRIS according to the AECR design as pioneered by the Berkeley group [3]. The new Al plasma chamber is being built by the ECRIS group of the university of Jyväskylä and is a close copy of theirs [4]. The ECRIS upgrade is discussed in more detail in a companion paper of these proceedings [5].

Some metal-ion beams have already been produced and delivered to the TRIμP facility with the current source using micro-ovens. These developments will be addressed in more detail below.

METAL-ION PRODUCTION

Many groups have developed a variety of techniques for the evaporation of metals in an ECRIS to produce metal-ion beams, see [6] and references therein. The preferred method depends on the physical and chemical properties of the specific metal and the requested charge state. In view of the first beams to be delivered to the TRIμP facility, i.e. ^{24}Mg^{6+} and ^{208}Pb^{27+} beams, we have chosen to use a micro-oven for the evaporation of these metals. Initial tests have been done with a GANIL-type micro-oven [7], which is shown in Fig. 1. The original oven is made from tantalum and has an outer diameter of 5 mm and length 50 mm. The small alumina crucible has an inner diameter of only 1.5 mm and length 31 mm, it is electrically heated with a tungsten filament. Crucible and filament are thermally and electrically insulated from the tantalum housing by an alumina shell. The oven assembly is mounted on a long hollow copper tube which is inserted into the ECRIS through the hollow inner conductor of the coaxial RF waveguide.

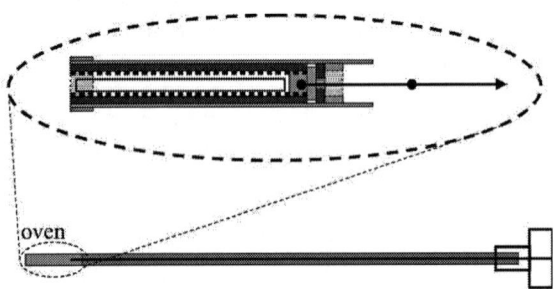

FIGURE 1. GANIL-type micro-oven for the KVI-ECRIS.

As an initial test we measured the maximum filament temperatures as a function of the filament current with an optical pyrometer. The results are shown in Fig. 2, and it can be seen that maximum temperatures up to 1500 °C can be reached. The experimental values could be reproduced with a simple

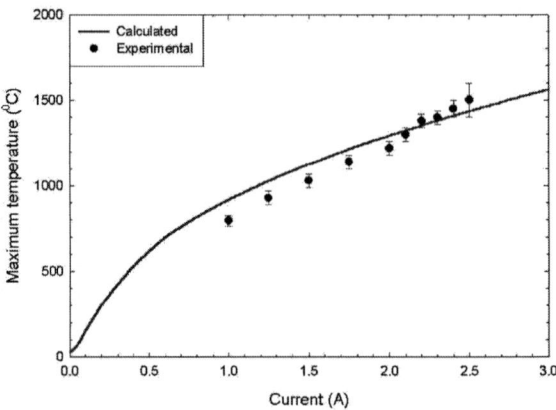

FIGURE 2. Maximum filament temperature as a function of the electrical current through the filament.

one-dimensional model of the heat transport through the filament taking into account conduction and radiation:

$$\frac{d^2T}{dx^2} - \frac{Kp}{\lambda A}\left(T^\omega - T_0^{\omega-\varepsilon}T^\varepsilon\right) + \frac{I^2}{\lambda \sigma A^2} = 0 \quad (1)$$

Here, x measures the distance along the filament, I is the current through the filament, λ the heat and σ the electrical conductivity, A and p the cross section and circumference of the filament, respectively. The radiative heat loss for a tungsten filament is taken from Langmuir and Taylor with the parameters $\omega = 5.332$, $\varepsilon = 0.87$ and $^{10}\log K = -16.2895$ [8]. The temperature dependency of both λ and σ is also taken into account. Eq. (1) is numerically solved and the maximum temperature, which occurs in the middle of the filament, is plotted in Fig. 2.

The micro-oven was used to produce ^{208}Pb^{27+}- and ^{24}Mg^{10+}-beams, with maximum extracted beam currents of 0.25 and 0.8 eμA respectively. A typical charge-state distribution for the Pb-beam is shown in Fig. 3. Although these initial results were encouraging, we had difficulties with the stability of the Mg-beam. Also, the time between oven refills during the Mg-runs was rather short, i.e. between 2 and 3 days. One reason for this is that we did not have a Ta heat shield in the plasma chamber, so that some fraction of the Mg was lost to the wall. After these first experiments with the GANIL-type oven we changed to the bigger GSI-type oven. This oven has an outer diameter of 14 mm and length 62 mm, a schematic drawing is shown in Fig. 4 [9]. The hollow inner conductor of the RF waveguide had to be enlarged to a diameter of 16 mm in order to fit this larger oven. We have had two successful runs with the larger oven, one with ^{24}Mg^{6+} and the .

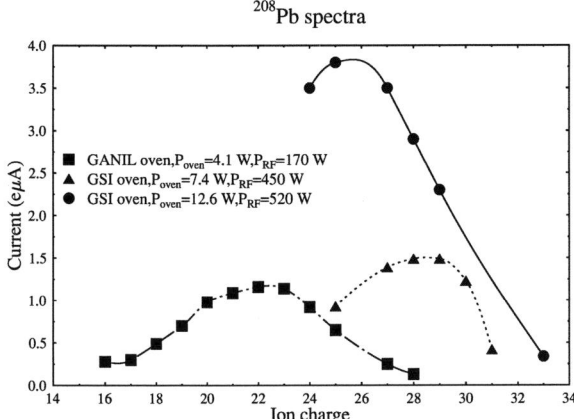

FIGURE 3. Charge-state distribution for ^{208}Pb. The extraction hole diameter for the GANIL-oven was 3.8 mm and for the GSI-oven 10 mm.

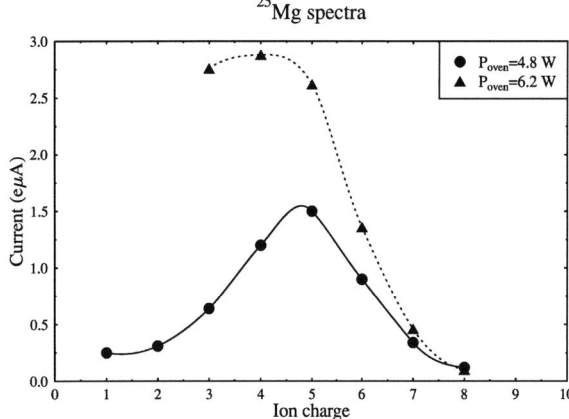

FIGURE 5. Charge-state distribution for ^{25}Mg obtained with the GSI-type oven. The extraction hole diameter was 10 mm.

other with ^{208}Pb^{27+}, both with regarding to intensity and stability. The maximum extracted beam current for the ^{24}Mg^{6+} beam was 1.4 eµA and for the ^{208}Pb^{27+} beam 3.5 eµA. The charge state distributions produced with the GSI-type oven are shown in Fig. 3 for the ^{208}Pb beam and in Fig. 5 for the ^{25}Mg beam, respectively.

injected microwaves and prevents Mg atoms from condensing on the cold plasma chamber wall. The Ta heat shield together with the GSI-type oven resulted in a very stable operation during the Mg run.

SUMMARY AND OUTLOOK

The new TRIµP program at KVI requires a much heavier demand on ECRIS operation both regarding availability, the development of new beams (in particular metal-ion beams) and a significant increase of delivered beam intensities on target. In this paper we have mainly focused on our efforts to produce metal-ion beams using the oven technique. In order to increase the extracted beam intensities we are upgrading the present CAPRICE-type ECRIS according to the AECR design. This is discussed in more detail in ref. [5].

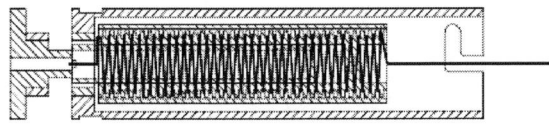

FIGURE 4. GSI-type micro-oven for the KVI-ECRIS.

In order to obtain a good long term stability with the Mg-beam we installed a thermally isolated tantalum heat shield in the plasma chamber following a GSI design. Our version has stainless steel endcaps, which are much easier to fabricate. The Ta shield is heated by the

ACKNOWLEDGMENTS

This work is part of the research programme of the "Stichting voor Fundamenteel Onderzoek der Materie" (FOM) with financial support from the "Nederlandse Organisatie voor Wetenschappelijk Onderzoek" (ZWO). It is supported by the Rijksuniversiteit Groningen (RuG) and the European Union through the Large-scale Facility program LIFE under contract number ERBFMGE-CT98-0125. The authors also gratefully acknowledge support and help from the H. Koivisto, K. Ranttila and J. Ärje from the university of Jyväskylä, R. Lang and K. Tinschert from GSI, R. Trassl from the university of Giessen and R. Vondrasek from Argonne National Laboratory.

REFERENCES

1. Drentje, A.G. et al., *Proc. of the 14th International Workshop on ECR Sources,* CERN – PS Division, 1999, pp. 90 -93.
2. Berg, G. P. et al., *Nucl. Instr. Meth. in Phys. Res. B* **204.**, 532-535 (2003).
3. Xie, Z. Q. and Lyneis, C. M., *Rev. Sci. Instrum.* **65**, 2947 (1994).
4. Koivisto, H. et al., *Nucl. Instr. Meth. in Phys.Res. B* **174**, 379 – 384 (2001).
5. Kremers, H.R. et al., ***these proceedings***.
6. Geller, R., *Electron Cyclotron Resonance Ion Sources and ECR plasmas*, IOP Publishing Ltd., London 1996.
7. Bourgarel, M.P. et al., *Rev. Sci. Instrum.* **63**, 2854 (1992).
8. Langmuir, I. and Taylor, J.B., *Phys. Rev.* **50**, 68 – 87 (1936).
9. Lang R., Bossler J., Schulte H., and Tinschert K., *Rev. Sci. Instrum.* **71**, 651 (2000).

II BEAM TRANSPORT

High current beam transport with Phoenix 28 GHz: experiment and simulation

T. Thuillier, J.C. Curdy, T. Lamy, A. Lachaize, A. Ponton, P. Sole, P. Sortais and J.L. Vieux-Rochaz

Laboratoire de Physique Subatomique et de Cosmologie, Grenoble, France

Abstract. High intensity emittance measurements achieved with Phoenix 28 GHz are presented for oxygen and lead beams. Discussion is proposed on the basis of the results of a simulation program.

INTRODUCTION

High intensity beam transport experiments have been performed with the PHOENIX 28 GHz Ion Source. This study became possible thanks to the use of a new sweep scanner emittancemeter developed at LPSC on the basis of the TRIUMF Vancouver one. We will describe the beam line and the new emittance meters, then the results of experiments done with oxygen and lead will be discussed. After the presentation of the simulation program, a tentative to fit the experimental data with the simulation will be proposed.

BEAM LINE OVERVIEW

The high current test bench of LPSC is equipped with the Phoenix 28GHz ECRIS [1], able to work up to a 60 kV extraction voltage. This test bench is sketched on Figure 1 for convenience. The ions are accelerated by means of a single gap (adjustable from 1 to 6 cm) between high voltage and ground. A Glaser lens placed at 450 mm from the source extraction hole, can deliver up to 1.4 T on axis, with a full width half-maximum ~120 mm. A 100 mm gap, 90° bending magnet, with a 550 mm radius is located 700 mm after the Glaser lens exit. The highest magnetic field achievable in the dipole is 0.33 T. The beam measurements are realized at the beam waist, 700 mm away from the dipole exit.

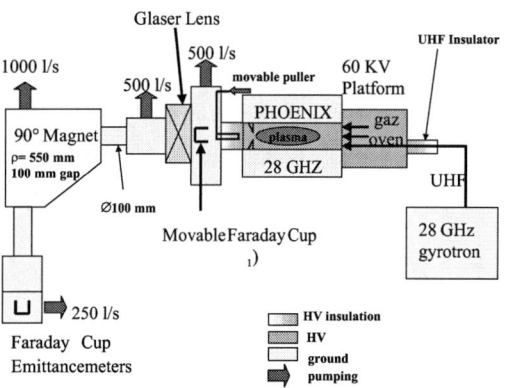

FIGURE 1. Overview of the LPSC high current test bench.

NEW EMITTTANCE SCANNERS

Two new compact emittance scanners have been successfully developed and installed on our high current test bench. These are Allison type emittance meters developed so far [2]. Their main specificity is a water cooled shielding to safely measure up to 500 Watt beams. A drawing to recall its basic principle is proposed on figure 2. An ion enters with an angle X' through a slit located at position X. An electric field E perpendicular to the axis of the emittance meter, established between two parallel plates set to a bipolar potential, permits to bring the ion to a second slit before its collection into a Faraday cup. The secondary electrons from the Faraday cup are reinjected by a repeller ring. The X' values are expressed as a function of E.

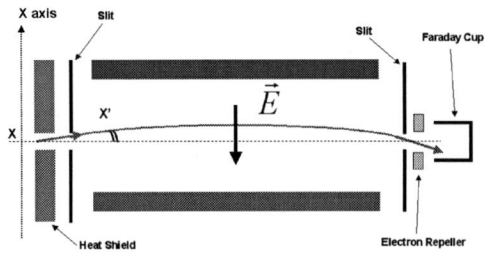

FIGURE 2. Emittance meter sketch.

The gap between the deflection plates is 3.5 mm, the effective length of the electric field is 64 mm allowing to study up to a 100 mrad divergence at a 60 kV ion extraction voltage. The overall dimension is 95 mm length, 59 mm width and 43 mm height. It can be used in CW or in pulsed mode enabling afterglow emittance measurements. The time to measure an emittance with 1000 phase space points is about 3 minutes. Improvements are under study to decrease this time down to one minute [3]. Below, any reference to emittance implicitly refers to a weighted root mean square (RMS) emittance calculated for one standard deviation:

$$\varepsilon = \sum_{i=1}^{N} w_i (X_i - X_m)(X'_i - X'_m) \quad (1)$$

where I_i is the current, measured in the emittance scanner faraday cup for phase position (X_i, X'_i); the weights are $w_i = I_i/\Sigma I_i$, X_m and X'_m are respectively the weighted mean position and divergence. The Normalized RMS emittance (NRMSE) is finally obtained by the multiplication of ε with the Lorentz factors: $\varepsilon_N = \varepsilon.\beta.\gamma$.

HIGH INTENSITY CW OXYGEN BEAM

This work has been performed in the framework of the SPIRAL II preliminary project study. The requirement is 1 mAe CW beam of A/Q=3 (mass number over charge state), with a NRMSE $\varepsilon_N <$ 0.4 π.mm.mrad at an extraction voltage of 60 kV. As a first step, a production of 1 mAe O^{6+} beam was expected. This intensity having already been extracted from ECRIS, the challenge was to manage the high extraction voltage and the emittance measurements for this 60 W beam. The 1 mAe O^{6+} beam has been successfully obtained with Phoenix 28 GHz both in CW and pulsed mode. The normalized measured emittances are reported in figure 3. The horizontal (H) NRMSE is 0.18 π.mm.mrad while the vertical one (V) is 0.26 π.mm.mrad. These results are promising since they fulfill the Spiral II preliminary project requirements.

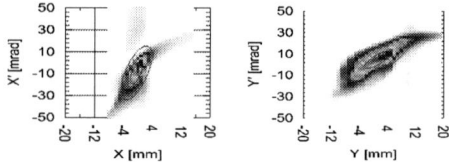

FIGURE 3. H (left) and V (right) NRMSE for the 1mAe O^{6+} beam extracted at 60 kV from Phoenix 28 Ghz.

The CW beam spectrum of oxygen analyzed in the faraday cup (FC) located after the bending magnet is shown Figure 4. Gaz mixing technique with Helium was used, its contribution is clearly seen on the A/Q= 4 peak. The drain current of the high voltage supply was 7.4 mAe. The different contributions to this current are the following: the Penning discharge (0.7 mAe), the plasma chamber cooling water resistive loss towards ground (0.3 mAe), the ionic current and the secondary electronic emission from the puller (6.4 mAe). The sum of the individual ionic currents in the analyzed spectrum reaches 5.5 mAe, so the global beam transport efficiency is at least ~85 %. This result was obtained with a very stable operation of the source at 1.8 kW of UHF power injected.

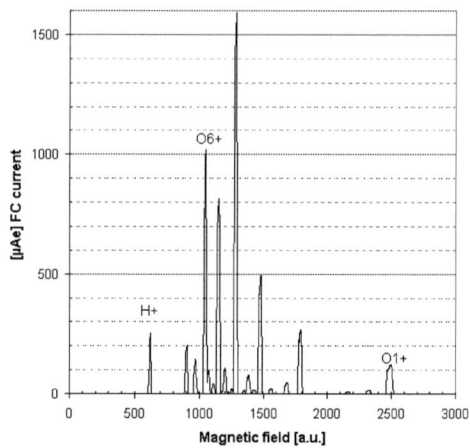

FIGURE 4. High current Oxygen spectrum analyzed after the bending magnet.

Under the conditions of the spectrum plotted in figure 4, a systematic emittance measurement has been performed for all the ionic species extracted from the source. These measurements are summarized figure 5.

An additional result for a lower total ionic current extracted at the same potential (2.2 mAe@60 kV) is represented on the same plot. In such a situation 0.45 mAe O^{6+} is extracted. It is interesting to note that the oxygen NRMSE increases with the total ionic current, this fact illustrates the emittance magnification due to the space charge effect. The two oxygen plots show that the NRMSE decreases for increasing charge states (> 4), independently from the intensity value. Another interesting point is the emittance increase with the mass decrease for a given charge state which is clearly illustrated by the He and H emittances values.

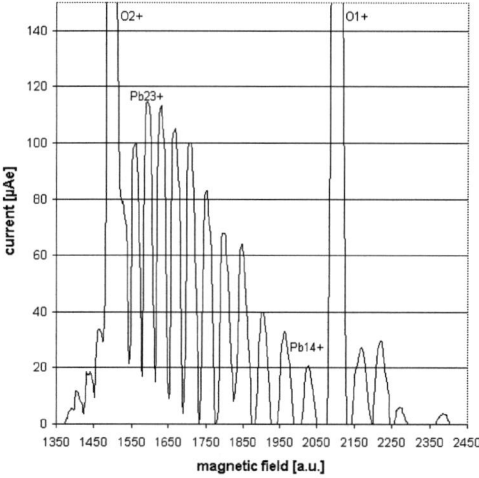

FIGURE 6. CW Lead spectrum analysed after the bending magnet.

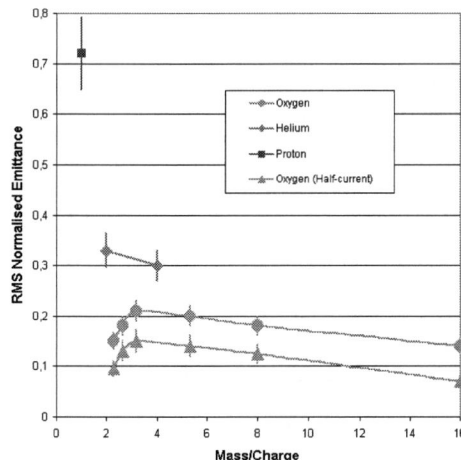

FIGURE 5. Normalized Horizontal RMS emittance evolution with the A/Q ratio. The triangles points have been measured for ~0.45 mAe of O^{6+} the other ones for 1 mAe.

HIGH INTENSITY CW LEAD BEAM

The production of high intensity metallic ion beams is of interest for accelerators. An oven has been installed on Phoenix to investigate CW production of lead at 28 GHz ECR frequency. The buffer gaz is oxygen. Former experiments in pulsed mode showed stable afterglows up to 500 µAe of Pb^{24+} [1]. Here the 100 µAe was easily reached for the 24+ charge state with a UHF power of 1.5 kW. See figure 6 for a typical spectrum of Lead extracted at 40 kV. The same level of current was achieved at lower voltage (25 kV), but the low resolution of our bending magnet could not discriminate charge states as properly as at 40 kV. The beam stability was very impressive and remained unchanged for hours. This high metallic current tends to demonstrate a very good atomic vapour capture by the plasma heated with the 28 Ghz microwave, certainly thanks to the highest electronic density available with respect to standard 14 Ghz ECR heating. Thus, a lead atom as ~4 times more chance to be first ionized while crossing the ECRIS plasma.

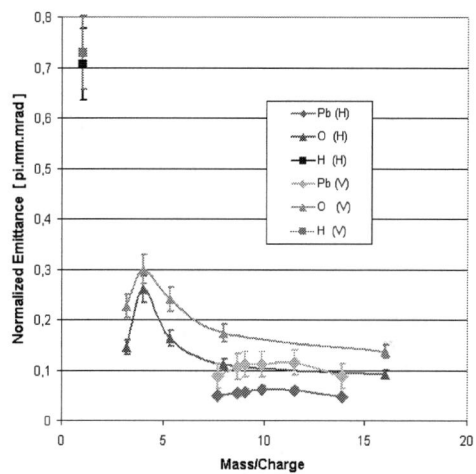

FIGURE 7. Horizontal and vertical Normalized RMS Emittance of the ionic species included in the Lead beam.

Like for the previously presented Oxygen experiment, systematic emittance measurement have been carried out. The H and V NRMS emittances are displayed in figure 7. A systematic higher V NRMSE with respect to H NRMSE is clearly visible for oxygen and lead. This phenomenon is discussed later. Once again, the emittances are better for heavy ions : lead emittances for charge state from 10 to 27 remain below 0.1 π.mm.mrad. The O^{5+} emittance decrease is coherent with the one seen in figure 5. The H and V lead 27+ NRMS emittances are displayed in figure 8.

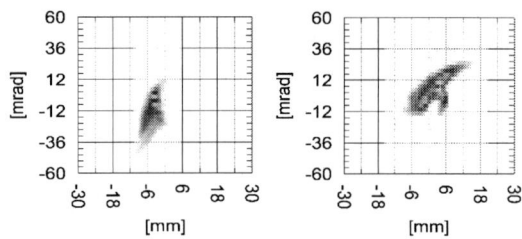

FIGURE 8. Horizontal (left) and Vertical (right) Pb^{27+} NRMS emittance measured at 40 kV. Their values are respectively 0.05 and 0.08 π.mm.mrad.

SIMULATION CODE

The code used to simulate the beams has been originally written by Xie and Antaya [4, 5], next developed at GANIL[6] and finally renewed at LPSC. It is written in fortran 77 and works on a linux server at LPSC. The optical elements of the beam line have been modelised with Poisson Superfish [7] and Radia [8]. The program propagates macro particles taking into account space charge effects. The number of ionic species to transport is a free parameter. Ions are generated at their initial low thermal energy. The ion spatial distribution is a free parameter and also their initial velocity dispersion. The propagation of macro particles is done with a simple, but fast, second order Runge-Kutta algorithm to solve the movement equations. The space charge effect is modelised using the simple Gauss theorem assuming an infinite beam pipe length. The beam formation at the exit of the plasma electrode is a crucial point. If one tries to accelerate ions assuming the full space charge effect in this area, one will realize that no beam can be extracted in this case (the beam explodes in any situation). So it means there is an active space charge compensation in this early stage. A simple model takes into account the presence of hot electrons in the first millimeters of the acceleration gap. These electrons screen space charge effects and allow beam formation. Hence, in the acceleration area, the neutralization model is :

$$f(z) = (1 - e^{\frac{-\Delta U}{kT}}), \forall z \leq z_E$$

where $\Delta U = U(0) - U(z)$ is the potential difference between the ECRIS plasma electrode (z=0) and the local position z (along the beam line axis). z_E is the location where $U(z_E)=0$ Volt, while kT is the electron mean temperature in the plasma. Next, after the ion acceleration area, the CW beam is in equilibrium with cold electrons provided by the beam pipe residual gaz. In this area, the neutralization function is simply :

$$f(z) = 1 - F, \forall z > z_E \quad (3)$$

where F is the mean beam rate of neutralization in the pipe, bounded between 0 and 1. The function $f(z)$ is then multiplied by the electrical space charge field before being applied to macro particles. The space charge effect is taken into account up to the entrance of the bending magnet where no specific model for space charge compensation is presently available in the simulation. Experiments performed on our test bench show that the space charge effect becomes negligible at the exit of the Glaser lens. This optics defocuses unwanted beams and thus minimizes their space charge effects on the ion beam selected through the bending magnet.

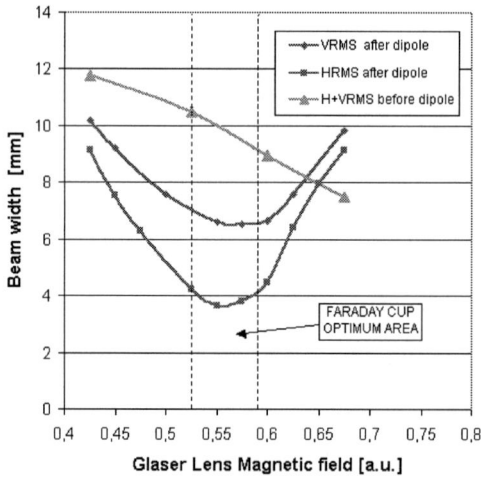

Figure 9. Simulated H and V beam width before and after the dipole for O^{5+} as a function of the Glaser lens magnetic field.

DISCUSSION

The difference of the emittance values between the horizontal and vertical planes has been observed in both oxygen and lead experiments (see figure 3 and 7) It appears to be a systematic effect. This difference has been investigated with the simulation program. Figure 9 shows the beam width evolution of O^{5+} as a function of the magnetic field in the Glaser lens. Before the dipole (triangle points) the spatial extension of the beam remains the same for the two planes H and V. This is normal since the beam optics has a symmetry of revolution. As expected, one can see that the beam size decreases when the magnetic field increases. The two other curves are the horizontal and vertical beam sizes after the bending magnet. It can be seen that the beam size goes down to a minimum for a given magnetic field of the Glaser lens. It is around this minimum that the beam is fully measured in the FC after the dipole. The magnetic field range where the size of FC is greater than the beam dimensions is delimited by dots lines. The emittance measurements are performed with such a Glaser lens tuning. The evolution of the H and V NRMS emittances with the Glaser magnetic field intensity is presented in figure 10. The FC measurement range has been reported for convenience. One can see that the simulation predicts that the V emittance is higher than the H one. The explanation of this result is as follow: when the beam is large at the magnet entrance (low magnetic field for H+VRMS in figure 9), it undergoes subsequent non linear effects in the bending magnet that mainly magnifies the V emittance. If the magnetic field of the Glaser lens is increased, the beam at the dipole entrance becomes smaller and convergent but tends to slightly diverge at its exit. Then the H and V emittances are low and have the same value, but the beam is not focused in the FC plane. This effect has been confirmed experimentally by measuring the NRMS emittance of a 1mAe O^{6+} beam at 0.2 π.mm.mrad in both H and V planes in the case of a higher magnetic field Glaser lens tuning.

strong constraint on the simulation.

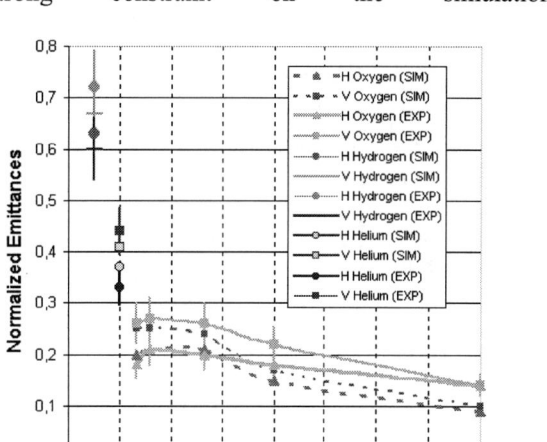

Figure 11. Comparison of simulated emittances with experimental values for the Oxygen beam tuning presented earlier.

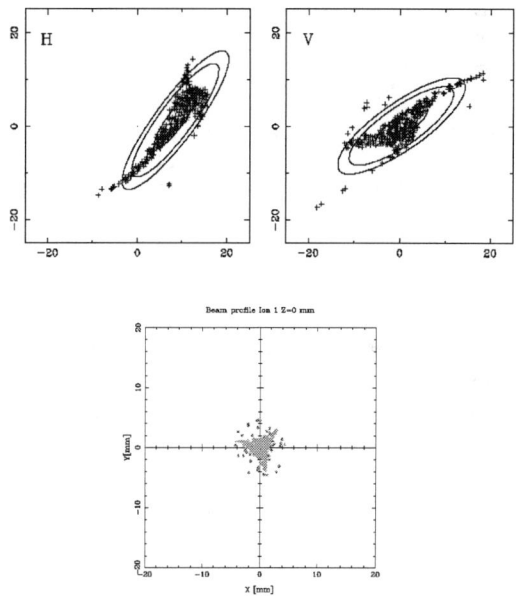

Figure 12. H and V simulated Lead 27+ emittance (TOP) and corresponding spatial distribution of ions at the source.

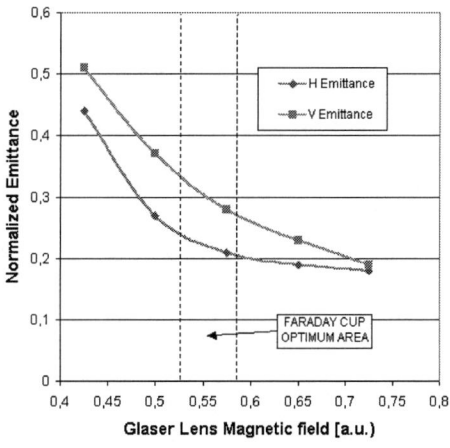

Figure 10. Evolution of H and V NRMS emittances with Glaser lens magnetic field. Simulation result.

The oxygen beams presented above have been simulated. The simulation fit consists in finding the input parameters that best reproduce all experimental emittances at the same time. This condition forms a

The results of the fit are summarized in figure 11. The initial ion temperature of the plasma is kT_i=5 eV and the plasma potential is fixed to 50 V. The neutralization factor as defined above is F=70% and the electron temperature is 2.5 keV. Low charge state ions (Z≤4) have a uniform spatial distribution over the whole plasma electrode surface (hole diameter = 10 mm in our experiment). Higher charge states (O^{5+}, O^{6+}, O^{7+}), which have a smaller experimental emittance (see figure 5), are distributed over smaller surfaces. The best fit is obtained when ions are extracted on a "plasma star" like surface. The mean extraction radii for O^{5+}, O^{6+} was found to be ~2.5 and 2 mm. Say differently, this result once again points out that the higher the charge state is, the closer the species are extracted from the plasma center [9].

At this time, only preliminary results are available from Lead experiment. On figure 8, a sort of triangle shape is visible on lead 27+ emittance, that recalls the plasma star topology. This point has been investigated with the simulation. Simulation shows that this aberration can only be seen after the bending magnet when the space charge effect is small. A large space charge tends to mix the ion trajectories extracted from different regions and then destroys the information about the initial spatial ion distribution at the source. If one assumes a fully neutralized beam, the size of the initial beam that fits data is only ~2-4 mm (see figure 12). In this case, the simulated emittances on figure 12 have the same kind of structure as in figure 8. Further developments are under study.

ACKNOWLEDGMENTS

The authors would like to warmly thank the ECRIS team from Vancouver who gave us crucial information to dare to develop a home made sweep scanner emittancemeter.

REFERENCES

1. T. Thuillier, J.-L. Bouly, J.C. Curdy, T. Lamy, P. Sole, P. Sortais, J.-L. Vieux-Rochaz, Proc. 8th European Particle Accelerator Conference, 3-7 June 2002, Paris, France.

2. P.W. Allison J. D. Sherman and D. B. Holtkamp, IEEE Trans. Nuc. Sci., Vol. NS-30, No. 4, 2204 (1983).

3. T. W. Debiak, Y. Ng, J. Sredniawski and W. Stasi , Proc. 18th Int. Linac Conf., Geneva, Switzerland, August 26-30, 1996.

4. T.A. Antaya and Z.Q. Xie, NSCL Report MSUCP-63, 1987.

5. T.A. Antaya, Inter. Workshop on non linear problems in accelerator physics, Berlin (Germany), April 1992.

6. P. Sortais, L. Maunory, A.C.C. Villary, R. Leroy, J. Mandin, J.Y. Pacquet and E. Robert, Proc. 13th Int. Workshop on ECRIS, 26-28 February 1997, Texas, U.S.A.

7. Poisson SuperFish software, Los Alamos National Laboratory, LAACG, http://laacg1.lanl.gov/laacg/services/possup.html

8. P. Elleaume, O. Chubar and J. Chavanne, presented at the PAC97 Conf. May 1997, Vancouver.

9. M.Cavenago, O.Kester, T.Lamy,and P.Sortais, Rev. Sci. Instrum. **73**, 537 (2002)

USE OF SIMULATIONS BASED ON EXPERIMENTAL DATA

P. Spädtke, K. Tinschert, R. Lang, R. Iannucci

GSI Darmstadt, Germany

Abstract. Compared to the simulation of classical high perveance extraction systems for high current ion sources, the extraction of typical ECRIS is more complicated because more parameters, partially unknown, are involved. To reduce the computational effort, it would be helpful to determine several of these unknown parameters affecting mainly the starting conditions for trajectories in space and in momentum. The influence of the magnetic field on the beam in the extraction region is studied. Here the magnetic flux density distribution is fixed, but the starting conditions of all particles can be modified within reasonable limits in the simulation to investigate this influence.

The GSI CAPRICE ECRIS was equipped with a movable accel-decel extraction system in order to investigate the influence of electric field gradient and space charge compensation as well as further effects of ion extraction. First results will be presented and will be compared with computer simulations. For the most simple case of Helium operation, simulation shows good agreement with experimental results, indicating the correctness of the applied model.

Introduction

To increase the extracted ion beam current the plasma generator must deliver the required particle density and the extractor must handle the space charge forces. To improve the design of the extraction system simulation codes can be used to decrease the necessary experimental effort. Here we tried to investigate the extraction system by experiment and simulation simultaneously. All experiments were made at the ECR Injector test setup (EIS) test bench, described in [1], for the simulation the 3D code KOBRA3-INP [2] was used.

Influence of the Hexapole

The plasma of an Electron Cyclotron Resonance Ion Source (ECRIS) for highly charged ions is confined in a magnetic structure defined by a solenoidal mirror field superimposed by a radial multipole cusp field. In most of the existing ECRIS a hexapole is used for the radial confinement. The traces on the plasma chamber and on the electrode surface (see Fig. 1-3), indicating the plasma loss areas have been observed by all ECRIS users, but this information was not used in the simulation of extraction, except in [3]. The reason for that might be the usage of 2D-simulation codes, where this information is neglected anyhow. The shape of the loss areas should be used as a relative measure for the plasma density distribution. Together with the total extracted current, important information about the start distribution in space of ions can be derived.

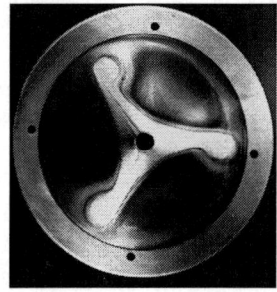
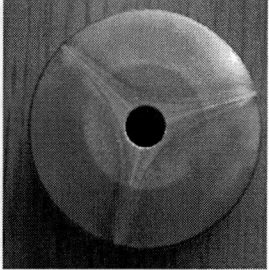

FIGURE 1. Sputter mark on the plasma electrode for first generation ECRIS and modern ECRIS. Left: 10 GHz ECR, ORNL [4]; Right: 18 GHz pure permanent magnet ECRIS, CEA [5].

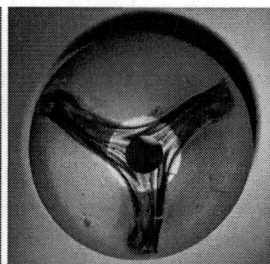

FIGURE 2. Left: sputter mark on the plasma electrode if the magnetic field was shielded 10 GHz CAPRICE, CEA [6]. Right: 14 GHz CAPRICE ECRIS, superposition of different source settings at different beam runs at GSI.

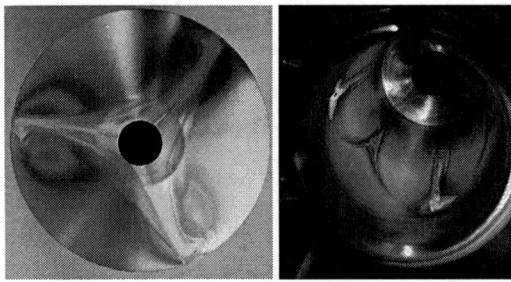

FIGURE 3. Noble gas operation with Ne and Xe at GSI. Left: extraction electrode, right: plasma chamber.

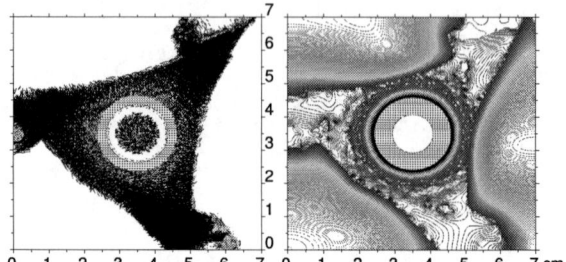

FIGURE 5. Left: trajectory distribution in the plane of extraction. The extracted beam is already separated from the ions which are lost to the plasma electrode and to the plasma chamber. Right: potential distribution in a plane perpendicular to beam extraction, close to the position of the extraction electrode, but still within the plasma. 100 potential lines between plasma electrode potential and the plasma potential itself are plotted.

The specific shape of the pattern shown in Fig. 1, left, Fig. 2, right, and Fig. 3, left, can be reproduced in the simulation if the stray field of the hexapole is included, which becomes relevant at larger radii (≈ 30 mm).

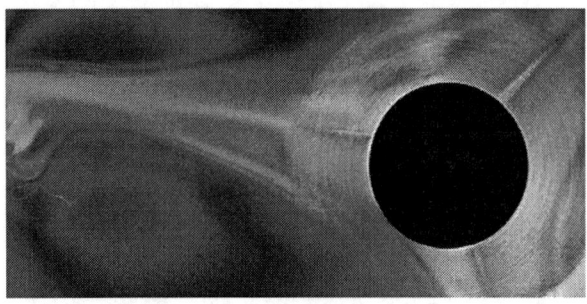

FIGURE 4. In a zoomed part of the plasma electrode a very thin groove can be seen. This groove seems to be a straight line in contrast to the full sputter mark, which shows an azimuthal deviation from this straight line.

Therefore, the first computer experiment was to start ions from a cylindrical volume defined by the diameter of the plasma chamber and a thickness of 10 mm located directly behind the extraction electrode inside the plasma chamber. The trajectories of these particles are mainly determined by the magnetic field because the plasma potential is assumed to be small and constant. The result of ray tracing shown in Fig. 5 explains the sputter marks on the plasma electrode as shown in Fig. 3, the thin groove in Fig. 4 is caused by ions generated close to the axis with specific starting conditions, derived below. The volume from which ions are able to reach the plasma electrode to be extracted can be used as an upper estimation of the real emitting area for the simulation. The problem remains how to describe this area analytically.

Assuming that the plasma is confined in a minimum B configuration, it is reasonable to start ions in the simulation only if they are launched at such a location where the magnetic flux density is less than a specific value B_{pl}, which is assumed to be the flux density separating plasma from vacuum.

Whereas at the minimum value of B (presumably on axis) the highest plasma density should be located, this density goes to zero for $B \geq B_{pl}$ in between the loss areas.

The correct B_{pl}-value will depend on the magnetic field distribution within the extraction region as well as on plasma heating parameters like rf power, electron density, and gas flux, which also control the charge state distribution described by a set of Saha equations. Different ion source operating conditions can be simulated by different B_{pl}-values, see Fig. 6. The resulting start distribution must be compared with the shape of the traces on the extraction electrode.

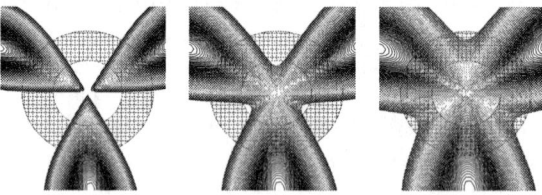

FIGURE 6. Different emission areas depending on source operating conditions in the plane of extraction with an aperture diameter of 10 mm (inner circle), using the absolute value of the magnetic flux density. From left to right increasing B_{pl}: 0.8 T, 0.9 T, and 1.0 T The minimum value is 0.55 T

The best correlation between the experimentally observed shape and theoretical assumptions was found using the radial component B_{rad} of the magnetic flux density only, see Fig. 7. The physical explanation is that the radial confinement is more important for the density distribution than the longitudinal confinement at that location. The flux density component in longitudinal direction might be used for the longitudinal distribution of the ions. Because of the strong influence of the magnetic field on the slow ions within the plasma, starting all ions from a single plane would limit the solution.

The influence of the hexapole on the extracted beam is negligible for the ray tracing, but it determines by us-

  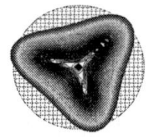

FIGURE 7. Different emission areas depending on source operating conditions in the plane of extraction with an aperture diameter of 10 mm, using the absolute value of the radial magnetic flux density component. From left to right increasing B_{pl}: 0.01 T, 0.033 T, and 0.1 T

ing a specific B_{pl} the start distribution of the ions, and this effect cannot be neglected. This was proven with the following simulations: a first simulation was made with hexapolar field, another without hexapolar field. In both cases the same starting conditions for all particles were used. These starting conditions were created taking the hexapole with B_{pl}=0.05 into account. In a third simulation a homogeneous start distribution was used. No difference in phase space for the first two cases is obtained. Different starting conditions however, do have a strong influence on the beam emittance, see Fig. 8.

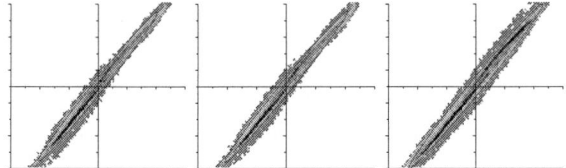

FIGURE 8. Influence of the hexapole on the emittance of the extracted beam: in the left figure ray tracing for the ions include the hexapolar field, the other two simulations were made without hexapolar field. In the left and the mid figure the starting conditions are defined with the hexapole, the right figure is assumed to be homogeneous in space. The same scale is applied to all three figures: +/- 30 mm and +/- 100 mrad. The emittance for all charge states is given 230 mm behind the plasma electrode.

The amount of ions I_{ex} which can be extracted depends on the total loss current on the plasma electrode I_{PE} and on the ratio of the current loss area at the plasma electrode A_{LA} and the size of the extraction hole A_{PE}:

$$I_{ex} = I_{PE} * MIN(1, A_{PE}/A_{LA}) \qquad (1)$$

Note that the ratio between A_{PE}/A_{LA} cannot become larger than 1, indicated by the minimum expression in Equ. 1. If the plasma loss area is increasing the plasma density in front of the extraction hole decreases, assuming that the plasma generation remains unchanged. Therefore, the extracted current becomes smaller. Reducing this loss area by a different source setting leads to an increased extracted current. As soon as A_{LA} at the extraction electrode becomes smaller than the size of the extraction hole the beam current will not increase anymore. However, if $A_{PE} \geq A_{LA}$ the beam emittance will increase due to nonlinear fields generated by the space charge of the extracted ions[3]. The higher the extracted current the more severe this effect will become. Even if the plasma generator would provide higher currents in such a mode, the disadvantages of aberrations will exceed the advantage of a higher current. In such a case the ion source would be tuned towards the correct operating mode for a good extraction: loss surface as large as the extraction hole to achieve a homogeneous distribution.

The remaining influence of the hexapole on the beam can be minimized by shielding the magnetic field component or by an active correction element. Magnetic shielding was used to modify the plasma density distribution in several ECRIS [6],[12], but no improvement was reported. The effect is shown in Fig. 2 left, where the experimental result (sputter marks on the extraction electrode) indicate a very large area with homogeneous current density. Most probably the extracted current was small because of the unfavorable ratio between the size of the extraction hole and the area of plasma loss, see Equ. 1.

A more flexible method to compensate the influence of the magnetic hexapole is promised by a correction hexapole, located directly behind the extraction. This proximity of the correction element with respect to the extraction is essential. In case of a space charge dominated drift the beam would homogenize itself due to the nonlinear fields at the expense of emittance growth. But even for the space charge compensated transport, lenses between the extraction and the location of compensation could introduce abberations which would make the original error irreversible.

The GSI CAPRICE extraction system was simulated here for 15 kV extraction voltage. The plasma is defined by the working gas Argon (4.2% Ar^{3+}, 7.3% Ar^{4+}, 11.3% Ar^{5+}, 14.1% Ar^{6+}, 16.3% Ar^{7+}, 17.2% Ar^{8+}) and Helium as auxiliary gas (27.7% He^+, 1.9% He^{2+}). The maximum magnetic flux density of the extraction side solenoid was 1.11 T on axis; the source is equipped with a hexapole with 1.0 T flux density at the pole tip. To generate the starting conditions the parameter $B_{pl} = 0.05$ T was used. For the definition of the starting velocity different models were investigated: assuming the same starting energy for all charge states, or assuming the presence of a potential drop, which leads to different energies for different charge states. In all simulations here a potential drop of $\Phi = 5V$ is assumed; the longitudinal starting energy is therefore $5 \cdot \zeta$[eV], with ζ denoting the charge state, and a transverse temperature of 1 eV. A larger rms-emittance for smaller mass to charge ratio is observed, see Fig. 9, even for similar starting conditions, but in contradiction to experimental experiences. This is a clear hint that the distribution parameter B_{pl} varies with the individual charge state.

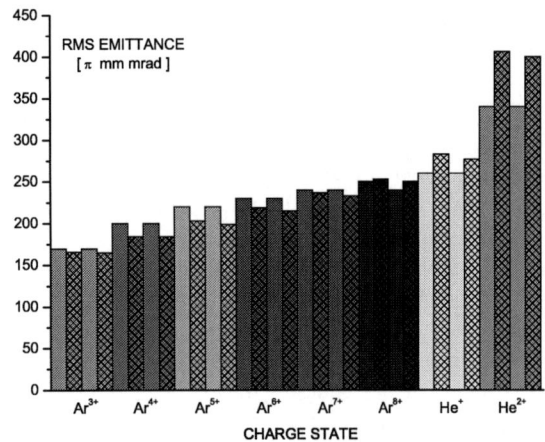

FIGURE 9. Difference in rms-emittance due to a constant potential drop (pure bars) or constant energy (hatched bars) for all charge states.

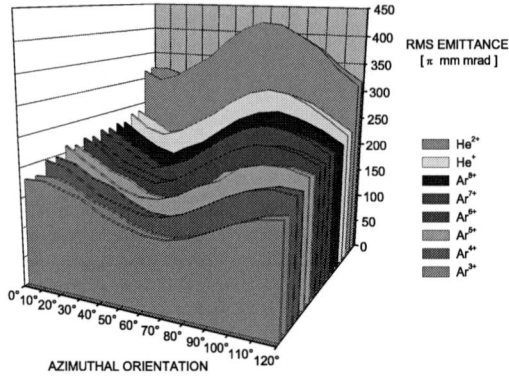

FIGURE 10. rms-emittance (both planes) for different charge states as a function of the azimuthal orientation of the correction hexapole. Magnetic flux density was chosen with 0.04 T @ 0.04 m. Constant starting energy of 1 eV for all charge states and 15 kV extraction voltage.

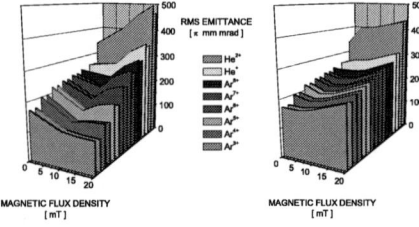

FIGURE 11. rms-emittance (both planes for each charge state) for different charge states as a function of hexapole flux density. Left: azimuthal orientation of $70°$, optimum for Ar^{3+}; Right: azimuthal orientation of $30°$, optimum for Ar^{7+}.

The correction must be applied in horizontal and vertical plane independently. An alternative for such a device, especially if available space is short, would be one hexapole, variable in flux density and in azimuthal orientation. For the simulation a hard edge model of a hexapole with up to 0.1 T flux density at the pole tip and an open aperture of 80 mm diameter and 100 mm length was assumed. The optimum for both values, flux density and azimuthal orientation differ for various mass to charge ratios, and more severely for different tuning of the source. The positive influence of such a correction element is shown in Fig. 10 and Fig. 11. Slower particles require a larger angular shift relative to the angular orientation of the hexapole of the source. One period ($120°$) was scanned in Fig. 10 with a constant hexapole flux density of 0.04 T at a radius of 40 mm. The decrease of the rms-emittance for the corrected beam can be explained by applying the correct nonlinear force to make the distribution linear again.

In Fig. 11 the rms-emittance of the Ar^{3+} charge state decreases from 170 mm mrad to 110 mm mrad with a setting of the correction hexapole of 0.08 T at 40 mm radius and an azimuthal location of $70°$, whereas for $30°$ orientation the improvement for Ar^{7+} is less pronounced. The positive action of the correction hexapole is shown for both transverse phase spaces in Fig. 12.

Another important topic not covered in detail here is the use of an electrode with a quasi-Pierce angle, promoting a flat plasma boundary to minimize radial electric fields and therewith a minimum of coupling between the strong longitudinal magnetic field and radial velocity component.

Accel-Decel Extraction and Matching

The necessity of an accel-decel extraction system, especially for higher currents and for low energies, was pronounced earlier [7], but was also demonstrated in an experiment [8] at the EIS [1] test bench at GSI, shown in Fig. 13, which was made to prepare a regularly planned beam time at the GSI accelerator [9].

The aim was to maximize the useable current in $^3He^+$. Enriched ^3He was used in that case. The linac requirement of 2.5 keV/u leads to an extraction voltage of only 7.5 kV. The extraction system was modified to an accel-decel electrode arrangement. The negatively biased screening electrode is mounted with a fixed gap width ($\approx$ 7.5 mm) to the ground electrode, see Fig. 14. Both electrodes are movable in longitudinal direction from 16 mm to 37 mm distance between the plasma electrode aperture and the front plane of the screening electrode. This adjustment can be made during operation by remote control. The effect of space charge compensation should be demonstrated in simulation based exactly on experimen-

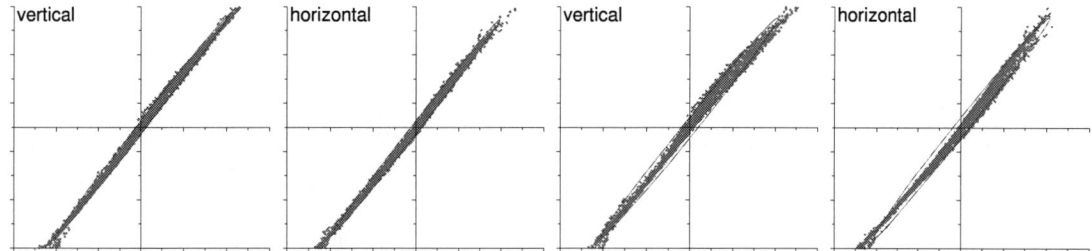

FIGURE 12. Emittance of Ar^{3+}. Left: azimuthal orientation of the hexapol 70 degree; Right: azimuthal orientation of the hexapol 120 degree. The real distribution and the rms-emittance ellipse are shown.

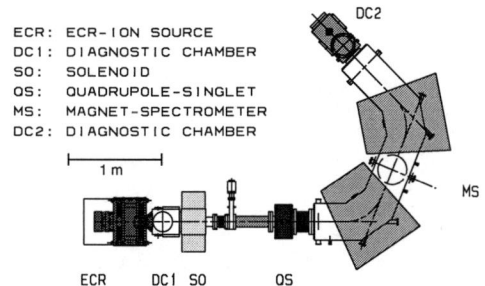

ECR: ECR-ION SOURCE
DC1: DIAGNOSTIC CHAMBER
SO: SOLENOID
QS: QUADRUPOLE-SINGLET
MS: MAGNET-SPECTROMETER
DC2: DIAGNOSTIC CHAMBER

FIGURE 13. EIS test bench.

tal data.

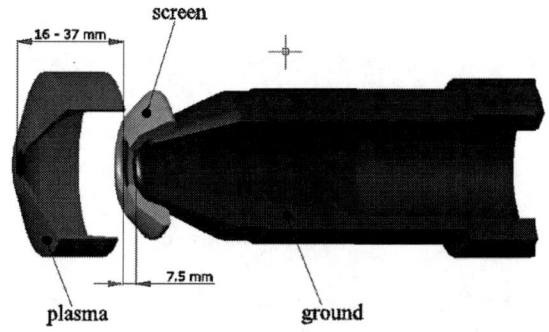

FIGURE 14. Geometry of the extraction system.

Mainly to simplify the charge state distribution, and to achieve a higher extraction current, helium is provided to the plasma only. The total extracted current was in the order of 5 mA. This information together with the assumption of a homogeneous plasma density distribution was used to obtain the first simulation result: the required plasma density for this current is so high that a convex plasma boundary is predicted by the simulation, see Fig. 15.

This is clearly not the best choice for a small emittance, but effective to increase the current within the comparatively large acceptance of our spectrometer, which provides a multiple of the linac acceptance. This result demonstrates that the source is always optimized

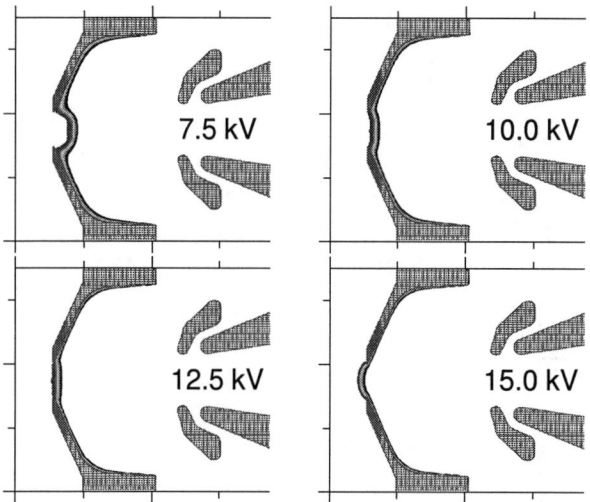

FIGURE 15. Shape of the plasma boundary for different extraction voltages and an extraction gap width of 36.5 mm, with constant source setting.

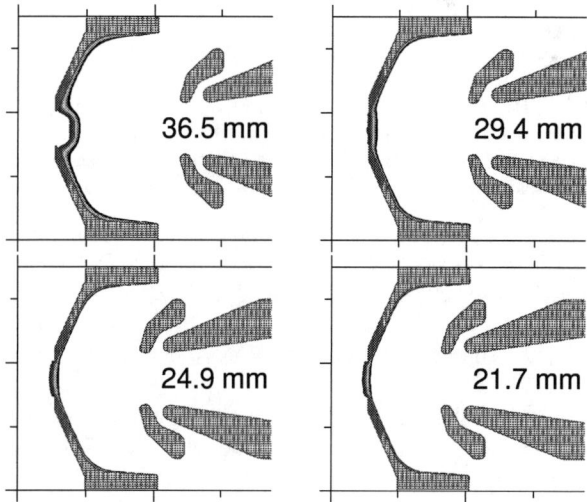

FIGURE 16. Shape of the plasma boundary for different extraction distances and an extraction voltage of 7.5 kV, with constant source setting.

with respect to the available acceptance as long as this is technically possible. When comparing different source types this fact has to be taken into account.

Simulations for different extraction voltages, shown in Fig. 15, demonstrate that the shape of the plasma boundary can be modified from convex to concave by increasing the extraction voltage. A flat plasma boundary should appear at about 12 kV. With this information the experiment was repeated, starting with the same source setting found in the previous experiment, this time with ^{4}He at both gas ports. Then the extraction voltage was increased without changing the plasma generator. The optic of the beam line was scaled according to the extraction voltage. The measured transmission shown in Fig. 17 shows the best value for 12.5 kV extraction voltage. Alternatively to the change of extraction voltage, the extraction gap width can be shortened, shown in Fig. 16. The experimental result in Fig. 17 shows the best transmission for an extraction gap width of 29.4 mm. For both cases the simulation predicts a flat plasma boundary. Beam transport for both cases is not equivalent, because of the different aspect ratio and the different beam energy.

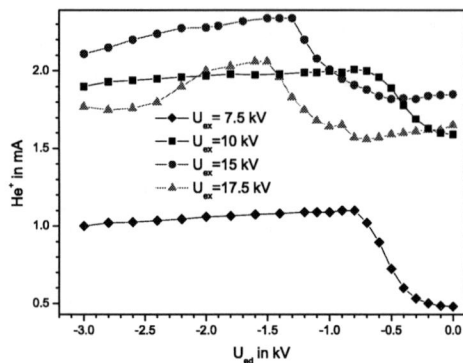

FIGURE 18. Influence of screening voltage on the extracted beam for different extraction voltages.

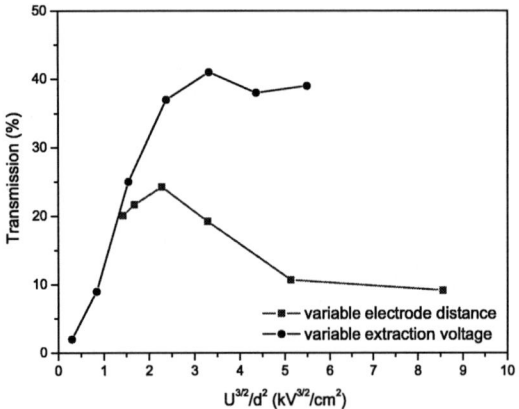

FIGURE 17. Experimental transmission at EIS as function of extraction voltage for 36.5 mm gap width and as function of gap width for an extraction voltage of 7.5 kV. The horizontal axis scales with $U_{ex}^{1.5}/d^2$.

Child's law was not to be demonstrated with that experiment, because the plasma density remained constant. Between 7.5 kV and 10 kV the analyzed current increases from 1.2 mA to 2.3 mA and stays constant for 15 kV, and it even decreases for 17.5 kV extraction voltage, see Fig. 18.

In this figure the analyzed current is plotted versus the negative screening voltage. The overall tendency of decreasing current with increasing screening voltage is due to the electro-static lens effect.

The influence of the negative screening voltage is as follows: when reaching a specific negative potential (in our case -0.5 kV, see Fig. 18), the analyzed current increases. Instead of being accelerated into the source plasma, the electrons will compensate the positive space charge created by the ions of the drifting beam.

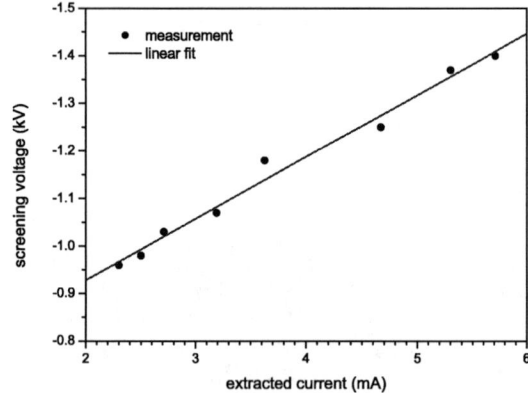

FIGURE 19. Necessary screening voltage for He as function of the total extracted current, measured for 15 kV extraction voltage and 36.5 mm gap width.

According to Fig. 19 there is a linear dependence of the minimum screening voltage for space charge compensation which can be described by the following equation for this specific geometry.

$$\Phi_{scr} = -670V - 129 \cdot I[mA]V \quad (2)$$

roughly independent from the extraction voltage. The transported current increases, indicating a compensated beam. In this experiment the gas flux was used to modify the current. The experimentally measured dependency of the necessary negative voltage has to be compared with the simulation result, shown in Fig. 20. The Laplace solution with a negative screening voltage of -670 V predicts still a negative voltage of -50 V on axis.

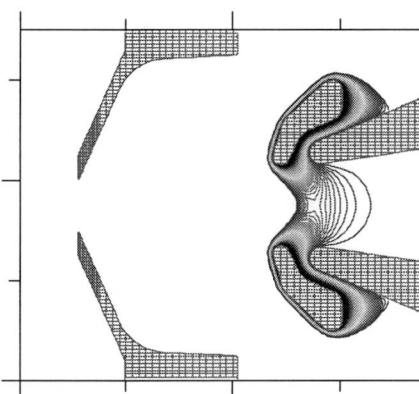

FIGURE 20. For -670 V on the electrode still -50 V are on axis for the Laplace case. 100 lines of constant potential are plotted in a window from 0 V to -670 V, with an increment of -6.7 V between two lines.

Above 10 kV extraction voltage the shape of the curves in Fig. 18 changes. A more negative voltage is required to obtain space charge compensation, which can be explained by the concentrated positive space charge on axis due to stronger focusing within the extraction system. This stronger focusing is predicted by simulation, shown in Fig. 21.

May be more important is that the intensity of visible light coming from the source is decreasing. This light is obtained by a CCD camera with an extreme tele lens (distance 3.3 m, focal length 800 mm, 1/2 inch CCD chip diameter), looking through the extraction hole into the plasma and at the gas injector tube, which is centrally mounted at the injection side. These are clear indications of screening the electrons of the extracted beam from the positive source potential. In case of operation with an oven, which is brought into the source on axis at the injection side (at the same position as the gas injection tube), the electrons which are accelerated from the beam to source potential are an additional, undesired source of oven heating. Because of the positive feedback of additional heating by the accelerated electron current with increasing density of neutrals and ions and therefore increasing electrons, this can provoke an unstable operation. Generally, a complete decoupling of ion beam extraction and plasma generation would be desired.

Clearly, this experimental condition favored an ion source setting with a large loss area, resulting in a homogenous distribution of ions at the location of the extraction electrode.

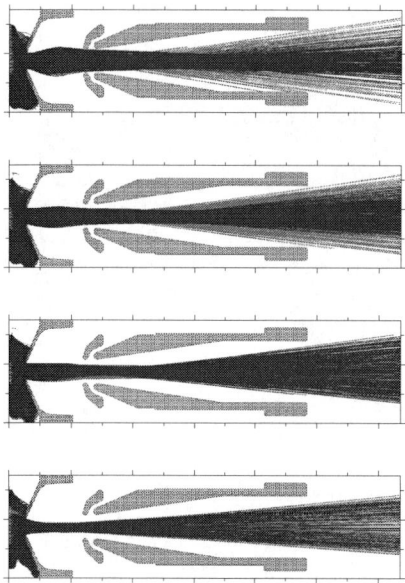

FIGURE 21. Trajectory plot for different extraction voltages. From top to botton: 7500 V, 10000 V, 12500 V and 15000 V extraction voltage. The plasma generator is unchanged for the different extraction voltages. He$^+$ is shown in red, He^{2+} in green. Note, that there are no cylinder symmetric starting conditions for the ions due to the hexapole.

Going further to higher currents for ECRIS [13] one should compare the situation with high current sources and with the simulation results regarding the optimum matching condition, given by a flat plasma boundary.

Sources for proton normalized current densities of 100 mA/cm^2 to 200 mA/cm^2 are typical for high current sources [10]. 35 kV seems to be a widely chosen extraction voltage for these sources for reasons of reliability, but the extraction distance would be in the range of 3...4 mm. In existing ECRIS the current density is roughly a factor of 100 less, but this reduces to a factor of 10 for the latest ECRIS operating at 18 GHz and 28 GHz. If similar extraction voltages are applied as in the case of high current sources an extraction distance of 10 mm seems to be reasonable. To keep the aspect ratio, a smaller extraction hole might be necessary. The plasma boundary should be as flat as possible. A movable plasma electrode as proposed in [11] could be helpful, because a position with homogeneous plasma density can be chosen.

For the decision on how the beam transport is to be planned, the question of space charge compensation arises. In case of an uncompensated transport the beam will double its diameter r_0 after a drift of L according:

$$L \propto r_0 \sqrt[4]{q/m}/\sqrt{P} \quad (3)$$

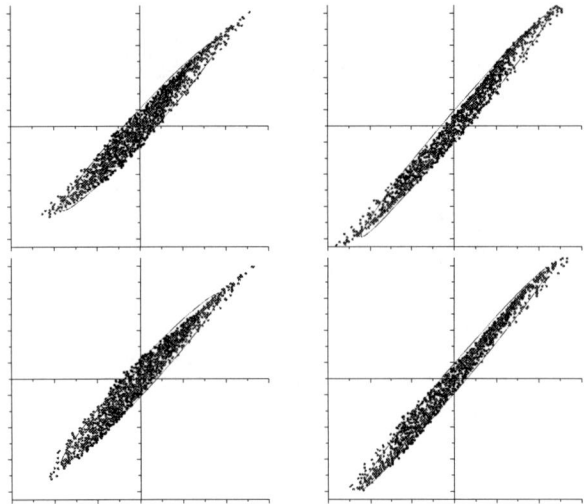

FIGURE 22. Influence of compensation behind the screening electrode on the beam divergence. Scaling is ± 150 mrad and ± 30 mm; left: compensated beam, right: uncompensated beam (3 mA at 12.5 keV). Top: vertical, bottom: horizontal.

where P denotes the perveance of the beam. A few examples given in Tab. 1 favor a space charge compensated transport.

TABLE 1. Characteristic length L for doubling the initial beam radius r_0 under the influence of space charge. The numbers are given for 1 mA electrical current and 10 mm initial beam diameter. U_{ex} is the extraction voltage, Φ_b the potential drop across the beam.

	U_{ex}[V]	Φ_b[V]	L[cm]
e	-15000	-0.715	430
p	15000	30	40
Ar$^+$	15000	189	16
Ar^{3+}	5000	189	9
Ar^{3+}	15000	109	20

In the simulation the difference between space charge compensated and uncompensated transport is shown in Fig. 22, where both beam emittances 230 mm behind the plasma electrode are compared. The maximum divergence angle of the rms-emittance, caused by a 3 mA He beam with 12.5 keV, increases from 100 mrad to 140 mrad already after 20 cm drift. This explains the reduced transmission for the uncompensated transport.

CONCLUSION

The specific operating conditions must be taken into account in the simulation. They can have a severe influence on the beam quality, especially if the source operation requires a strong confinement with high plasma density. The radial flux density can be used in the simulation to define the starting conditions of ions. The shape and location of the plasma boundary was simulated and compared with experimental results, which agree quantitatively. Computer simulation can be used not only to reproduce the reality but also to increase the understanding of physical processes.

Good matching of plasma density and applied field strength is required to achieve a flat plasma boundary, which gave the best transmission.

A rotatable hexapol correction is proposed directly behind the extraction system, which could improve the rms-emittance by a remarkable amount of 40% leading to an enhanced transmission in our high charge state injector at GSI.

For a space charge compensated transport an accel-decel extraction is required. Such a system was taken into operation at the EIS test bench, and its functionality was proven. This system showed such a functionality that it will be used for the next scheduled ^{48}Ca beam time [14] to de-couple oven heating from beam extraction.

REFERENCES

1. K. Tinschert et al., Rev. Sci. Instrum. 69 (2), p. 709, (1998).
2. KOBRA3-INP, INP, Junkernstr. 99, 65205 Wiesbaden.
3. P. Spädtke, ECRIS'02 Proceedings, Jyväskylä, Finland, p. 77, (2002).
4. Fred W. Meyer (ORNL), private communication.
5. D. Hitz et al., An all-permanent magnet ECR ion source for the ORNL MIRF upgrade project, these proceedings.
6. D. Hitz (CEA), private communication.
7. Y. Torii, M. Shimada, I. Watanabe, Rev. Sci. Instrum. 63 (4), p. 2559, (1992).
8. K. Tinschert et al., Experiments on Beam Extraction from the CAPRICE ECRIS, these proceedings.
9. T. Haberer, beam development for cancer therapy, SBIO May 2004, beam time schedule at GSI.
10. The Physics and Technology of Ion Sourcea, edited by I.G. Brown, John Wiley & Sons, New York, (1989).
11. S. Biri and A. Valek, ECRIS'02 Proceedings, Jyväskylä, Finland, p. 49, (2002).
12. S. Druck, P.A. Schmelzbach, Emittance Measurements at the PSI ECR Heavy Ion Source, Paul Scherrer Institut, Annual Report 2001.
13. Ion Sources for Intense Beams of Heavy Ions, A joint research activity in the Frame Programme 6 of the EU.
14. M. Schädel, F. Heßberger, beam development for the SHE program, U208/U209, beam time schedule at GSI.

The WARP Code: Modeling High Intensity Ion Beams

David P. Grote, Alex Friedman

LLNL, Livermore, CA, USA

Jean-Luc Vay

LBNL, Berkeley, CA, USA

Irving Haber

University of Maryland, Collage Park, MD, USA

Abstract. The Warp code, developed for heavy-ion driven inertial fusion energy studies, is used to model high intensity ion (and electron) beams. Significant capability has been incorporated in Warp, allowing nearly all sections of an accelerator to be modeled, beginning with the source. Warp has as its core an explicit, three-dimensional, particle-in-cell model. Alongside this is a rich set of tools for describing the applied fields of the accelerator lattice, and embedded conducting surfaces (which are captured at sub-grid resolution). Also incorporated are models with reduced dimensionality: an axisymmetric model and a transverse "slice" model. The code takes advantage of modern programming techniques, including object orientation, parallelism, and scripting (via Python). It is at the forefront in the use of the computational technique of adaptive mesh refinement, which has been particularly successful in the area of diode and injector modeling, both steady-state and time-dependent. In the presentation, some of the major aspects of Warp will be overviewed, especially those that could be useful in modeling ECR sources. Warp has been benchmarked against both theory and experiment. Recent results will be presented showing good agreement of Warp with experimental results from the STS500 injector test stand. Additional information can be found on the web page http://hif.lbl.gov/theory/WARP_summary.html.

INTRODUCTION

The Warp code was originally developed to model the high current, high brightness beams that are required heavy-ion driven inertial confinement fusion (HIF).[1,2] HIF offers a path to fusion as an energy source. It relies on having ion beams focused down onto the small fusion target, driving it to ignition. In order to provide the required energy, the ion beams must be high current, but have low enough emittance (or temperature) to be focusable. These beams are "space-charge dominated" – the self-field effects are significantly larger than the thermal effects. The beams act as non-neutral plasmas. An ideal method to simulate these beams is the particle-in-cell (PIC) method from plasma physics. This method fills the phase-space with representative particles and couples them by solving Maxwell's equation on a grid.

The Warp code begins with the PIC method and extends it by incorporating a description of the applied fields of the accelerator lattice. The PIC method is implemented in axisymmetric mode, transverse slice mode, and in full 3-D mode. Due to the relatively low energy per nucleon of the beams in HIF, only an electrostatic, Poisson, solver has been implemented. The solver allows internal boundary conditions – extensive tools have been developed for their specification. The particle advance is 2nd order leap-frog, and for the coupling to the grid, linear, or cloud-in-cell interpolation is done. Multi-species can be modeled, such as multiple charge states and multiple ions. For electrons, an advanced integrator is being developed that allows large time-steps compared to the electron cyclotron frequency.[3] The lattice description allows a range of field descriptions, from uniform, pure multipole components, to axially varying, mixed components, to gridded field data.

The natural mode of operation of the PIC method (and thus of Warp) is to be time-dependent, which is well suited for the modeling of space-charge dominated beams. A consequence is that the fields from the lattice are applied directly to the particles, rather than via mapping methods, as is usual in the modeling of emittance dominated beams.

WARP OVERVIEW

Combined in Warp are many different pieces that cover a wide variety of scenarios, covering various dimensionality, levels of problem description, and kinds of physics. All pieces of the code have been adapted to run in parallel-processing environments.

Warp3D

The original package of Warp was the 3-D package, which models the beam in full three-dimensional physical space and three-dimensional velocity space. The self-fields are calculated on a Cartesian mesh laid down in the frame of the beam. The mesh can move with the beam or remain static. In a bend, warped-Cartesian coordinates are used, which are cylindrical coordinates, with the angle theta replacing the axial coordinate z. A single mesh can contain areas with and without bends. In a bend, the coordinate system follows a defined physical centerline of the bend, which does not necessarily coincide with the trajectories of any particles (which depend only on applied and self fields). The coordinates of the particles, however, are stored relative to the warped coordinates – a particle which does follow the bend centerline will have $x = 0$.

A number of field solvers (Poisson solvers) are available. The first is an FFT based solver. Bends can be included by moving the curvature related terms to the right hand side, treating them as sources, and iterating to convergence. Simple internal boundaries can be included using the capacity matrix method. For larger, more complicated conductors however, the matrix becomes very large and is costly to generate.

For this reason, an iterative multigrid solver was developed. This solver can include arbitrary internal boundaries. At the internal boundaries, cut-cell or embedded boundary conditions are used to maintain second order convergence of the solver. Extensive tools have been developed to specify the conductors, allowing combinations of basic geometric objects, such as cylinders and tori, and more complicated objects, such as those describable as surfaces of revolution. In bends, the curvature terms are directly included in the iteration. An adaptive mesh refinement (AMR) capability in three-dimensions is in development. Two and four-fold transverse symmetries can be taken advantage of for efficiency.

The basic time advance for the Warp3D is fully time-dependent. Various approximations can be used to gain efficiencies, however. For example, "quasi time-dependence" can be used – the particle advance is fully time-dependent, but the self-field calculation is done only periodically. A further approximation is an iterative steady-state mode, where a single bunch of particles is tracked through the system, accumulating the charge density. The self-fields are recalculated with the accumulated density and the iteration repeated. This is a standard method in many gun codes. It can sometimes converge to a bi-stable state, however. The quasi time-dependent does not suffer from this problem.

WarpRZ

Beams can be modeled assuming axisymmetry – variation along the azimuth is ignored. Warp actually follows the particles in full 3-D space, but the charge density is mapped to and the self-fields mapped from the r-z plane. The Poisson solver uses the multigrid method, and includes internal boundaries using the same cut-cell methods as in the 3-D solver. The adaptive mesh refinement methods in the RZ solver are more developed.[4]

WarpXY

The third model implemented is a transverse slice model which effectively models a steady flow. A thin transverse slice of the beam is followed through the lattice, ignoring any z-dependence of the self-fields. Each time step, the particles are advanced to the same z position - they all have the same z-step size. The particles can have a variation in their axial velocity, and z-dependent and z-directed applied fields are included. Each particle has its own time-step size, which is adjusted inversely to the axial velocity to keep the z step size constant. Each step, as the axial velocity changes, the advance is iterated to update the time-step size of each particle. In bends, the slice moves in steps of the angle theta around the bend - particles at large x in the bend move further each step (the time-step size is adjusted accordingly). There are two Poisson solvers implemented, an FFT based solver with optional capacity matrices, and a multigrid/AMR based solver.

The Lattice

The lattice description is used to set the applied fields and geometry of bends. The fields can be specified at several levels of description. Any elements can be overlapped. The lowest level is the axially uniform, hard edge approximation. Any multipole component can be applied, for example solenoid, dipole, quadrupole, sextapole, *etc*. Accelerating fields can be applied as well. When the fields are applied to the particles, "residence corrections" are used, where, upon entering or exiting the element, the applied field is scaled by the fraction of the time-step spent inside the element. With the corrections, 2nd order accuracy is maintained in the advance.

The next level of description is to use axially varying multipole components. The z variation of the coefficients of the components is tabulated. Currently, linear interpolation is done between data points. Any component or combination of components can be applied, including both fundamentals and their axial derivatives. In a bend, the transverse center follows the curvature.

A further detailed description is to use fully tabulated field data. For magnetic fields, the three components of the field are each specified in three-dimensional Cartesian grids. For electric fields, the potential is specified on a single three-dimensional grid, and finite differences are done on a per particle basis. In both cases, tri-linear interpolation is done between grid points. As with the other descriptions, in a bend, the grid follows the curvature.

For electrostatic elements, the fields can be applied by directly including the conductors as boundary conditions in the self-field calculation. For example, with interdigitated electric quadrupoles, the geometry as described via the lattice can be included. In a diode, the voltage drop can be modeled by including the anode and cathode plates.

The bend elements are different than the others since they are only specifying geometry – no fields are applied. For any elements that overlap a bend, the center of the element follows the curvature. Currently, Warp only supports bends in one plane, the *z-x* plane. Also, note that two bends can overlap each other. Figure 1 shows an example lattice that includes bends.

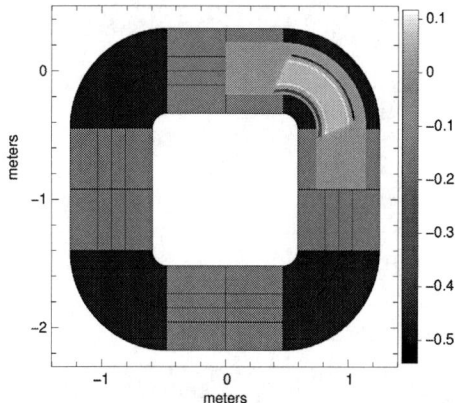

Figure 1: This shows an example lattice – a storage ring experiment at MSU containing only 4 bends. The magnetic field is gridded. The blue shows the extent of the bend. The green shows where the gridded field is (though it is covered over in the bend). The color scale is the B_y field component, in Tesla.

Injection

A significant capability of Warp is its ability to model particle injection. Fixed-current, space-charge limited injection, and secondary emission can be modeled. Injection from a plasma source, using the standard approximation of electrons with the Boltzmann distribution, is in development.[5] The emission of particles can be from curved surfaces. Some examples are shown in Figures 2, 3 and 4.

Unlike many gun codes that launch particles from a virtual surface in front of the true emission surface, Warp launches particles directly from the true emission surface. This offers several advantages: for sources immersed in a magnetic field, particles are advanced correctly in that field from birth; for time-dependent problems, particles can spend a significant time traversing the virtual region in front of the source, and in order to correctly model the head of the beam, the detailed motion in that region must be captured. As part of this, a capability was added to model this region using one-dimensional mesh refinement along lines normal to the emitting surface. The refinement is non-linear, following the Child-Langmuir density scaling. Refinement factors as high as 10,000 are used regularly [4].

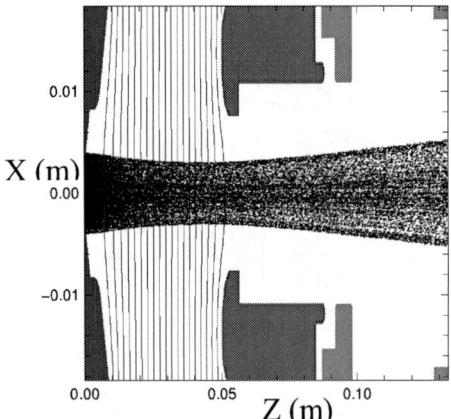

Figure 2: The extraction region of the VENUS ECR source. The nearly vertical lines are evenly spaced contours of constant potential.

Figure 3: The same region as shown in figure 2, but rendered in 3-d.

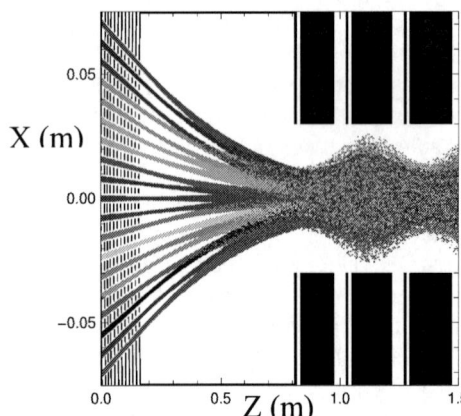

Figure 4: From a multiple beamlet merging injector. This shows the slice at y=0. Here, 119 beamlets are independently injected and accelerated and then merged into a single beam that flows into a transport channel.

The modeling of secondary emission of particles is under development. With this capability, a simulation can for example include emission of electrons when an ion or electron strikes a surface. The motion of these particles are tracked self-consistently.

Python interface

The user interface to Warp is the modern scripting language Python.[6] This is a fully object oriented language that is well developed and is used extensively throughout the world. While the core of the code is written in modern Fortran, Python is the interface for data input, steering, and post-processing. Python gives the user great control over the problem description and how the simulation is carried out. The authors of Warp do not have to foresee all possible modes of operation, diagnostics, post-processing, *etc.* that the users may need. The users input file becomes the "main" routine. Python also gives interactive access, allowing such things as rapid problem setup and debugging, and interactive experimentation to help in aiding the understanding of the problem of interest.

CONCLUSIONS

Warp was originally developed to study the high-current, high-brightness beams required for the HIF approach to fusion energy. It was designed to be flexible, including various degrees of approximation and dimensionality. Warp should work well for ECR ion sources. Complex conductor geometries can be modeled and bends included. Multiple species can be injected and followed. A plasma source model is in development. This covers much of the capability required for ECR source modeling.

REFERENCES

1. A. Friedman, D. P. Grote, I. Haber, "Three Dimensional Particle Simulation of Heavy Ion Fusion Beams," *Phys. Fluids B.* **4**, 2203 (1992)
2. D. P. Grote, A. Friedman, G. Craig, I. Haber, W. M. Sharp, "Progress Toward Source-to-Target Simulations," *Nulc. Instrum. Methods Phys. Res. A*, 464, p. 563 (2001).
3. R. Cohen, *et. al.*, "Electron-Cloud Simulation and Theory for High-Current Heavy-Ion Beams," to be published in *Phys. Rev. ST Accel. and Beams*.
4. J. L. Vay, et. al., "Application of adaptive mesh refinement to particle-in-cell simulations of plasmas and beams," *Phys. Plasma* **11**, 2928 (2004)
5. http://ww.science.doe.gov/sbir/awards_abstracts/sbirsttr/cycle22/phase1/024.htm
6. http://www.python.org

III DIAGNOSTICS AND TRANSPORT

Plasma Potential Measurements With A New Instrument

O. Tarvainen*, P. Suominen*, T. Ropponen*, H. Koivisto*, R.C. Vondrasek**, R.H. Scott**

*University of Jyväskylä, Department of Physics (JYFL), Jyväskylä, Finland
**Argonne National Laboratory, Illinois, USA

Abstract. An efficient and fast instrument to measure the plasma potential of ECR ion sources (ECRIS) has been developed at the Department of Physics, University of Jyväskylä (JYFL). The operating principle of the instrument is to measure the energy of the ion beam by applying a decelerating voltage to a mesh located in the beam line after mass analysis. The plasma potential is determined by measuring the current at the grounded electrode situated behind the mesh as a function of this adjustable voltage. The measurements were performed with ECR ion sources at JYFL (6.4 and 14 GHz) and at Argonne National Laboratory (14 GHz). The plasma potential was measured as a function of different source parameters such as microwave power, gas feed rate (with different gases), voltage of the biased disk and magnetic field strength. The effects of gas mixing and double-frequency heating were also studied. The energy of the ions extracted from an ECRIS plasma comes from the source potential, plasma potential and the thermal energy of the ions. In order to distinguish the effect of the ion temperature on the measured curve simple computer simulations were performed. With the aid of the simulation and assuming a certain potential profile and Maxwellian velocity (energy) distribution of the ions, it was seen that the ion temperature should affect the shape of the measured curve in the region where the adjustable deceleration voltage is close to the value of the plasma potential. In the measurements it was observed that the shape of the curve in this region changed dramatically when gas mixing was used. However, the effect was typical only for low charge states of the heavier element while the curves measured with higher charge states remained almost unchanged. The effect of gas mixing on the ion temperature will be discussed based upon the obtained results.

INTRODUCTION

A positive plasma potential with respect to the source potential is characteristic of ECRIS plasmas. The plasma potential is created because electrons and ions have different diffusion rates out of the plasma due to the higher mobility of electrons[1]. Low-energetic electrons are especially easily scattered into the loss cone because the electron-ion collision frequency v_{ei} depends on the electron temperature T_e as $v_{ei} \propto T_e^{-3/2}$. The imbalance of diffusing charge is compensated by the positive plasma potential retarding the loss of electrons and repelling ions[2]. The value of the plasma potential is determined by the equilibrium of escaping charge and therefore reflects the performance of the ion source. It has been observed that the value of the potential is on the order of tens of volts[3]. A low plasma potential is usually characteristic of a well-performing ion source[1]. In order to understand the plasma processes of an ECRIS an instrument to measure the plasma potential has been developed at the Department of Physics, University of Jyväskylä (JYFL). With the aid of the device the plasma potential can be determined in a single measurement without disturbing the plasma or changing the source parameters.

EXPERIMENTAL RESULTS

The instrument used for the plasma potential measurements and the method of determining the plasma potential from the obtained data has been described in ref. 4. The data curves can be divided into four parts (A-D, see fig. 1). In region A the measured current decreases only slightly. In region B the current decreases linearly until it starts to saturate in region C before attaining a certain small value in region D. The shape of the measured curve in region C seems to depend on the ion temperature, which was the greatest motivation to develop the simulation code described later in this paper. The measurement error related to the instrument (not to the plasma instabilities or the ion temperature) was estimated by measuring the plasma potential corresponding to certain source settings several times without changing the tuning of the source. The ion optics of the beamline was changed in order to ensure that different focusing does not affect the measured value of the plasma potential. According to the results the error related to the plasma potential values is about 0.5 V for the ANL ECR2 and JYFL 14 GHZ ECRIS, whilst it is about 1 V for JYFL 6.4 GHz ECRIS. For clarity, error bars are not presented in the following chapters.

The plasma potential of the JYFL 6.4 GHz ECRIS has been measured with different elements. Several gas feed rates and microwave power of up to 350 W were used with the magnetic field configuration and the voltage of the biased disk kept constant. It was observed that the potential tends to increase as the microwave power or the gas feed rate is increased[4]. These tendencies are probably due to the increased plasma density and density gradients, which lead to faster escape of the electrons from the plasma[3]. A higher plasma potential is consequently needed to balance the loss rates of electrons and ions. The measured plasma potentials were somewhat higher in the case of heavier ions, which is probably due to their lower mobility compared to lighter ions. The range of plasma potentials obtained with different gases is presented in table 1.

TABLE 1. Range of the Plasma Potentials Measured with the JYFL 6.4 GHz ECRIS with Different Gases.

Plasma	Potential Range [V]	Gas Feed Rate [cm^3/h]
Hydrogen (^{1}H)	22 – 33	0.41 – 0.81
Deuterium (^{2}H)	19 – 39	0.40 – 0.82
Helium (He)	36 – 47	0.59 – 1.25
Nitrogen (N)	36 – 46	0.12 – 0.22
Oxygen (O)	32 – 55	0.09 – 0.22
Neon (Ne)	43 – 60	0.08 – 0.17
Argon (Ar)	40 – 65	0.14 – 0.23
Krypton (Kr)	47 - 62	0.16 – 0 20

The measurements presented in ref. 4 showed that the plasma potential of an oxygen plasma decreases 5-10 volts as the negative voltage of the biased disk is increased. This indicates that the biased disk affects the loss rates of electrons and ions (at least locally). The voltage of the disk pushes electrons towards plasma and pulls ions out of it close to the disk resulting in the decrease of the plasma potential due to additional compensation of the loss rates. It has also been observed[4] that the strength of the axial magnetic field affects the plasma potential only slightly. A possible explanation is that the thermal electron population responsible for the plasma potential build up is not affected by magnetic mirrors as much as the hot electron population.

In the case of the JYFL 6.4 GHz ECRIS there is practically no difference between the plasma potential values obtained with different charge states. Figure 4 shows the normalized voltage-current curves measured with the JYFL 6.4 GHz ECRIS with charge states O^{6+}, O^{5+} and O^{3+} (tuned for O^{5+}). The shape of the curve is almost independent of the charge state in the case of the JYFL 6.4 GHz ECRIS. The values of the plasma potential determined from the curves shown in figure 1 are 39.6 V, 40.2 and 39.4 volts measured with O^{6+}, O^{5+} and O^{3+} ion beams, respectively.

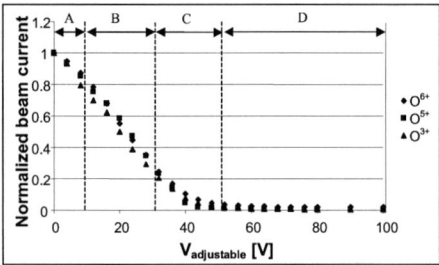

FIGURE 1. Measurement result obtained with O^{6+}, O^{5+} and O^{3+} ion beams with the JYFL 6.4 GHz ECRIS

The measurements with the JYFL 14 GHz ECRIS and with ECR2 (14 GHz) at ANL were initiated by measuring the plasma potential with different charge states of oxygen. It was observed that unlike the JYFL 6.4 GHz ECRIS the value of the potential was dependent on the charge state. Some results obtained with different ion sources are shown in table 2.

TABLE 2. Plasma Potentials of Different Ion Sources.

Ion Source	ANL ECR2 (14 GHz)		JYFL 14 GHz		JYFL 6.4 GHz
Power	333 W	485 W	350 W	497 W	150 W
	Plasma Potential [V] / Beam Current [µA]				
O^{7+}	7.7 / 7	6.5 / 21	17.3 / 53	20.1 / 69	49 / 5
O^{6+}	8.7 / 58	8.2 / 100	16.5 / 186	20.4 / 213	47 / 75
O^{5+}	9.5 / 79	8.0 / 90	19.2 / 109	24.4 / 117	47 / 133
O^{4+}	11.8 / 92	10.2 / 75	18.3 / 76	27.7 / 81	49 / 185
O^{3+}	18.9 / 69	12.3 / 49	20.6 / 40	28.2 / 30	48 / 102
O^{2+}	23.7 / 53	19.3 / 40	20.2 / 6	27.6 / 5	48 / 68
O^{+}	19.9 / 6	24.0 / 10	- / -	- / -	49 / 44

The plasma potentials obtained with 14 GHz ion sources are significantly lower than those obtained with the 6.4 GHz source. The lowest plasma potential values were obtained with the ECR2 at ANL. The only significant difference between the ECR2 at ANL and the JYFL 14 GHz ECRIS is the strength of the radial multipole field, being 0.95 T and 0.85 T respectively (on the wall of the plasma chamber). In the case of the JYFL 14 GHz ECRIS the plasma potential tended to increase as the microwave power was increased but with the ANL ECR2 the potential dropped as the power came up. The dependence of the shape of the measured curve on the charge state can be clearly seen from figure 2, presenting normalized curves measured with different charge states (O^{6+}, O^{3+} and O^+) of oxygen with the ANL ECR2 (14 GHz / 485 W).

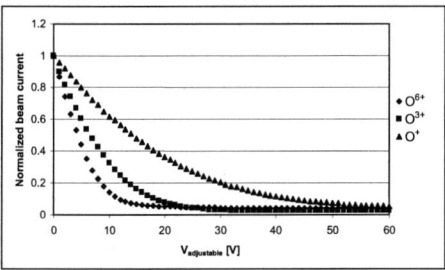

FIGURE 2. Normalized plasma potential curves measured with the ECR2 at Argonne National Laboratory.

The plasma potential values determined as described in ref. 4 were 8.2 V (O^{6+}), 12.3 V (O^{3+}) and 24.0 V (O^+). This means that the measurement includes the thermal energy of the ions in addition to the plasma potential, which has to be the same for all charge states. The shape of the measured curve and thereby the value of the plasma potential is probably affected more by the ion temperature than in the case of the JYFL 6.4 GHz ECRIS. This is because the ratio of ion temperature and the plasma potential (about 10 V) is higher in the case of the 14 GHz ion source. According to ref. 5, the ion temperature increases as the electron density of the plasma increases. With higher frequencies one can achieve higher electron densities, favorable for the production of highly charged ions, because the "cutoff" electron density for the wave penetration becomes higher as the frequency increases. However, as the electron density increases due to higher frequency or higher microwave power, the ion temperature should also increase. If the ion temperature is high compared to plasma potential, the results which best represent the true plasma potential are obtained with high charge states. If we, for example, considered an ion originating from plasma potential of 10 volts with a thermal energy of 7 eV, the decelerating voltage needed to stop O^{7+} ions is 11 volts being 17 volts for O^+ ions.

In the measurements performed with the JYFL 14 GHz ECRIS it was observed that carbon contamination on the wall of the plasma chamber has a significant effect on the plasma potential as well as on the performance of the source. In the case of severe carbon contamination the plasma potential was twofold greater than the potential measured with clean source. In addition to higher plasma potentials it was observed that the shape of the measured plasma potential curve changed dramatically due to the contamination. It seems that secondary electrons affect the profile of the plasma potential or the ion spatial distribution so that less ions originate from low potential. The plasma potential build up process is assumed to be related to ambipolar diffusion[3] and the diffusion rate of charged particles is proportional to density gradients. Secondary electrons emitted from the walls of the plasma chamber (flowing towards the plasma from the exterior of it) probably decrease the gradient of the electron density, which leads to lower plasma potential despite the probable increase in electron density. It has been observed that coating the copper wall of the plasma chamber with aluminum oxide i.e. increasing the secondary electron emission reduces the plasma potential and makes it nearly independent of microwave power[3]. The carbon contamination may explain the difference between the behavior (power vs. plasma potential) and plasma potentials measured with the JYFL 14 GHz ECRIS and the ANL ECR2 because the former is often used for the production of metal ion beams with the MIVOC-method. As the power is increased, more carbon comes off from the walls of the plasma chamber and affects the plasma potential.

It has been demonstrated that the performance of the ANL ECR2 ion source has been improved through the use of two frequency heating[6], However, the physical processes leading to enhanced production of high charge states due to the secondary frequency are still largely unknown. This gave the motivation to study the effect of two frequency heating on the plasma potential. At first, the plasma potential was measured with a power of approximately 330 W with different frequencies (klystron / TWTA). The source was tuned with 14 GHz and the adjustable source settings were kept constant during the measurements with certain total power. Then the power of the 14 GHz klystron was increased up to 626 W and the plasma potential was subsequently measured. After this, the output power of the klystron was lowered to approximately 320 W and the output power of the TWTA was chosen so that the total power launched into the plasma chamber was 600-630 W. The same frequencies as in earlier measurements (330 W) were used for the TWTA. The source settings were kept constant after tuning the source for 14 GHz only. Some results obtained with different frequencies and microwave powers measured with different charge states of oxygen ion beams are shown in table 3.

TABLE 3. Plasma Potentials and Beam Currents Measured with ANL ECR2 in Different Electron Heating Modes.

RF-frequency [GHz]	11.08	12.18	12.48	14	14	14 / 11.06	14 / 12.19	14 / 12.48
RF-Power [W]	323	344	330	333	626	326 / 300	318 / 283	309 / 304
				Plasma Potential [V] / Beam Current [µA]				
O^{7+}	7.9 / 4	9.8 / 13	10.1 / 17	7.7 / 7	6.5 / 42	5.9 / 64	6.4 / 61	5.9 / 40
O^{6+}	9.3 / 32	10.8 / 72	12.4 / 84	8.7 / 58	7.7 / 166	7.2 / 188	7.3 / 190	7.9 / 148
O^{5+}	10.1 / 45	11.5 / 79	13.1 / 83	9.5 / 79	8.1 / 135	7.4 / 127	7.7 / 125	7.9 / 112
O^{4+}	10.9 / 58	12.8 / 87	16.1 / 82	11.8 / 92	7.9 / 110	7.6 / 89	7.1 / 80	8.0 / 93
O^{3+}	12.8 / 58	19.0 / 72	19.4 / 64	18.9 / 69	8.6 / 58	8.7 / 54	8.2 / 51	9.9 / 63
O^{2+}	18.3 / 60	27.8 / 65	23.6 / 50	23.7 / 53	11.5 / 50	9.9 / 44	9.9 / 31	10.9 / 51
O^{+}	20.3 / 4	23.4 / 13	18.2 / 7	19.9 / 6	14.3 / 14	14.0 / 17	14.8 / 17	14.5 / 23

The results obtained with different frequencies in single frequency mode show that there seems to be a slight dependence between the plasma potential (measured with high charge states) and the microwave frequency. The plasma potential seems to increase as the frequency of the TWTA is increased. However, the plasma potentials measured with 14 GHz with approximately the same output power are somewhat lower. No satisfactory explanation for the difference between the operation with the klystron and the TWTA was found. The plasma potential measured with O^{7+} in double frequency mode is about 10 % lower than in single frequency mode with the same total power in the cases of secondary frequency being 11.06 or 12.48 GHz. This means that that the proportion of the electrons escaping from the plasma out of the whole electron population may be slightly affected by the secondary frequency. Our observation does not exclude the possibility that the electron density is higher in either mode. The shape of the measured curves was almost similar in single and double frequency modes. However, the shape of the potential profile may be different compared to single frequency heating (see R. C. Vondrasek et al. in these proceedings).

The effect of the gas mixing[7] on the plasma potential of the ANL ECR2 was studied with 14 GHz microwave frequency as follows. First, oxygen was fed into the plasma chamber and a microwave power of 570 W was used. The plasma potential of the oxygen plasma was subsequently measured with different charge states (series #1). After this, helium was added into the existing plasma without changing the tune of the ion source. The plasma potential was once again measured (series #2) with oxygen ion beams before tuning the source so that the extracted current of O^{7+} ion beam was at maximum. Finally, the plasma potential was also measured in this source configuration (series #3). The plasma potentials and extracted currents of different ion beams obtained during the gas mixing measurements are presented in table 4. For comparison table 4 also includes plasma potentials and currents obtained in a measurement with an oxygen plasma resulting in approximately the same drain current as during the gas mixing measurements (series #4, 523 W).

TABLE 4. Gas Mixing versus Plasma Potential.

Series #	1	2	3	4
Plasma	O	O + He	O + He	O
I_{drain} [mA]	0.76	1.43	1.45	1.34
	Plasma Potential [V] / Beam Current [µA]			
O^{7+}	7.1 / 24	5.9 / 47	6.2 / 50	7.0 / 40
O^{6+}	7.1 / 92	6.7 / 182	7.5 / 194	7.9 / 155
O^{5+}	8.5 / 70	7.7 / 103	7.4 / 91	8.3 / 125
O^{4+} (He^+)	9.1 / 59	8.0 / 132	8.2 / 132	8.1 / 105
O^{3+}	11.3 / 38	7.8 / 30	7.5 / 19	8.3 / 60
O^{2+}	19.7 / 31	8.1 / 21	7.3 / 12	12.0 / 52
O^{+}	23.5 / 16	9.9 / 9	8.0 / 4	16.0 / 21

The results show that adding helium into the oxygen plasma affects the plasma potential. The plasma potential is slightly lower in the case of gas mixing because the ions of the lighter element carry the charge out from the plasma more effectively due to their higher mobility. The effect of the lighter element is especially clear with low charge states. The difference in the drain current cannot explain this because the plasma potential values (low charge states) obtained in series #4 were higher than with oxygen-helium plasma. The effect of the gas mixing on the shape of the measured plasma potential curves of low charge states was dramatic while the shape of the curves measured with high charge states did not change. The charge state dependence of the measured curves was observed to almost vanish due to the gas mixing. Figure 3 shows the measured plasma potential curves of O^{7+} and O^+ ion beams in the case of oxygen (series #1) and oxygen-helium (series #3) plasmas. The curves are normalized with respect to the maximum current of each measurement. The effect of the gas mixing on the curves measured with O^{2+} and O^{3+} ion beams was almost as clear as in the case of O^+. The results indicate that the gas mixing affects the temperature of the low charged ions of the heavier element more than the temperature of the highly charged ions of the heavier element.

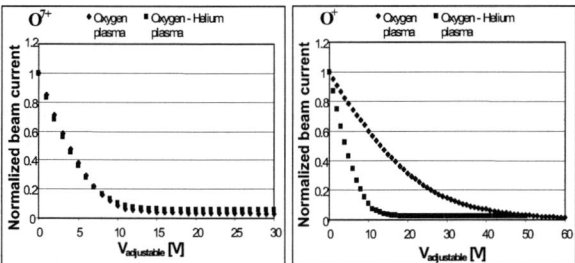

FIGURE 3. Normalized plasma potential curves measured with O^{7+} and O^+ ion beams.

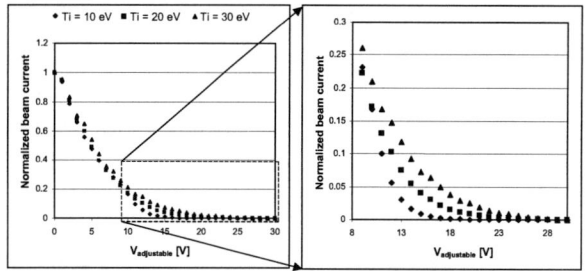

FIGURE 4. Simulated plasma potential curves of O^{7+} ion beam with different ion temperatures.

SIMULATION RESULTS

In order to study the effect of ion temperature on the measured plasma potential curves (plasma potential + T_i) a simple computer simulation code was written with the Mathematica 4.1 program. The input parameters of the code are the adjustable plasma potential profile (a negative potential dip can be included), the temperature of the ions, the adjustable spatial distribution of the ions relative to the potential profile and the ion mass and charge. By spatial distribution we mean the relative amount of ions originating from a certain potential. The velocity distribution of the ions in an ECRIS plasma was assumed to be Maxwellian.

With the aid of the simulation it was seen that the temperature of the ions indeed determines the shape of the measured voltage-current curve in the saturation region whereas the shape of the curve at low stopping voltages depends mainly on the potential profile and the spatial distribution of the ions. The potential profile and the spatial distribution of the ions were chosen so that the simulated curves matched the measured curves with voltages lower than the plasma potential determined from the measured curve. The negative dip was not included in the potential profile. As an example, figure 4 shows simulation results of O^{7+} ion beam with three different ion temperatures (T_i). In this case, the plasma potential profile was chosen to match a plasma potential curve measured with the ANL ECR2 in single frequency mode with a frequency of 12.48 GHz and power of 330 W (arbitrary choice). The plasma potential value determined from the measured curve is 9.5 V, which was used as the maximum of the potential profile in the simulation.

Changing the temperature used in the simulation clearly affects the saturation of the plasma potential curve. In the case of figure 4 the correlation between the measured result and the simulation with an ion temperature of 20 eV was found to be 0.9999. Theoretically estimated ion temperatures in ECRIS plasma are only a few electron volts[5]. However, in reference 2 ion temperatures of ~20 eV measured with a single cell quadrupole mirror (constance B) for plasma parameters close to an actual ECRIS are reported. These temperatures were determined by measuring the Doppler broadening of the emission lines for different charge states of oxygen. Some ion temperatures of different charge states of oxygen ions determined by comparing the simulated and measured plasma potential curves are presented in table 5. The value of the maximum plasma potential for each column was determined with the aid of the measured curve of O^{7+} and the same value was used for the other charge states. The error related to the ion temperatures was estimated by comparing the correlation between simulation results and measured curves. It was observed that the error is about ± 2 eV for high charge states and about ± 1 eV for low charge states (O^+ and O^{2+}).

Table 5 shows that the temperature of the extracted ions is usually between 10 and 20 eV depending on the charge state. The simulated temperatures and the dependence of the temperature on the charge state are consistent with the experimental result of ref. 2. The temperature of the low charge state ions (O^+, O^{2+}) tends to be lower than the temperature of highly charged ions (O^{7+}, O^{6+}), which is probably due to the shorter confinement time of low charge states having no time to gain energy in collisions with electrons. It is a well-known fact that the electron density n_e of ECRIS plasma increases with microwave power. This should produce higher ion temperature[5]. However, according to our results the ion temperature is practically independent of the microwave power. This indicates that the average energy of thermal electrons does not change significantly and that the highly charged ions and thermal electrons are close to thermal equilibrium.

TABLE 5. Ion Temperatures and Beam Currents with Different Ion Sources and Source Settings.

Ion Source	ANL ECR2							JYFL 14 GHz	
RF-frequency [GHz]	11.08	12.48	14	14	11.06 / 14	14	14	14	14
RF-power [W]	323	330	333	626	300 / 326	570	570	350	630
Plasma	O	O	O	O	O	O	O + He	O	O
Ion Temperature [eV] / Beam Current [µA]									
O^{7+}	15 / 4	17 / 17	16 / 7	14 / 42	14 / 64	14 / 24	14 / 50	21 / 53	23 / 101
O^{6+}	15 / 32	24 / 84	16 / 58	15 / 166	16 / 188	14 / 92	16 / 194	21 / 186	21 / 278
O^{5+}	14 / 45	22 / 83	17 / 79	14 / 135	15 / 127	14 / 70	12 / 91	19 / 109	20 / 136
O^{4+} (He^+)	12 / 58	23 / 82	17 / 92	11 / 110	11 / 89	13 / 59	11 / 132	17 / 76	19 / 83
O^{3+}	12 / 58	23 / 64	22 / 69	9 / 58	10 / 54	14 / 38	6 / 19	13 / 40	17 / 27
O^{2+}	12 / 60	19 / 50	20 / 53	9 / 50	8 / 44	17 / 31	4 / 12	10 / 6	8 / 4
O^+	8 / 4	9 / 7	9 / 6	6 / 14	6 / 17	11 / 16	2 / 4	- / -	- / -

There seems to be a small difference in ion temperatures between the ANL ECR2 and the JYFL 14 GHz ECRIS, the ion temperatures in the latter being slightly higher for high charge states. A possible explanation is the residual carbon contamination of the JYFL 14 GHz ECRIS affecting the ion temperatures in addition to plasma potential. For example the temperature of extracted O^{7+} ions was determined to be 27 eV with a microwave power of 400 W after the source (JYFL 14 GHz ECRIS) was severely contaminated. The ion temperatures of the JYFL 6.4 GHz ECRIS were observed to be similar to those of 14 GHz ion sources. More information about the electron densities and other plasma parameters of different ion sources would be needed in order to explain this.

Various explanations for the increasing performance of ECRIS's due to gas mixing have been suggested (see ref. 8). Our results strongly imply that the ion cooling process due to the ion-ion collisions plays a dominant role. According to the simulation results and experiments the effect of the gas mixing on the ion temperature is clear. The temperature of low charge states of oxygen ions (heavier element) decreased significantly although the temperature of highly charged oxygen ions did not change as helium was added into the existing plasma. This indicates that adding a lighter element into the plasma gives the low charged ions of the heavier element a better chance to be confined longer and therefore to be further ionized before escaping the plasma. The thermal energy is transported from the plasma by the ions of lighter element. The temperature of the highly charged ions is higher because their confinement time is longer and they approach thermal equilibrium with the cold electron population having a temperature of some tens of electron volts. The amount of high charge states in the plasma and consequently the extracted currents increase as the gas mixing is used due to the reduced losses of low charged ions of the heavier element. The effect of the gas mixing on the plasma potential and ion temperature will be studied in more detail in near future in conjunction with emittance measurements with the JYFL 14 GHz ECRIS.

ACKNOWLEDGMENTS

This work has been supported by the Academy of Finland under the Finnish Centre of Excellence Programme 2000-2005 (Project No. 44875, Nuclear and Condensed Matter Programme at JYFL). This work was also partially supported by the U.S. Department of Energy, Nuclear Physics Division, Contract Number W-31-109-ENG-38.

REFERENCES

1. A. G. Drentje, Rev. Sci. Instrum., 74 (5), 2003, p. 2631.
2. C. C. Petty, D. L. Goodman, D. K. Smith and D. L. Smatlak, Journal de Physique (Paris), 50 (C1), 1989, p. 783.
3. Z. Q. Xie and C. M. Lyneis, Rev. of Sci. Instrum., 65 (9), 1994, p. 2947.
4. O. Tarvainen, P. Suominen and H. Koivisto, will be published in Rev. Sci. Instrum., 75 (10), 2004.
5. G. Melin, A. G. Drentje, A. Girard and D. Hitz, Journal of Applied Physics, 86 (9), 1999, p. 4772.
6. R. C. Vondrasek, R. H. Scott and R. C. Pardo, Proceedings of the 15th International Workshop on ECR Ion Sources, Jyväskylä, 2002, p. 63.
7. A. G. Drentje, Nucl. Instrum. and Methods in Physics Research, B9, 1985, p. 526.
8. A. G. Drentje, A. Girard, D. Hitz and G. Melin, Rev. Sci. Instrum., 71 (2), 2000, p. 623.

Developments and Plasma Studies at the ATOMKI-ECRIS

S. Biri[1], A. Valek[1], E. Takács[2], B. Radics[2], J. Pálinkás[2], J. Karácsony[3], L. Kenéz[3], A. Kitagawa[4], M. Muramatsu[4]

[1]*Institue of Nuclear Research (ATOMKI), Debrecen, Hungary*, [2]*University of Debrecen, Debrecen, Hungary,*
[3]*Babes-Bolyai University, Cluj-Napoca, Romania,* [4]*Nat. Inst. Rad. Sci. Sci. (NIRS), Chiba, Japan*

Abstract. The 14.5 GHz ECR ion source of the ATOMKI is a stand-alone device producing highly charged ion beams for ion-surface experiments and a variety of low charged plasmas and beams for plasma physics studies and for practical applications. In the past two years we performed plasma diagnostics measurements using Langmuir-probes and X-ray camera. Langmuir-probe results allowed estimating the plasma potential close to the resonance zone. The studying of X-ray pictures of Xe-Ar plasmas helps understanding the gas-mixing phenomena. A mixture plasma of fullerene and ferrocene was generated and FeC_{60} hybrid molecules were detected in the extracted beam.

INTRODUCTION

In the ATOMKI a 14.5 GHz electron cyclotron resonance ion source (ECR, ECRIS) operates as a stand-alone device to produce variously stripped plasmas and low-energy ion beams [1,2]. So far it delivered H, He, N, O, Ar, Kr, Xe (from gases) and C, C_n, C_{60}, F, Fe, Ni, Zn and Pb (from solids) plasmas and beams. During the last few years several improvements have been carried out on the ECRIS. The mirror ratio of the magnetic trap was increased by inserting additional soft iron plugs at the injection side. A movable, biased electrode was installed to tune the plasma potential and to effect on the extraction conditions. More recently this electrode was equipped with water cooling while it is still biased and movable on the high voltage platform. By this we expected to decrease the outgassing and so that we can apply higher microwave powers. The result of the first tests of the disc cooling is however, strange. It remarkably improved the intensities and charges for the case of xenon (we obtained Xe^{27+}), but showed almost no effect on argon. The installation of a water cooling system for the puller electrode and placing soft iron plugs at the extraction side, are also under way. In Figure 1 the cross sectional view of the ion source is shown.

One of the specialties of our source it is equipped with three vacuum motion feedthroughs. One moves the first group of electrodes of the extraction system (puller and pre-lens). Another one is normally used for the biased electrode which in our case has a disc shape or sometimes star (or triangle) shape. For oven experiences the biased electrode is changed by the oven so the oven in whole can also be moved and biased. The third feedthrough is used for Langmuir-probe diagnostics (not shown in the figure).

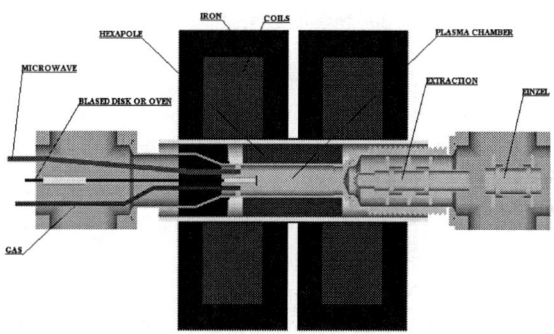

FIGURE 1. Layout of the ATOMKI ECR ion source. The internal diameter of the plasma chamber is 58 mm.

PLASMA POTENTIAL

The measurement of space potential in laboratory plasmas is often necessary because information on electric fields, potential distribution etc. is frequently needed. The potential plays substantial role ECR ion sources which combine magnetic mirror confinement of electrons with electrostatic confinement of the highly charged ions. Conventional Langmuir probes in ECR plasmas have been used to measure the plasma density [3,4]. To determine the plasma potential emissive Langmuir probes can be used. An emissive probe is essentially a biased hot wire probe, heated sufficiently to allow thermal emission of electrons. When the probe is biased negatively with respect to the local plasma potential, the emitted electrons can escape into the plasma and appear as an effective emission ion current. When the probe is biased positive to the local plasma potential, the emitted electrons are reflected back to the probe and the emission current disappears. There are several ways to determine the plasma potential from emission probes characteristics. In our experiments we applied the simplest way. The U-shape emission probe was moved into the required position of the plasma chamber. First we developed the necessary heating power [5]. Then the probe potential was set to a value (we call as wall potential) where the total current to the probe is zero. This potential in the cases of sufficiently strong emission is very close to the plasma potential. Its changes caused by plasma modifications can be used to study the effect of external settings to the plasma. Further in this paper we simply call it as plasma potential.

Our first attempts at the ATOMKI-ECRIS and at the NIRS-KEI2-ECRIS [6] with classical U-shape emission probes (Current-Heated Emission Probes, CHEP) have been unsuccessful so far due to the limited lifetime of these types of probes. Nevertheless the advantage of these probes that they can be used in most places within the plasma chamber. We continue our efforts to find a better making and operation technique of CHEPs.

Then we decided to apply simple Langmuir probes heated to a high emission rate by the plasma itself. Obviously such probes can be used only in the hottest plasma region that is in the resonance zone (RZ). This is a spatial limit but the lifetime of these Plasma-Heated Emission Probes (PHEP) is long and if their size is small enough the plasma is almost undisturbed. We used 2.5 mm long and 0.4 mm thick cylindrical tungsten probe positioned to one of the estimated hottest plasma spots (as in Figure 3, later on). The holder of the probe was fit into thin ceramics tube.

We studied the variation of the plasma potential in the RZ with respect to external tuning factors as extraction voltage on/off, biased disc voltage and gas mixing. The ion source was tuned for Ar^{8+} or Ar^{11+} current. The magnetic field was less than the optimal to form a larger resonance zone and to position the resonance zone exactly to the probe. Figure 2 summarizes the results. For additional information the Ar^{8+} ion current (in arbitrary units) is also shown in the figure. It belongs to the Extraction ON case.

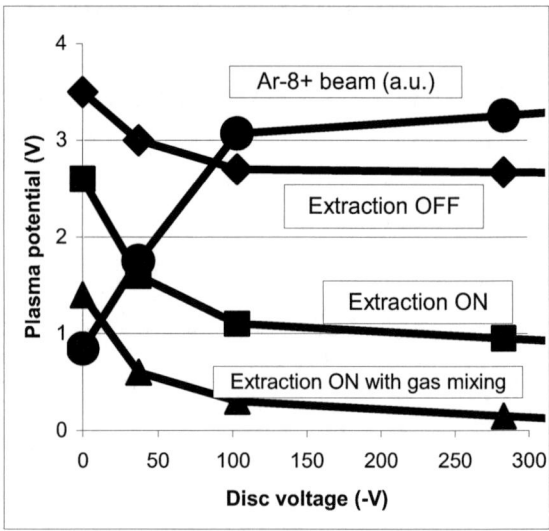

FIGURE 2. The plasma potential at the resonance zone.

From the figure several conclusions can be drawn. The plasma potential decreases by both techniques (biased disc voltage and gas mixing) which increased the Ar^{q+} (q=8 and higher) currents. Switching the extraction voltage off increased the plasma potential indicating a less-charged plasma. This means the charge state distribution (CSD) of the extracted beam does not mirror back completely the CSD in the plasma.

X-RAY PICTURES

X-ray plasma images were made at the ATOMKI-ECRIS in 2003 using a pinhole and a high resolution CCD camera. This method had good spatial resolution as well as the capability of post-processed filtering of the images. During the measurements low and high charge state Ar, Xe and Fe plasmas were produced with simultaneous beam extraction. Full-size and selected region images were recorded and analyzed. The details of the measurement and some results of the prompt analysis were already published [7]. Since then no new measurement could be carried out however the analysis of the tremendous measured data was continued. Based on the CERN ROOT software package a method was developed to process and analyze the data. We concentrated to full-size images of xenon and xenon-argon mixture plasmas. In these experiments the ECRIS was tuned for Xe^{3+} ions applying low microwave power (40 W). In the X-ray spectra strong xenon emission lines appeared between 4 keV and 5 keV originating from L shell transitions (from n=3 to n=2 bound state transitions) in different charge states of xenon. We developed a new post-processing technique to remove all of those parts from the X-ray pictures which can not originate from the xenon plasma as line emission. The details of this method were recently published [8].

Figure 3 shows a spatial distribution map of xenon in the plasma (upper image) and the case when argon gas is added to (bottom image). It is important to note that since the lifetimes of the transitions involved are usually short, line emission only occurs at places where energetic electrons to excite the ions are also present. The x-ray emission from xenon ions is the strongest where the overlap between the electron and the ion clouds is large. Our observation is that the high intensity xenon regions are more localized (more concentrated around high electron density regions) in the mixture case.

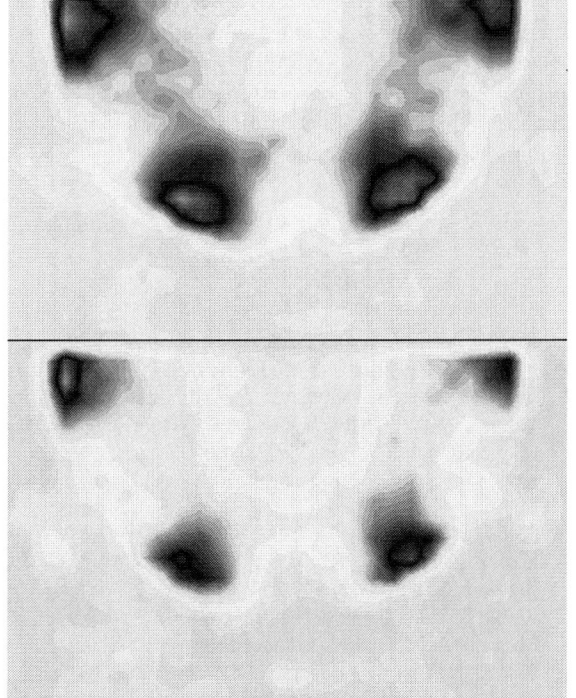

FIGURE 3. Comparison of spatial distribution of Xe ions in pure Xe plasma (up) and in Xe-Ar mixed plasma (bottom)

The indication for xenon ions being in higher charge states when mixing gas is added, is showed by the charge state distribution of the extracted beam and by the shift of the xenon L peak position. In the mixture case there is strong shift of the line position as a function of the brightness of the spot. This appears only to a lesser extent in the pure xenon case.

The temperature of the ion cloud in the plasma is basically determined by the resulted depth of the potential dip of the electron and the ion clouds (generally around a few eV). The strong heating by the energetic electrons by coulomb collisions is balanced by the self cooling of the ion cloud due to the so called evaporation process controlled by the depth of the local ion trapping. Ions with energies higher than the potential dip leave the region removing energy from the local ion cloud providing the cooling effect. If additional cooling is provided by a lighter mixing gas ions can stay trapped longer and the cloud not only becomes colder but the ions will become further ionized.

The above observations and conclusions help a better understanding of the experimentally well-known fact that evaporative cooling of the ion cloud takes place in ECR sources. This results in a more localized and higher charged xenon ion cloud when the lighter argon gas is added. We must note however that these X-ray images were taken in case of low-charged plasmas. We consider important to repeat these type of measurements for the case of higher charged plasmas and we have plans to perform it in near future.

IRON-FULLERENE PLASMA

In many laboratories new materials useful for nanotechnology and medical applications are searched and studied. In Atomki one of our future goals is to produce endohedral fullerene molecules (e.g. $N@C_{60}$, $Fe@C_{60}$) in large quantity. If this comes true, it will be possible to make building blocks for nano-parts, an ultra-contrast medium of MRI, and a magnetic nano-particle for treatment of cancer.

For this experiment some modifications were carried out on the ECRIS similarly to [9]. The waveguide of the 14.5 GHz microwave generator was splitted and divided in order to couple very low powers (in the range of 1 watt or less) into the plasma. The iron plug at the injection side was removed. A low temperature oven was installed in the place of the biased disc. The oven can be biased and is on-line movable on the axes of the plasma chamber. The temperature was monitorized by thermocouple probe. Usually 200…300 mg fullerene powder was placed into the stainless steel crucible of the oven. A simple

MIVOC chamber was connected to one of the two gas feeding lines as close to the plasma chamber as possible. The chamber was filled with 200 mg ferrocene powder. The evaporation rate was controlled by careful external heating. The extraction voltage had to be kept as low as 600 V, because of the low mass-energy product of our bending magnet. So the extracted beam intensities were low.

First we developed independently the rough working conditions for single-charged dense iron and fullerene plasmas. Then a clean fullerene plasma was made. The temperature of the oven was about 450 C which corresponds to approx. 1 mbar local fullerene vapour pressure range. The bending magnet was set to the C_{60} peak (M=720) and about 50...100 nA intensity of single-charged fullerene peak was obtained. Then the magnet was set to the position of the expected Fe@C_{60} or FeC_{60} peak (M=776) and the ferrocene valve was opened. This usually caused strong instabilities in the plasma, all the fullerene peaks decreased and many new ones from the ferrocene appeared. A very difficult and long tuning was then followed. While the first ionization potential of the C_{60} and Fe are low (7.8 and 16.2 eV, respectively), obviously different working conditions are necessary to break the ferrocene and to ionize/excite the fullerene molecules. Finally we found reproducible setting to get a high peak around M=776. The key point of the tuning was the microwave power and the magnetic field strength at the extraction side. The typical power was several watts (sometimes several tens of watts) and the extraction magnetic peak had to be decreased down to the lowest possible value. At this value the resonance zone was very close to the extraction slit.

A typical C_{60}+Fe beam spectrum can be seen in Figure 4 together with an original (clean) C_{60} spectrum. The centre of the new big peak on the right side is at M=776 which corresponds to FeC_{60} and/or Fe@C_{60} molecules. However the peak is wide and shows some structure. We think it may contain impurities attached to the C_{58}, C_{59}, C_{60} and FeC_{60} molecules. These impurities can be H, C and C_xH_x fragments (x=1...5). Unfortunately the resolution of our bending magnet does not allow separating these components. Of course we can not tell at that moment the iron atoms are inside or outside the carbon cage, neither.

As a conclusion our experiment demonstrated that the ECR ion source generally can be used to produce mixed iron-fullerene plasma and FeC_{60} molecules both in the plasma and in the beam. However, most probably other method (e.g. sputtering, electron bombardment) will be necessary to make a more clean iron component of the plasma.

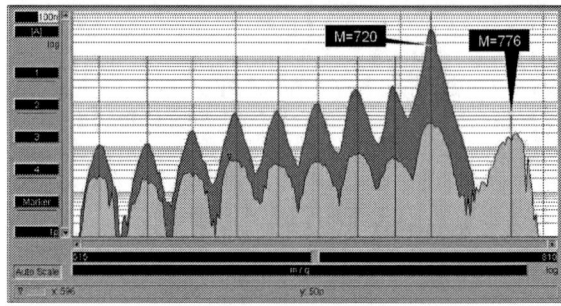

FIGURE 4. Fullerene beam spectrum (upper curve) optimized for C_{60}^+ (M=720). Lower curve: fullerene+ferrocene spectrum optimized for the peak at M=776.

ACKNOWLEDGEMENTS

This work was supported by OTKA grants (T42729 and T046454) and by a Hungarian-Japanese intergovernmental scientific-technological co-operation (OM-TET-JAP-3/00). S. Biri is a grantee of the Bolyai János Scholarship. J.Karácsony was supported by the Domus Hungarica Scientarium et Artium Fellowship.

REFERENCES

1. Biri S., Vámosi J., Valek A., Kormány Z., Takács E., Pálinkás J. Nucl. Instr. Methods B124 (1997) 427-430.
2. Biri S., Valek A., Kitagawa A., Muramatsu M., Proc. 15. Int. Ws. ECRIS. Univ. Jyvaskyla, Finland, 2002 (JYFL Research Report 4/2002), pp. 49-52.
3. Kenéz L., Biri S., Karácsony J., Valek A., Nakagawa T., Stiebing K. E., Mironov V., Rev. Sci. Instrum. 73 (2002) 617-619.
4. Kenéz L., Biri S., Karácsony J., Valek., Nucl. Instr. Methods B187 (2002) 249-258.
5. Kenéz L. et al., paper under preparation.
6. Muramatsu M. et al, this proceedings.
7. Biri S., Takács E., Hudson L.T., Valek A., Radics B., Imrek J., Juhász B., Suta T., Szabó Cs., Pálinkás J., Rev. Sci. Instrum. 75 (2004) 1420-1422.
8. Takács E., Radics B., Szabó C.I., Biri S., Hudson L.T., Imrek J., Juhász B., Suta T., Valek A., Pálinkás J., 12th Int. Conf. on the Physics of the Highly Charged Ions (HCI'04), Vilnius, Lithuania, 6-10 September 2004.
9. Biri S., Valek A., Kenéz L., Jánossy A., Kitagawa A., Rev Sci. Instrum. 73 (2002) 881-883.

Effect of the plasma electrode position on the beam intensity and emittance of the RIKEN 18 GHz ECRIS

Y. Higurashi, T. Nakagawa, M. Kidera, T. Aihara*, M. Kase and Y. Yano

RIKEN, Hirosawa 2-1, Wako, Saitama 351-0198, Japan

*SHI, Accelerator Service Ltd, Kita-shinagawa 5-9-11, Shinjuku-ku, Tokyo 141-0001, Japan

Abstract. We measured the beam intensity of Ar, Kr and Xe ions as a function of plasma electrode position. The beam intensities of lower charge state heavy ions increased when moving the plasma electrode toward the ECR zone. The RF power absorbed in the plasma chamber was slightly dependent on the gas pressure and 70% of injected RF power was absorbed. The emittance of ion source was gradually increased with increasing the drain current, which may be due to the space charge effect.

INTRODUCTION

To improve the performance of RIKEN 18 GHz ECRIS, we intensively studied the effect of the plasma electrode position on the beam intensity of heavy ions during past several years.[1,2] In previous experiments, we investigated the effect of the plasma electrode position on the beam intensity of $Ar^{8+,9+}$ and increased their beam intensities up to order of mA.[3] However, these effects on the beam intensities of heavier ions (Kr and Xe ions) are still not clear. Furthermore, such intense beam should affect on the beam quality (emittance, brightness and so on).[4]

RF power is one of the important parameters to evaluate the performance of the ion source. In our case, we used the RF power subtracted from the injected RF power to reflected one. However, this value is strongly dependent on the distance between the RF power supply and ion source, and the number of components used in the RF transmission line. To evaluate the performance of the ion source, it is better to use the RF power absorbed in the plasma chamber. For this reason, we tried to measure the RF power absorbed in the plasma chamber.

In this paper, we report the effect of the plasma electrode position on the beam intensity of Ar, Kr and Xe ions and the emittance of these ions. We also report the measurements of the RF power absorbed in the plasma chamber.

EXPERIMENTAL RESULTS AND DISCUSSION

Plasma electrode position

Figure 1 shows the position of movable plasma electrode and beam extraction side. The diameter of the plasma electrode hole was 10 mm. The distance between the extraction electrode and the plasma electrode was ~15 mm. The hole of the extraction electrode was 12 mm.

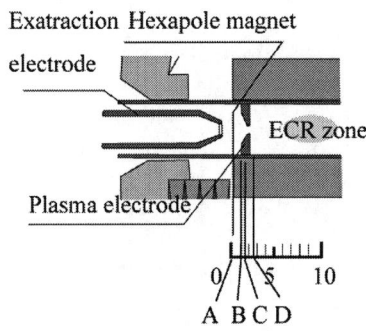

FIGURE 1. Cross sectional view of the movable plasma electrode and beam extraction side of the RIKEN 18 GHz ECRIS.[5]

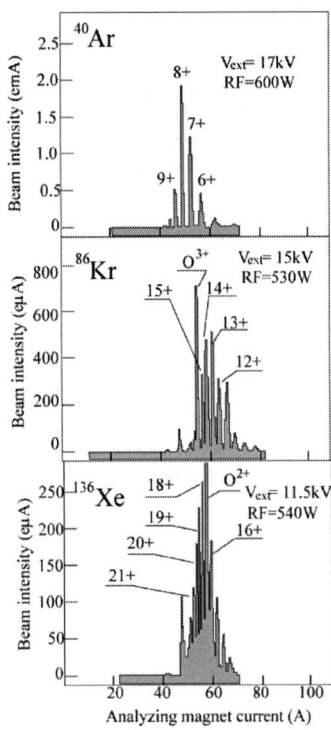

FIGURE 2. Charge distribution of the Ar, Kr and Xe ions

In a previous paper,[3] the plasma electrode position for maximizing the beam intensities of the $Ar^{8+,9+}$ was found to be at electrode position C (see Fig. 1). For investigating this effect on the beam intensity of Kr and Xe ions, we measured the beam intensities as a function of plasma electrode position. We have chosen four positions (position A(edge of the hexapole magnet), B, C, and D (2.5 cm far from the edge of the hexapole magnet) as shown in Fig. 1. The other parameters (magnetic field, gas pressure, biased disc position, and negative bias voltage of the disc) were tuned to maximize the beam intensity.

Figure 2 shows the charge distributions of Ar, Kr and Xe ions produced from RIKEN 18 GHz ECRIS at the plasma electrode position C. The ion source was tuned to produce Ar^{8+}, Kr^{13+} and Xe^{18+} ion beams. Beam intensities of $Xe^{18,20+}$ increased from 0.2 to 0.3 emA, when the plasma electrode was moved from position A to C. At plasma electrode position C, beam intensities of 0.6 mA of Kr^{13+}, 0.5 emA of Kr^{14+} and 0.3 emA of $Xe^{18,20+}$ were extracted at the RF power of ~700W.

Figure 3 shows the beam intensity of Ar, Kr and Xe ions at the plasma electrode position A and C. We observed that the beam intensity of medium charge state Kr and Xe ions strongly enhanced at the plasma electrode position C, which is the same tendency as Ar ion production. Figure 4 shows the beam intensity of Ar, Kr and Xe ions as a function of plasma electrode position. It is clearly seen that beam intensities of lower charge state heavy ions increased when moving the plasma electrode toward the ECR zone.

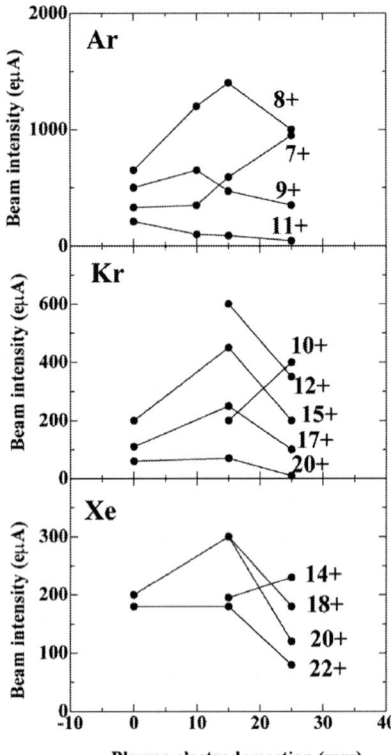

FIGURE 3. Beam intensity of Ar, Kr and Xe ions at the plasma electrode position A and C.

FIGURE 4. Beam intensity of Ar, Kr and Xe ions as a function of plasma electrode position

Figure 5 shows the beam intensities of Ar, Kr and Xe ions as a function of RF power at the plasma electrode position C. The ion source was tuned with 600 W of RF power and then the RF power was changed without changing any other parameters. It should be stressed that the beam intensity is not saturated at the maximum RF power in this experiment. This means that higher beam intensities can be realized at higher RF power.

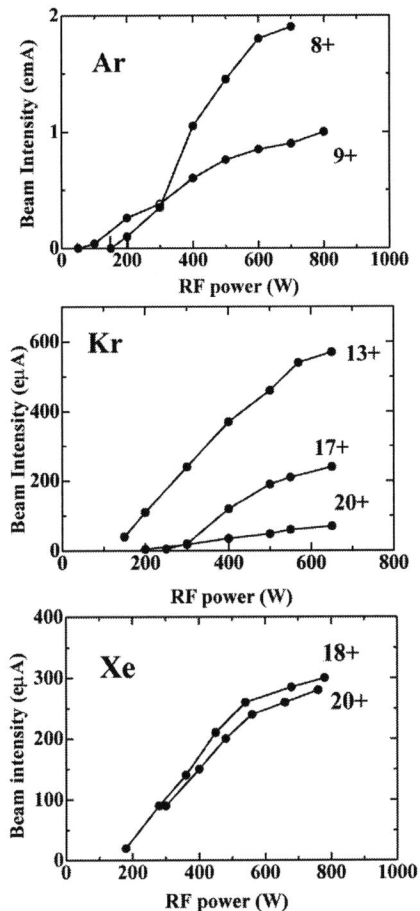

FIGURE 5. Beam intensities of Ar, Kr and Xe ions as a function of RF power at plasma electrode position C

RF power absorbed in the plasma chamber

Figure 6 shows the schematic diagram of the RF power supply and transmission lines. The microwaves emitted from the klystron are injected through several components into the plasma chamber as shown in Fig. 6. Input and reflected RF powers were measured by the input and reflected RF power monitor. In order to measure the absorbed RF power in the plasma chamber, we measured the temperature deference between input and output cooling water of the plasma chamber. The flow rate of the cooling water was 2.2 l/min.

Figure 7 shows the absorbed RF power in the plasma chamber as a function of input RF power. In order to minimize the effect of extraction current on the increase (or decrease) of the temperature of the cooling water, we did not supply the extraction voltage. To investigate the effect of gas pressure, we measured the RF power at the gas pressure of 6×10^{-7} to 2×10^{-6} torrs. We observed that the absorbed RF power in the plasma chamber was independent on the gas pressure as shown in Fig. 7 and the 70% of input RF power was absorbed in the plasma chamber.

The total length of waveguide (15.8 mm in width 7.9 mm in height) is ~4m. It is estimated that the ~90W of RF power is absorbed in the waveguide at the input RF power of 500W.[6] We also used the flexible waveguide (total length of 20 cm) which has the absorbed RF power of 3.3W/cm. It means that the 66W of RF power have to be absorbed in it. From these results, the total absorbed RF power in the wave guide is estimated to be 161W. The absorbed RF power in the plasma chamber was ~280 W. In transmission line, we used about 10 wave guide flanges. The RF power may be absorbed in these flanges.

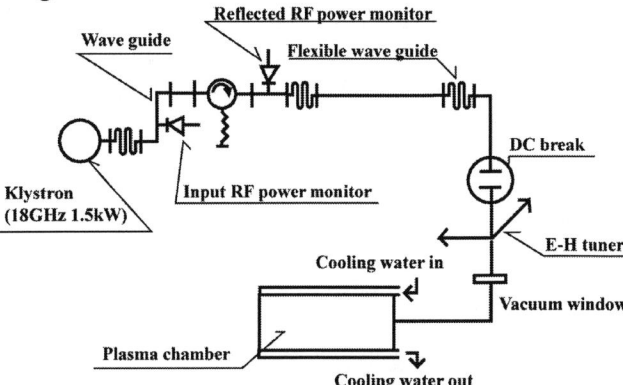

FIGURE 6. Schematic drawing of the RF power supply and transmission lines

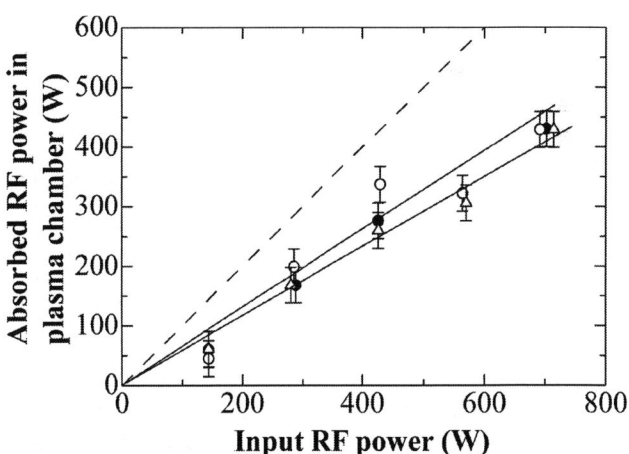

FIGURE 7. Absorbed RF power in the plasma chamber as a function of input RF power

Emittance measurement for intense beams

Measurements of the emittance in both horizontal and vertical were performed in wide range of the drain current of RIKEN 18 GHz ECRIS. Figure

8 shows the normalized 99% emittance of Ar^{8+} as a function of drain current. It is clearly seen that the emittance gradually increased from 0.5 to 1.1 πmm mrad with increasing the drain current from 4 to 15 mA. At the highest drain current, the beam intensity of Ar^{8+} was 1.5 emA at the extraction voltage of 17kV. In this case, the unnormalized emittance was 420 πmm mrad, which is 3 times as large as the acceptance of our RFQ linac (Acceptance is 150 πmm mrad.).[7] To accelerate the full beam of 1.5mA Ar^{8+} in this condition, we surly need the extraction voltage of 60 kV. At lower drain current, 90 % of Ar^{8+} beam will be accepted by the RFQ linac at the extraction voltage of 20kV. The emittance of Xe^{20+} were 0.45 and 0.47 πmm mrad for the drain current of 3.5 and 1.5 mA, respectively. It seems that the size of emittance is almost constant at lower drain current. In this test experiment, other effects (beam instability and so on) may be larger than the space charge effect. To understand these phenomena, we need further investigation.

99% emittance of Ar^{8+} (beam intensity of 1.5 mA) was 420 πmm mrad at the extraction voltage of 17 kV and drain current of 15 mA.

REFERENCES

1. Y.Higurashi, T.Nakagawa, M.Kidera, T.Aihara, M.Kase and Y.Yano Rev. Sci. Instrum. 73, 598 (200s)
2. Y.Higurashi, T.Nakagawa, M.Kidera, T.Aihara, M.Kase and Y.Yano Jpn. J. Appl. Phys. Vol. 42 (2003) pp.3656
3. Y.Higurashi, T.Nakagawa, M.Kidera, T.Aihara, M.Kase and Y.Yano Nuclear Instruments and Methods in Physcs Research – A 510 (2003) 206-210
4. S.Gamino et.al. Rev. Sci. Instrum. 75, 1637(2003)
5. T.Nakagawa and Y.Yano, Rev. Sci. Instrum. 71, 637 (2000) and references therein
6. Handbook of ION SOURCES (CRC, New York. 1995) Edited by B.wolf.
7. O.kamigaito et.al. Rev. Sci. Instrum. 70, 4523 (1999)

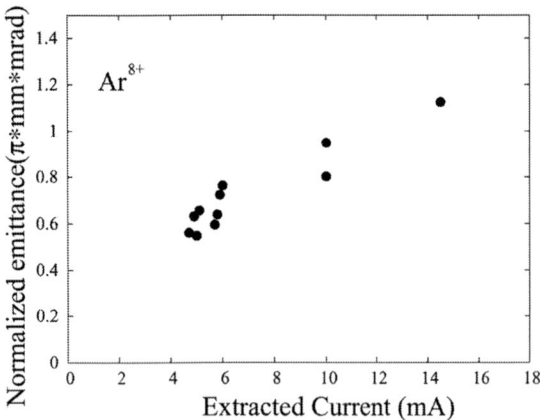

FIGURE 8. Normalized 99% emittance of Ar^{8+} as a function of drain current

Conclusions

We measured the beam intensity of Ar, Kr and Xe ions as a function of plasma electrode position. The beam intensities of lower charge state heavy ions increased when moving the plasma electrode toward the ECR zone. Using this method, we obtained 2mA of Ar^{8+}, 0.6mA of Kr^{13+} and 0.3mA of Xe^{20+}. The RF power absorbed in the plasma chamber was measured. It was slightly dependent on the gas pressure and 70% of injected RF power was absorbed. The emittance of ion source was gradually increased with increasing the drain current. It may be the space charge effect. The

Radioactive Beams from ^{252}Cf Fission Using a Gas Catcher and an ECR Charge Breeder at ATLAS

Guy Savard, Richard C. Pardo, E. Frank Moore, Adam A. Hecht, and Sam Baker

Argonne National Laboratory, 9700 S. Cass Avenue, Argonne, IL 60439

Abstract. A proposed upgrade to the radioactive beam capability of the ATLAS facility has been proposed using ^{252}Cf fission fragments thermalized and collected into a low-energy particle beam using a helium gas catcher. In order to reaccelerate these beams the ATLAS ECR-I will be reconfigured as a charge breeder source. A 1Ci ^{252}Cf source is expected to provide sufficient yield to deliver beams of up to ~10^6 far from stability ions per second on target. A brief facility description and the expected performance information are provided in this report.

INTRODUCTION

After over a century of research in nuclear physics, the information needed to address the current outstanding questions of the field is obtained with greater and greater difficulty by the experiments that can be carried out with stable beams. Often these questions cannot be addressed at all with these beams. Experiments with stable beams naturally tend to explore the proton-rich side of the valley of stability and the nearby neutron-rich region of the isotope landscape.

Important questions to be addressed in nuclear physics that need neutron-rich radioactive beams include:

a. modification of the nuclear structure in neutron-rich systems such as shell-structure quenching, single particle structure near n-rich magic nuclei, and pairing interaction in weakly bound systems;
b. collective behavior in neutron-rich systems and
c. information on the astrophysical r-process path including nuclear masses, lifetimes, and ß-delayed neutron branching ratios.

Most existing facilities can only probe the proton-rich side well and dabble on the periphery of the much larger neutron-rich region. Facilities that will reach further into this region and provide interesting beams into the far neutron-rich region, such as the Rare Isotope Accelerator (RIA) [1], are still many years from operation.

In the interim period, an interesting, transitional facility based on fission fragments of ^{252}Cf can allow a large class of important measurements prior to the realization of these much larger facilities. Figure 1 shows the distribution of fission fragments from ^{252}Cf [2]. The distribution covers a wide region of the neutron-rich side populating some of the important nuclei for addressing the issues mentioned above. In addition the mass distribution of ^{252}Cf is quite complimentary to that of proton or neutron-induced fission of ^{235}U [2].

At Argonne National Laboratory we propose to make use of this unique fragment distribution to provide the nuclei which will then be thermalized in a helium gas-catcher system scaled from a prototype design being developed for the RIA facility. The remainder of this paper gives an overview of the proposed facility, describes its main features, discusses the challenges in implementing such a facility, and describes the expected performance.

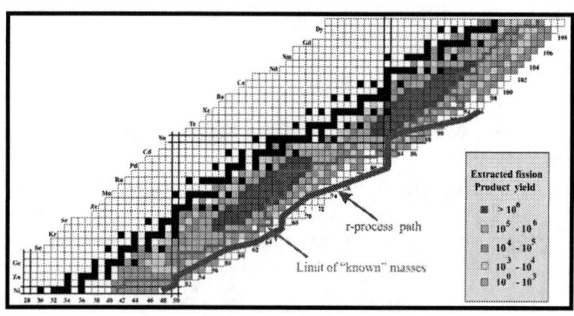

Fig. 1. Distribution of fission products from the spontaneous fission of ^{252}Cf. The color code gives the extracted low-energy ion beam intensity (s^{-1}) for a 1 Curie source.

TECHNICAL PLAN

The proposed facility consists of six major components:

a. A one Curie ^{252}Cf fission source mounted on a strong backing but open in the forward direction except for a thin gold layer.
b. A helium gas catcher and RFQ system to thermalized the fission products and collect them into a singly- or doubly-charged ion beam with very low emittance and energy spread.
c. A sophisticated mass analysis system tailored to this excellent quality beam with a mass resolution of 1:20,000.
d. An ECR charge-breeder ion source for stripping the 1+ ions to a charge state suitable for further acceleration.
e. The ATLAS superconducting linac for acceleration to the necessary beam energy.
f. Additional diagnostics systems in ATLAS to provide the necessary information for beam tuning and delivery.

A schematic overview of the planned facility and its relationship to the existing ATLAS linac is shown in figure 2. The entire assembly of fission source, gas catcher and ECR ion source will be mounted on high-voltage platforms in order to provide the ions with the necessary velocity(0.086c) for injection into the linac.

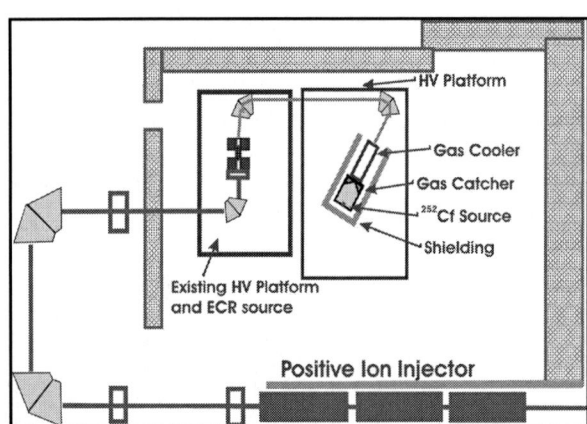

Fig. 2. Schematic overview of the proposed ^{252}Cf fission fragment beam facility. An existing ECR ion source will be modified for charge breeding and then inject these beams into the ATLAS linac.

The ^{252}Cf source will be deposited on a tantalum backing to provide sufficient mechanical strength. A thin gold foil will provide isolation from the rest of the vacuum system and will also serve as an energy degrader to match the fission fragment range in the helium gas to the available volume. The source will be installed in a vacuum assembly which will mate to the gas catcher system by remote control. The assembly will be enclosed in a neutron and gamma shield consisting of 0.75 m polyethylene and heavy metal.

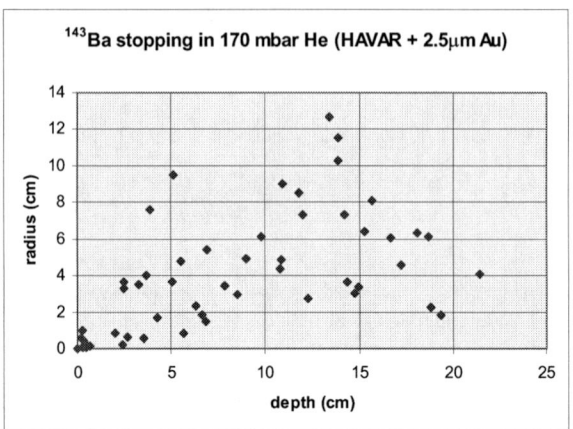

Fig 3. Distribution of stopping positions for ^{143}Ba fission fragments in 170 mbar He.

The use of helium gas catchers to slow down and thermalize ions produced in nuclear reactions as well as fission fragments has been demonstrated over the past few years and is in routine use at ATLAS [3]. In addition this technology is key to the RIA project and a prototype gas catcher system suitable for RIA has been developed [4]. The ion distribution stopped in such a gas catcher for ^{143}Ba is shown in figure 3 as calculated by the program SRIM[5].

A shortened version of the RIA gas catcher system will be used for this project. The gas catcher volume has a radius of 13 cm and length of ~0.5 m. Once thermalized the ions remain charged in the gas and are pushed toward the extraction end by DC fields, are focused toward the nozzle and kept away from the electrodes by an RF cone and finally extracted by gas flow into a radio frequency quadrupole (RFQ) focusing channel collecting ions extracted from the nozzle of the gas catcher. The RFQ guides and cools the ions while the helium extracted with the ions is pumped away. This results in extracted ion beams with excellent energy spread and transverse emittance - $\delta E \sim 1$ eV and $\varepsilon \sim 3\pi$mm mrad at 50 keV [6]. The assembled RIA prototype gas catcher is shown in figure 4.

Good beam purity is desirable is all cases, but is very difficult to achieve due to the extremely small mass differences for differing isobars. This is most serious near the valley of stability where good separation requires a mass resolution of 1:50,000. As one moves further away from the stable region the mass differences increase and, for most cases, a mass resolution of 1:20,000 is adequate or at least provides significant discrimination from the unwanted isotopes. With the expected beam quality, it is possible to achieve a mass resolution of 1:20,000 using a scaled-down version of a mass spectrometer designed for RIA [7] without energy correction. The geometry of such a layout is indicated in figure 2 with a total bend of 120 degrees. The mass analysis must be performed at a beam energy approximately 50 kV higher than provided by the ECR source/fission fragment bias and will be accomplished by floating the spectrometer at this voltage relative to the platform.

Fig. 4. The assembled RIA prototype gas catcher as seen from the nozzle/RF cone end. A shorter version of this device will be used for this project.

To reach Coulomb barrier energies (E/A ≥ 5 MeV/u), the ATLAS linac must be provided with ions whose charge-to-mass ratio (q/M) is at least 0.15. Thus the 1+ ions provided by the gas catcher system must be stripped to higher charge states. This task will be accomplished by transporting the ions to an ECR ion source operating as a charge breeder [8], stopping the ions in the ECR source and stripping the ions, as in normal ECR ion source operation. An existing ATLAS ECR ion source, known as ECR-I, will be modified into a charge breeder for this purpose. This source operates at 10 GHz RF frequency, but is capable of accepting 14 GHz RF as well. Both frequencies will be implemented for this project in an effort to improve source performance, especially with regard to total efficiency. Work by Sortais and others [9] has shown charge breeding efficiencies of around 5% for solid materials and as high as 12% for some gases. A view of ECR-I modified as a charge breeder is shown in Figure 5.

Shielding design and radiological monitoring and protection are important issues for this proposed facility. We expect to perform hands-on maintenance to most of the facility and to work near the source during beam development. The entire facility will be housed in a large high bay area that also contains the positive ion injector and support utilities and facilities. The initial mass separation takes place on the first high voltage platform so as to contain as much of the unwanted activity as possible to that platform. We have developed a shielding concept which will allow remote control during installation of the source as well as storage of the source during maintenance requiring access to the gas catcher interior or other portions of the nearby beam transport system.

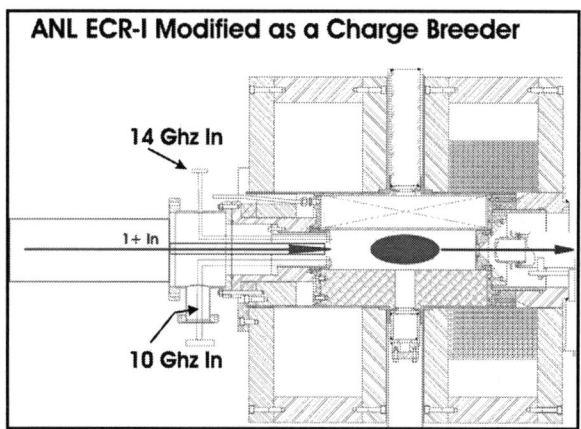

Fig. 5 10 & 14 GHz ECR-I ion source modified on the injection side for charge breeding operation.

Shielding from the fission-generated neutrons requires about 0.75 m of polyethylene as well as a small amount of lead shielding to attenuate the γ-ray flux. This level of shielding will reduce the on-contact radiation field to approximately 1 mrem/hr. To create 4π shielding, the associated beamline must also be heavily shielded. In addition to neutron shielding, significant quantities of decaying fission fragments will be accumulated on the gas catcher walls, magnet chamber and other beamline walls. Shielding of this radiation is simple during operation, but it does present a challenge for maintenance activities both from potential direct exposure and limiting any possibility for spreadable contamination. Temporary tent enclosures and the use of attachable glove boxes appear to be good solutions for this problem. The shielding concept is shown in Figure 6.

FACILITY PERFORMANCE

The fission mass distribution from ^{252}Cf is shown in figure 1. The yields identified in that figure assume approximately 2% total system efficiency for delivery to an experiment target location. Some specific additional examples are listed in Table I. The efficiencies assumed in this estimate include:

a. 50% of fission fragments enter the gas catcher,
b. 45% of fission fragments entering the gas catcher emerge as a beam of 1+ ions,
c. 10% ECR stripping efficiency into one charge state,
d. 80% bunching efficiency and acceleration in linac,
e. 90% beam transport efficiency.

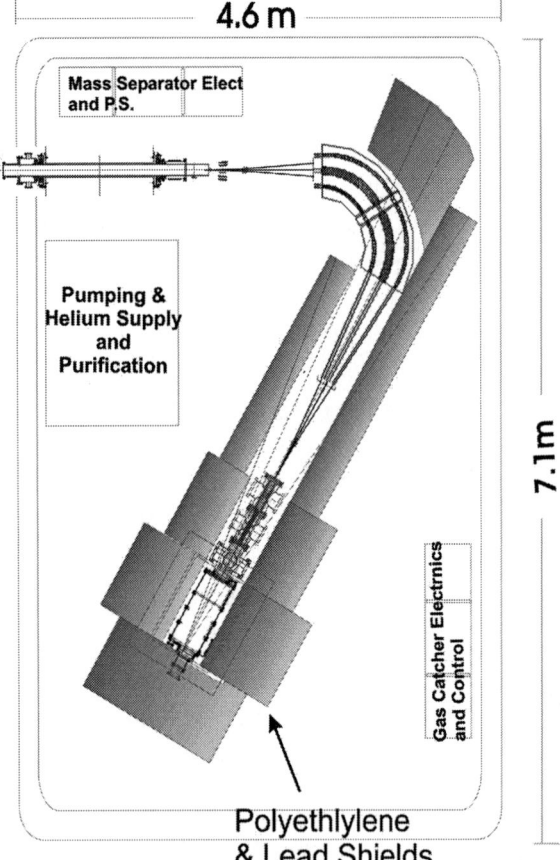

Fig. 6 ^{252}Cf fission source and gas catcher beam transport shielding plans on the HV platform.

Improvements in a number of areas will be required to achieve these results.

Table I
Examples of expected maximum beam intensities for 1Ci ^{252}Cf fission source using efficiencies discussed in test

Isotope	Half-life (s)	Yield (s^{-1})
^{143}Ba	14.3	8x10^5
^{145}Ba	4.0	4x10^5
^{130}Sn	187	7x10^4
^{132}Sn	40	2x10^4
^{110}Mo	2.8	4x10^3
^{111}Mo	0.5	3x10^2

SUMMARY AND STATUS

A ^{252}Cf fission source delivers a unique class of radioactive species that can provide tools to address important physics questions during the era leading up to RIA. Unique technologies and expertise available at ATLAS can provide the necessary capabilities in a timely manner and at low cost (approximately $2.5M). The proposed upgrade has great synergy to RIA on both the technical and physics fronts and will provide a very smooth transition toward RIA research. This upgrade will keep ATLAS and the US competitive in radioactive beam physics until RIA becomes available.

The project is currently in the design and proposal writing phase. A complete proposal will be submitted to DOE in the fall of 2004. This work is supported by the U.S. Department of Energy, Office of Nuclear Physics, under contract W-31-109-ENG-38.

REFERENCES

1. J.A. Nolen, Proceeding of XXI International LINAC Conference, MO302, Gyeongju, Korea, August 19-23, 2002.
2. T.R. England and B.F. Rider, Los Alamos National Laboratory, LA-UR-94-3106; ENDF-349 (1993).
3. J.A. Clark, G. Savard, K.S. Sharma, J. Vaz, J.C. Wang, Z. Zhou, A. Heinz, B. Blank, F. Buchinger, J.E. Crawford, S. Gulick, J.K.P. Lee, A.F. Levand, D. Seweryniak, G.D. Sprouse and W. Trimble, Phys. Rev. Lett. 92 (2004) 192501.
4. G. Savard, A. Heinz, J.P. Greene, D. Seweryniak, Z. Zhou, J. Clark, K.S. Sharma, J. Vaz, J.C. Wang, C. Boudreau and the S258 collaboration, Nucl. Instr. & Meth. **B204** (2003) 582.
5. J. F. Ziegler, J. P. Biersack and U. Littmark, **The Stopping and Range of Ions in Solids**, Pergamon Press, New York, 1985.
6. F. Herfurth, Nucl. Instr. & Meth. **B204** (2003) 587.
7. Portillo, M.; Nolen, J. A.; Barlow, T. A., Proceedings 2001 Particle Accelerator Conference, Chicago, IL, 6/18/2001, IEEE, **4**, 3015-17, (2001)
8. R. Geller, C. Tamburella, and J.L. Belmont, Rev, Sci. Instr., **67**, #3, 1281(1995).
9. T. Lamy, J.-C. Curdy, R. Geller, C. Peaucelle, P. Sole, P. Sortais, T. Thuillier, D. Voulot, K. Jayamanna, M. Olivo, P. Schmor, D. Yuan, Proceedings of the 2003 European Particle Accelerator Conference, Paris, France, June 3-7, 2003, ISSN 1684-761X, page 1724.

IV DIAGNOSTICS AND APPLICATIONS

High Resolution He-like Argon And Sulfur Spectra From The PSI ECRIT

M. Trassinelli[*], S. Biri[†], S. Boucard[*], D.S. Covita[**], D. Gotta[‡], B. Leoni[§], A. Hirtl[¶], P. Indelicato[*], E.-O. Le Bigot[*], J.M.F. dos Santos[**], L.M. Simons[§], L. Stingelin[§], J.F.C.A. Veloso[‖††], A. Wasser[§] and J. Zmeskal[¶]

[*]*Laboratoire Kastler Brossel, Université Pierre et Marie Curie, Paris, France*
[†]*Insitute of Nuclear Research (ATOMKI), Debrecen, Hungary*
[**]*Physics Department, University of Coimbra, Portugal*
[‡]*Institut für Kernphysik, Forschungszentrum Jülich, Jülich, Germany*
[§]*Paul Scherrer Institut, Villigen PSI, Switzerland*
[¶]*Österreichisch Akademie der Wissenschaften, Wien, Austria*
[‖]*Physics Department, University of Coimbra*
[††]*Physics Department, University of Aveiro, Portugal*

Abstract. We present new results on the X-ray spectroscopy of multicharged argon, sulfur and chlorine obtained with the Electron Cyclotron Resonance Ion Trap (ECRIT) in operation at the Paul Scherrer Institut (Villigen, Switzerland). We used a Johann-type Bragg spectrometer with a spherically-bent crystal, with an energy resolution of about 0.4 eV. The ECRIT itself is of a hybrid type, with a superconducting split coil magnet, special iron inserts which provides the mirror field, and a permanent magnetic hexapole. The high frequency was provided by a 6.4 GHz microwave emitter.

We obtained high intensity X-ray spectra of multicharged F-like to He-like argon, sulfur and chlorine with one 1s hole. In particular, we observed the $1s2s\ ^3S_1 \rightarrow 1s^2\ ^1S_0\ M1$ and $1s2p\ ^3P_2 \rightarrow 1s^2\ ^1S_0\ M2$ transitions in He-like argon, sulfur and chlorine with unprecedented statistics and resolution. The energies of the observed lines are being determined with good accuracy using the He-like M1 line as a reference.

We surveyed the He-like M1 transition intensity as a function of the ECRIT working conditions. In particular we observed the M1 intensity dependency on the coil current and on the injected microwave power.

INTRODUCTION

The Electron Cyclotron Resonance Ion Trap (ECRIT) of the Paul Scherrer Institut (PSI) has been set up with the goal of producing narrow X-ray lines that would yield the response function of the Bragg crystal spectrometer with high accuracy [1]. Such a measurement is crucial for the ongoing pionic hydrogen experiment at PSI [2, 3, 4]. For low- to medium-Z atoms, the E1 X-rays from hydrogen-like ions and the M1 X-rays from helium-like ions have an energy of a few keV, with a natural width which is negligible compared to the expected resolution of the Bragg crystal spectrometer. The ion kinetic energy in an ECRIS is small: it is on the level of less than 1 eV [5, 6]. Because of this, a Doppler broadening of less than 40 meV can be expected in transitions in the 3 keV energy range in He-like argon.

In 2002, the $1s2s\ ^3S_1 \rightarrow 1s^2\ ^1S_0\ M1$ X-ray transition in He-like argon has been used to study the response function of the Bragg crystal spectrometer with quartz [10-1] and silicon [111] crystals [7]. In spring 2004, the characterization of the spectrometer was completed with the test of our whole set of crystals (quartz [10-1], quartz [100] and silicon [111]) using the M1 transition line from He-like argon, chlorine and sulfur. In addition, we obtained X-ray spectra of highly-charged ions of these elements with unprecedented statistics and resolution.

EXPERIMENTAL SET-UP

The experimental set-up is almost identical to the 2002 run set-up described in Ref. [7]. It is composed of the ECRIT source and of a Bragg crystal spectrometer coupled to a position sensitive detector (see fig. 1).

The ECRIT consists of a pair of a superconducting split coil magnets (which, together with special iron inserts, provides the mirror field configuration), of an AECR-U style permanent hexapole magnet, and of a 6.4 GHz power regulated emitter. The mirror field pa-

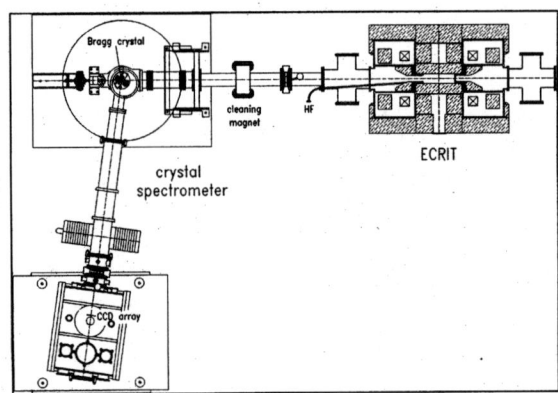

FIGURE 1. Set-up of the experiment.

rameters provide one of the highest mirror ratios for ECR sources, with a value of 4.3 over the length of the plasma chamber.

By using a cryopump and reducing the surface of iron insertion pieces the reference pressure (i.e. without plasma) was reduced from $1.7 \cdot 10^{-7}$ mbar (2002) to $3 \cdot 10^{-8}$ mbar Gas filling was supplied radially through the gaps in the open structure hexapole. The gas composition was routinely controlled and stabilized with a quadrupole mass spectrometer.

A Bragg crystal spectrometer of reflection type (Johann configuration) was installed at a distance of 2200 mm from the plasma center. Silicon and quartz crystal have used for the 2004 run, with different Bragg angle values. The crystals are 100 mm diameter circular plates, with a thickness of 0.2–0.3 mm, and are spherically bent with a curvature radius of 2982.4 ± 0.6 mm, by optical attachment to high-quality quartz spherical lenses.

The detector is an array of 6 CCDs of 600×600 pixels each [8], with an energy resolution of 140 eV at 3 keV. The pixel size at working temperature ($-100°C$) has been recently measured to be 39.9943 ± 0.0035 μm [9]. The granularity of the detector was decisive in discriminating charged particle events against X-rays possessing different topologies. The CCD chips and the associated electronics were protected against light and high-frequency (HF) power by a 30 μm thick Beryllium window installed in the vacuum tube between the crystal spectrometer and the ECRIT.

ECRIT at work

During spring 2004, we injected different kinds of gases in the ECRIT in order to obtain X-ray spectra from highly-charged argon, chlorine and sulfur. For this propose, we used a gas mixture of O_2 and, respectively, Ar, $CHClF_2$ and SO_2. Using the experience acquired in the previous run, we adjusted a mixing ratio of around 1:9, with a total pressure in the plasma chamber of $3 - 4 \cdot 10^{-7}$ mbar. In order to recognize the different charge states, we used as an initial reference the $K\alpha$ or $K\beta$ lines of the neutral gas, which are easily recognizable: they are the brightest when only a few watts of HF power are injected. Using known energy intervals, we were then able to move the spectrometer to the region of the nearby $1s2s\ ^3S_1 \rightarrow 1s^2\ ^1S_0$ M1 transition in the He-like ion, and to observe it. We then optimized the different ECRIT and spectrometer parameters in order to maximize the line intensity in the detector. The typical illumination time of the CCD chips before the readout is 1 min. Due to the large intensity of the X-rays, this allows for a sizeable probability of double hits for pixels near the line peak. In order to reduce this effect and improve the peak-to-background ratio, a densimet® collimator was inserted at a distance of 150 mm from the center of the plasma, leaving an aperture of 28 mm(h) × 4 mm(v) or 28 mm(h) × 1 mm(v), depending on the configuration. During the different runs the (double hits) to (single hits) ratio always stayed below 5%, which is small enough to properly handle double hits in the final analysis, and to neglect triple hit processes.

During the optimization of the apparatus, we studied the M1 line intensity as a function of the injected HF power. As expected, we observed a strong dependence between the M1 intensity and the HF power (see fig. 2). In contrast, we noted an unexpected behavior of the maxima of the curves, whose HF intensity does not increase with the ionization energy of the He-like ion.

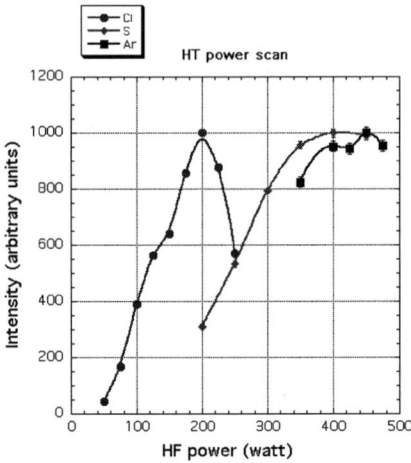

FIGURE 2. Injected HT power scan for argon, chlorine and sulfur versus the He-like M1 intensity (in arbitrary units).

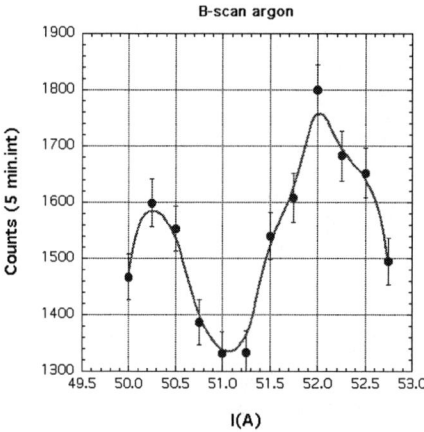

FIGURE 3. He-like argon M1 line intensity dependency versus the superconducting coils current. 55 A corresponds to a magnetic field value of 2.35 kG at the center of the ECRIT.

During the ECRIT parameters adjustment, we observed a non-trivial dependency of the M1 intensity against the longitudinal magnetic field. As shown in fig. 3, we observe the presence of two distinct maxima in the M1 intensity-coil current relationship.

Atomic Spectra

One of the most important goals of the spring 2004 run was the high-precision measurement of the X-ray spectra of argon, chlorine and sulfur. With 1–2 hours maximum acquisition time, we obtained high-statistics spectra of He-, Li- and Be-like ionic states of these elements. A crystal spectrometer like ours can only measure energy differences between atomic transitions. A reference is thus needed. Due to the lack of high quality reference line in neutral atom X-ray spectra (see, e.g., [10]) we used as a reference He-like $1s2s\ ^3S_1 \rightarrow 1s^2\ ^1S_0$ $M1$ transition. All the transition energies provided in the present work are based on the M1 theoretical transition energy calculated with a multi-configuration Dirac-Fock code [11, 12]. The peaks in the spectra were fitted with a simulated spectrometer response function which was convoluted to a Gaussian. The response function was obtained through a Monte Carlo X-ray tracking simulation based on the theoretical reflection function of the crystal obtained with the XOP code [13]. The reliability of the simulation had previously been tested during the crystal response function study [7]. The results obtained in highly-charged argon and sulfur spectroscopy have an unprecedented precision of the order of 10 meV and they agree with the previous experimental values and theoretical predictions (see tables 1, 2).

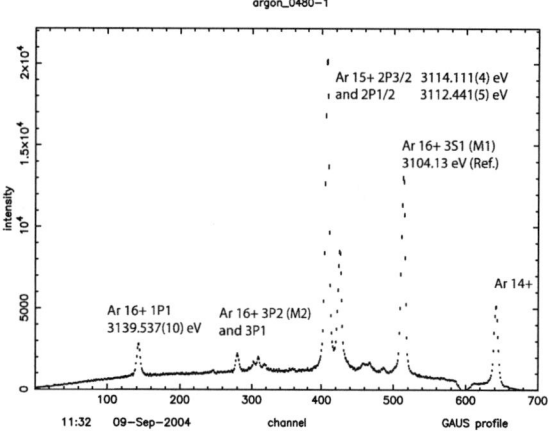

FIGURE 4. He-like and Li-like argon spectrum using quartz [10-1] crystal. HF power injected $P = 400\ W$. argon partial pressure $p_{Ar} = 5 \cdot 10^{-9} mbar$, total pressure $p_{tot} = 5 \cdot 10^{-7} mbar$

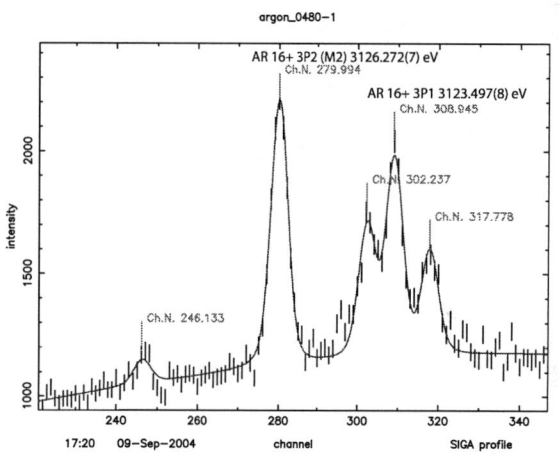

FIGURE 5. Detail of the spectrum in figure 4 around the $1s2p\ ^3P_n \rightarrow 1s^2\ ^1S_0$ transitions

OUTLOOK AND CONCLUSION

In this paper, we demonstrated once more the possibility of using high-precision X-ray spectroscopy with a Bragg spectrometer in the study of ECR ion sources. Moreover, we presented new results on X-ray transition energies in highly-charged argon, chlorine and sulfur.

TABLE 1. He-like argon energy transitions in eV

	$1s2p\ ^1P_1 \to$ $\to 1s^2\ ^1S_0$	$1s2p\ ^1P_1 \to$ $\to 1s2p\ ^3P_1$
Costa [11] (th.)	3139.57	16.05
Plante [14] (th.)	3139.6236	16.0484
Lindgren [15] (th.)		16.048
Deslattes [16] (exp.)	3139.553(36)	16.031(72)
This Work (preliminary results)	3139.537(10)	16.040(17)

TABLE 2. He-like sulfur energy transitions in eV

	$1s2p\ ^1P_1 \to$ $\to 1s^2\ ^1S_0$	$1s2p\ ^1P_1 \to$ $\to 1s2p\ ^3P_1$
MCDF [12] (th.)	2460.6169	13.4875
Plante [14] (th.)	2460.6707	13.4857
Schleinkofer [17] (exp.)	2460.67(9)	13.62(20)
This Work (preliminary results)	2460.608(9)	13.483(16)

The next steps will consist in finishing the data analysis of lower-charge states (Li- and Be-like) of argon and sulfur, in analyzing the chlorine spectra and in studying the injected HF power dependency of the satellite transition in the He-like ions spectra.

ACKNOWLEDGMENTS

The technical support of L. Stohwasser, H. Schneider and D. Stückler was essential in obtaining the results described here. The advice and help of D. Hitz and K. Stiebing during the preparation of the ECRIT experiment is warmly acknowledged.

REFERENCES

1. Gotta, D., *Prog. Part. Nucl. Phys.*, **52**, 133–195 (2004).
2. Biri, S., Simons, L., and Hitz, D., *Rev. Sci. Instrum.*, **71**, 1116–18 (2000).
3. Anagnostopoulos, D. F., Biri, S., Boisbourdain, V., Demeter, M., Borchert, G., Egger, J. P., Fuhrmann, H., Gotta, D., Gruber, A., Hennebach, M., Indelicato, P., Liu, Y. W., Manil, B., Markushin, V. E., Marton, H., Nelms, N., Rusi El Hassanii, A., Simons, L. M., Stingelin, L., Wasser, A., Wells, A., and Zmeskal, J., *Nucl. Instrum. Methods B*, **205**, 9–14 (2003).
4. Pionic Hydrogen Collaboration, *PSI experiment proposal R-98.01* (1998), URL http://pihydrogen.web.psi.ch.
5. Bernard, C., Ph.D. thesis, Université J. Fourier, Lyon (1996).
6. Sadeghi, N., Nakano, T., Trevor, D. J., and Gottscho, R. A., *J. Appl. Phys.*, **70**, 2552 (1991).
7. Anagnostopoulos, D., Biri, S., Fuhrmann, H., Gotta, D., Gruber, A., Indelicato, P., Leoni, B., Simons, L. M., Stingelin, L., Wasser, A., and Zmeskal, J., eprint: physics/0408081 (2004).
8. Nelms, N., Anagnostopoulos, D. F., Ayranov, O., Borchert, G., Egger, J. P., Gotta, D., Hennebach, M., Indelicato, P., Leoni, B., Liu, Y. W., Manil, B., Simons, L. M., and Wells, A., *Nucl. Instrum. Meth. A*, **484**, 419–31 (2002).
9. Trassinelli, M., "Precision Spectroscopy Of Pionic Atoms: From Pion Mass Evauation To Tests Of Chiral Perturbation Theory," in *DAΦNE 2004 proceeding, Frascati Physics Series*, to be published, eprint: physics/0409066.
10. Anagnostopoulos, D. F., Gotta, D., Indelicato, P., and Simons, L. M., *Phys. Rev. Lett.*, **91**, 240801 (2003).
11. Costa, A. M., Martins, M. C., Parente, F., Santos, J. P., and Indelicato, P., *At. Data Nucl. Data Tables*, **79**, 223–39 (2001).
12. Indelicato, P., private communication (2004).
13. Sanchez del Rio, M., and Dejus, J., "XOP: Recent development," in *SPIE proceedings*, 1998, p. 3448.
14. Plante, D., Johnson, W., and Sapirstein, J., *Phys. Rev. A*, **49**, 3519–3530 (1994).
15. Lindgren, I., Åsén, B., Salomonson, S., and Mårtensson-Pendrill, A. M., *Phys. Rev. A*, **64**, 062505 (5) (2001).
16. Deslattes, R., Beyer, H., and Folkmann, F., *J. Phys. B: At. Mol. Opt. Phys.*, **17**, L689–L694 (1984).
17. Schleinkofer, L., Bell, F., Betz, H., Trolman, G., and Rothermel, J., *Phys. Scr.*, **25**, 917–923 (1982).

Novel Technique for Trace Element Analysis using the ECRIS and Heavy Ion Linear Accelerator (ECRIS-AMS)

M. Kidera[*], T. Nakagawa[*], K. Takahashi[*], S. Enomoto[*], R. Hirunuma[*], K. Igarashi[*], M. Fujimaki[*], E. Ikezawa[*], O. Kamigaito[*], M. Kase[*], and Y. Yano[*]

Cyclotron Center, RIKEN, 2-1 Hirosawa, Wako, Saitama 351-0198, Japan

Abstract. We have developed the new analytical system which consists of electron cyclotron resonance ion source (ECRIS) and heavy ion linear accelerator. ECRIS-AMS (Accelerator Mass Spectrometry using Electron Cyclotron Resonance Ion Source) has several advantages described below. 1) The production of positive ions in the ECRIS is not influenced by ionization selectivity. 2) We can analyze many trace elements simultaneously in the material with very low background. 3) We can minimize the spectroscopic interference by using the high temperature ECR plasma. Using this system, we have measured elemental compositions in rock reference samples (JB-2). From our experimental results, it is considered that the further development and establishment of this method will play an important role in the trace element analysis. For this application, we just need small heavy ion linear accelerator which has acceleration energy of ~ 1MeV/u. In this contribution, we will present the procedure of analysis in detail and several experimental results for trace element analysis in the materials.

INTRODUCTION

Recent years the techniques for trace element analysis and accelerator mass spectrometry have been rapidly improved by great effort of many engineers and scientist and the detection limit has become lower and lower. However, these methods have several disadvantages. To overcome these disadvantages, new technique for trace elements analysis by using ECRIS and heavy ion linear accelerator were developed.

The conventional mass spectrometric method which entails the use of an ion source and a mass analyzing system is widely used in various field.[1] This method is often unable to detect some elements when they have the same mass as that of some molecules. For minimizing the influence of molecules with the same mass, the key issue is to identify the atomic and mass numbers of various elements. For this purpose, an accelerator mass spectrometry (AMS) which consists of a negative ion source and an electrostatic tandem accelerator has historically been used as an ultra sensitive device, particularly for measuring the isotopic ratio involving rare radio isotopes of interest for dating applications.[2] Negative ionization of the element strongly depends on the electron affinity. For the production of negative ions from elements of low electron affinity, great effort was exerted in the sample preparation. In many cases, results of the measurement have large errors due to the effect of various treatments of the elements. To measure a new element, it is necessary to make a test experiment many times. Furthermore it is impossible to produce negative ions from rare gases, such as Ar and Kr. These problems are disadvantages of the method, although a tandem accelerator is used for AMS.

As an extension of the AMS for detecting trace elements in various materials and for overcoming the disadvantages of the present AMS, we propose a novel method using the ECRIS and heavy ion linear accelerator[3] (ECRIS-AMS). The productions of positive ions in ECR plasma do not have problems such as the difficulty in production of negative ions, which is due to the electron affinity, because the electrons in ECR plasma have a high kinetic energy and easily ionize any atom. Furthermore, when using this system, the acceleration as well as transport depends only on the mass-to-charge ratio (A/q) of heavy ions. This indicates that many trace elements can be analyzed simultaneously. Thus, this method allows us to analyze about 90% of the elements when choosing only ions of A/q = 3, 4, and 5. It is possible to carry out a very low background experiment, because the effect of molecules will be minimized using the accelerator. Part of the experimentally results are described in Refs. [4] and [5].

In this paper, we explain our new experimental method for trace element analysis which uses ECRIS and a heavy ion linear accelerator. In section 2, the experimental method is described. In subsection of section 2, we present the typical experimental results obtained using this system.

EXPERIMENT BY USING ECRIS-AMS

Process of Experiment

Figure 1 shows the schematic drawing of the experimental setup for the trace element analysis. The multi charged ions produced from the ECRIS are selected using a bending magnet (A) placed downstream of the ECRIS. These ions are accelerated by the RFQ linear accelerator[6] and the RIKEN heavy ion linear accelerator (RILAC) (B). Then, the charge state of ions is exchanged using a charge-stripper C-foil (C). After passing through the foil, the charge state of ions is exchanged for another charge state which is strongly dependent on the velocity and atomic number of ions.[7] These ions are analyzed using a bending magnet (D). The analyzed particles are detected and identified using a ΔE-E counter telescopes which consists of ionization chambers.

To ionize the trace elements in a sample, we inserted it into the plasma of the ion source directly. For example, the solid rod is inserted into the plasma and then heated to obtain sufficient vapor pressure. For production of the intense beam of highly charged ions with A/q = 3, 4, and 5, we need a high-performance ion source. For this purpose, we used the RIKEN 18GHz-ECRIS which is highly suitable for producing the intense beam of highly charged heavy ions. The performance of this ECRIS is described in Ref.[8].

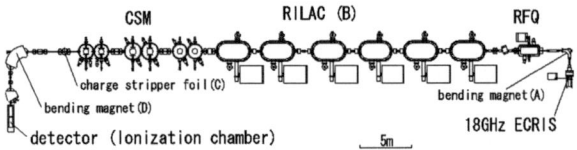

FIGURE 1. Whole view of the systems of ECRIS-AMS.

Measurement of Trace Elements in the Cinnabar

The concentration of composite elements of cinnabar in the ancient Tenjinyama tomb was measured using the system.[9] A powdered sample was put into a Ta tube of 4 mm in diameter and 0.2 mm in thickness, and was heated by ECR plasma. The particles with A/q = 5 analyzed by the first analyzing magnet were accelerated to 2.0 MeV/u with RILAC and the RFQ linear accelerator. The parameters of each magnet were fixed to detect particles of A/q = 5. Nine kinds of elements were identified. All the identified elements are the multiples of 5.

Performance Evaluation of the System

In order to investigate the performance of this system, the trace element in a reference sample was measured. Sample was a rock reference sample JB-2 provided by the AIST (National Institute of Advanced Industrial Science and Technology).

The rock reference sample was sintered after mixing with pure Al_2O_3 powder. The sintered sample was made into the rod (4 mm in height, 4 mm in width and 40 mm in length), and was ionized by the insertion method. Beam energy was 1.4 MeV/u. A/q = 5 was selected in this experiment. Standard sample gas which consists of rare gases was used. The standard gas was made by mixing small amount of Ne, Ar, Kr, and Xe into pure helium gas. This gas can be used as standard gas, when we selected the ions which has A/q = 2, 3, 4, and 5.

Figure 2 shows the ΔE-Total E spectrum of trace elements in Al_2O_3 rod used as a base material. Open circles are the calculated results. Italic (or green) symbols are elements detected as the elements from the standard gas, other elements were measured as contamination in the base material Al_2O_3 or in the standard gas or on the surface of the plasma chamber.

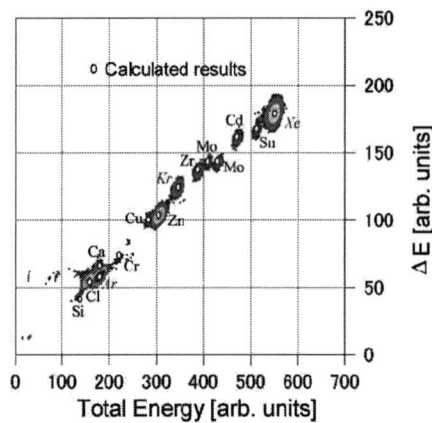

FIGURE 2. ΔE-Total E spectrum of trace elements of Al_2O_3 rod used as a base material.

10 elements were assigned by using the ECRIS-AMS. We found that many kinds of trace elements exist in the Al_2O_3 rod used base material. Moreover in this experiment, the large amounts of ^{70}Zn ions were detected.

Figure 3 shows the ΔE-Total E spectrum of trace elements of Al_2O_3 containing the rock reference sample. Open circles are the calculation results. Bold (or red) symbols are the detected trace elements in the JB-2. In this experiment, 15 elements were identified among 23 elements in the rock reference sample which must be detected in this experimental condition.

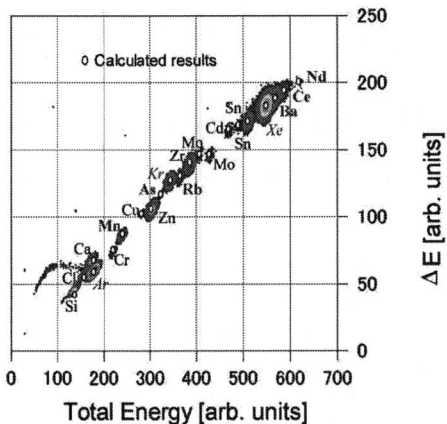

FIGURE 3. ΔE-Total E spectrum of trace elements of Al_2O_3 rod containing the rock reference sample (JB-2).

DISCUSSION

By measurement of the trace element in the cinnabar, it was able to check that there was no principle problem in this system and method.

Direct insertion of a sample mixed the inorganic substance with ceramics into the plasma is suitable method for ECRIS-AMS, because such a sample does not affect the plasma condition. However, in order to measure the trace element in inorganic substance, the highly pure ceramics powder is required.

In the experiment of the rock reference sample, large amount of ^{70}Zn were detected. Before this experiment, ^{70}Zn ions were generated in the ion source for the search of the super heavy element for a long time (about 3 month), so that ^{70}Zn exists in somewhere of plasma chamber. Even if the plasma chamber wall was cleaned up, ^{70}Zn might be remained on it. In order to perform the trace element analysis, it is necessary to clean the plasma chamber wall more.

In order to carry out experiment of quantitive-analysis measurement, we have to perform precise measurement of ionization efficiency, maintenance of the standard sample for using relative comparison, etc.

SUMMARY AND CONCLUSION

We demonstrate a novel trace element analysis using the ECRIS and heavy ion linear accelerator. We can analyze many trace elements simultaneously in materials. We successfully identify many kinds of trace elements without any sample preparation of the cinnabar. In order to carry out the quantitive measurement, we have to take following points into consideration. 1) We have to clean the plasma chamber wall to minimize the contamination. 2) We have to use the highly pure ceramics or base material. 3) We have to prepare a standard gas or sample used as standard elements for carrying out relative comparison. 4) It is necessary to establish the ionization method which suited the condition and physical-properties-character of a sample to measure.

REFERENCES

1. C. Vandecasteele and C. B. Block: in Modern Methods for Trace Element Determination, (John Wiley & Sons, Chichester, 1997).
2. For example, in Proceedings of the 8th International Conference on Accelerator Mass Spectrometry.
3. M. Odera et al., *Nucl. Instrum. and Methods* **227**, 187 (1984).
4. T. Nakagawa and M. Kidera, *J. Mass Spectrom. Soc. Jpn.* **48 (2)**, 169 (2000).
5. M. Kidera et al., *Nucl. Instrum. and Methods* **B172**, 316 (2000).
6. Kamigaito et al., *Rev. Sci. Instrum.* **4523**, 70 (1999).
7. K. Shima, N. Kuno, M. Yamanouchi and H. Tawara, NIFS-DATA-10, (1991).
8. T. Nakagawa et al., *Nucl. Instrum. Methods Phys. Res. A* **396**, 9 (1997).
9. M. Kidera et al., *Anal. Sci.* **17**, i17-20 (2001)

Photoionization of Multiply Charged Ions at the Advanced Light Source

A. S. Schlachter[1], A. L. D. Kilcoyne[1], A. Aguilar[2], M. F. Gharaibeh[2], E. D. Emmons[2], S. W. J. Scully[2], R. A. Phaneuf[2], A. Müller[3], S. Schippers[3], I. Alvarez[4], C. Cisneros[4], G. Hinojosa[4], and B. M. McLaughlin[5]

[1] Advanced Light Source, Lawrence Berkeley National Laboratory, Berkeley CA 94720, USA
[2] Department of Physics, MS 220, University of Nevada, Reno NV 89557, USA
[3] Institut für Atom- und Molekülphysik, Justus-Liebig-Universität, 35392 Giessen, Germany
[4] Universidad Nacional Autonoma de Mexico, Cuernavaca, Mexico
[5] Queen's University Belfast, BT71NN, UK

Abstract. Photoionization of multiply charged ions is studied using the merged-beams technique at the Advanced Light Source. An ion beam is created using a compact 10-GHz all-permanent-magnet ECR ion source and is accelerated with a small accelerator. The ion beam is merged with a photon beam from an undulator to allow interaction over an extended path. Absolute photoionization cross sections have been measured for a variety of ions along both isoelectronic and isonuclear sequences.

INTRODUCTION

Photoionization of ions is a fundamental process of importance in many high-temperature environments, such as in stars and nebulae, and in interstellar space. It is also an important process in hot dense laboratory plasmas, and of potential importance for modeling a plasma source for future lithography applications. There have been extensive theoretical calculations of cross sections for photoionization of ions, but few experiments to serve as benchmarks.

Absolute cross sections for photoionization of ions have been measured for many species of ions. Experiments are performed using a merged-beams end station at the Advanced Light Source (ALS). An ion beam from an electron-cyclotron-resonance (ECR) ion source is merged with an ultrabright beam of x rays from an undulator. Results are reported for several of the ion species studied to date, including measurement of lifetimes of K-shell vacancies in atomic carbon, observation of a truncated giant resonance in photoionization of Ti^{3+}, and absolute photoionization cross-section measurements in Sc^{2+}, metastable O^+, and other ions.

EXPERIMENTS

Experiments have been conducted at undulator beamline 10.0 at the ALS, using the ion-photon merged-beams end station [1,2] shown schematically in Fig 1. The technique was developed by Lyon and coworkers [3], and has been recently employed by Kjeldsen et al [4] and others [5]. An energy selected photon beam is merged over a path length of approximately one meter with an ion beam produced by an ECR ion source and a small accelerator.

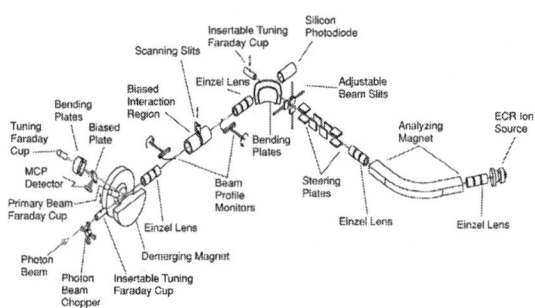

FIGURE 1. Schematic diagram of ion-photon merged-beams end station.

Ion beams are produced with a 10-GHz all-permanent-magnet ECR ion source based on a design by Trassl et al [6] shown schematically in Fig. 2. The ion source produces a high-intensity ion beam from either a gas or a metal evaporated in an oven inserted into the ion source. The ion beam in the interaction region is nearly parallel, has a typical diameter of 1-2 mm, and has a current in the 1-100 nA range.

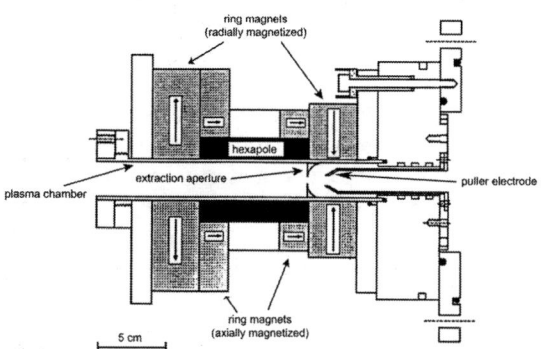

FIGURE 2. Schematic diagram of all-permanent-magnet ECR ion source [6].

The photon beam from an undulator is energy selected by a spherical-grating monochromator, and has a spectral resolving power in the range 1000 to 34,000 determined by the width of the entrance and exit slits of the monochromator. Typical photon flux is of the order of 10^{13} photons per second (an order of magnitude less at energies above 250 eV) for photons in the range 17 to 340 eV at modest resolving power, and correspondingly less with increasing spectral resolution.

The ion beam and the counter-propagating photon beam are merged, and a potential is applied to a central 29.4-cm-long interaction region. This potential energy-labels ions produced in the interaction region, thus providing a known interaction length. Downstream is an analyzing magnet which demerges the beams, separating ions which have changed charge from the primary ion beam. The current in the primary ion beam is measured in a Faraday cup, while the ions which have changed charge are counted using a single-particle detector. The photon beam is mechanically chopped to allow separation of ions produced by collisions in background gas from photoions.

Two-dimensional intensity distributions of both beams are measured with rotating-wire beam-profile monitors and with a translating-slit scanner. Photon intensity is measured with a calibrated photodiode. The efficiency of the single-particle detector is measured in a separate experiment. All quantities required for measurement of an absolute cross section are thus measured.

PHOTOIONIZATION OF METASTABLE O^+ IONS

An ECR ion source can produce a mixture of ground-state and metastable ions, making the analysis of photoionization spectra challenging. Absolute experimental measurements and theoretical calculations have been published [2] for cross sections for photoionization of O^+ ions from the $^2P^o$ and $^2D^o$ metastable levels and from the $^4S^o$ ground state in the photon energy range 30-35.5 eV. The measurements, reported by Covington et al [2] and by Aguilar et al [7] reveal a rich spectrum of resonances (Fig. 3) arising from 2p-ns and 2p-nd transitions originating from ions in metastable states.

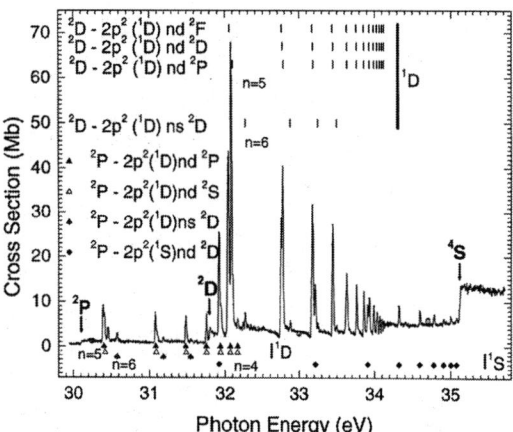

FIGURE 3. Experimental cross section for photoionization as a function of photon energy for an admixture of 57% metastable and 43% ground-state O^+ ions [2]. The energy resolution is 17 meV.

Comparison with theory shows that predicted resonance structure associated with the photoexcitation of autoionizing states is sensitive to the choice of basis functions.

PHOTOIONIZATION OF Sc^{2+} IONS

Cross sections for the photoionization of Sc^{2+} ions have been measured using the merged-beams apparatus at ALS as reported by Schippers et al [8]. The photon energy range 23-68 eV encompasses the direct 3d and 3p photoionization thresholds. The experimental cross section is dominated by autoionizing resonances due to 3p excitations predominantly decaying by Coster-Kronig and super-Coster-Kronig transitions.

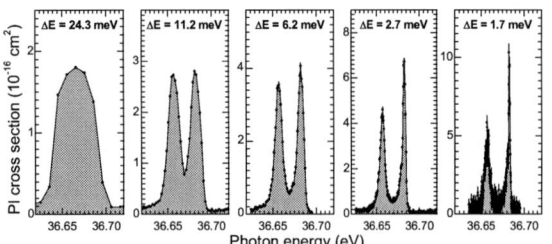

FIGURE 4. Influence of the experimental photon energy resolution on the measured photoionization cross section for a group of three resonances in photoionization of Sc^{2+} ions [8].

This system was studied with varying spectral resolution, shown in Fig 4: individual resonances were measured with a spectral resolution as high as 1.2 meV, corresponding to a resolving power of nearly 34,000. Although the effective ion target density is many orders of magnitude smaller, this resolution is comparable to that normally obtained in experiments with gas targets at ALS, illustrating the sensitivity of the merged-beams technique.

PHOTOIONIZATION OF C^{2+} IONS

Photoionization of an admixture of ^1S ground-state and $^3P^o$ metastable-state C^{2+} ions has been studied [9].

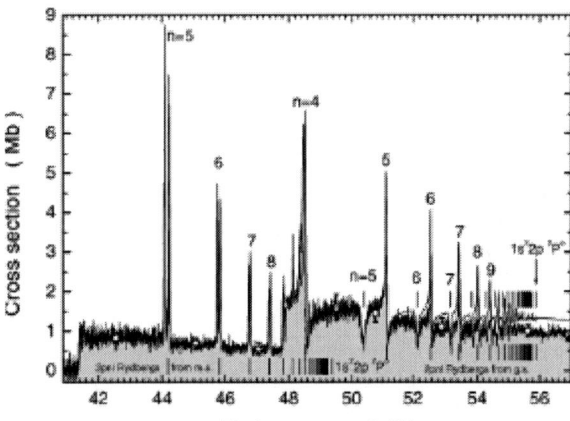

FIGURE 5. Measured and calculated cross sections for photoionization of C^{2+} ions [9]. Theory is shown by the curve; experimental results are shown as individual data points.

Theoretical cross sections were used to estimate that the primary ion beam consisted of 40% metastable ions. Figure 5 compares the experimental results with the results of an R-matrix theoretical calculation. The results are almost indistinguishable except at higher energies near the Rydberg-series limits.

LIFETIME OF A K-SHELL VACANCY IN ATOMIC CARBON

Lifetimes for K-shell vacancy states produced by removing a K-shell electron from atomic carbon have been determined by measurement of the natural linewidth of the 1s-2p photoexcited states of C^+ ions [10]. The K-shell vacancy states produced by direct K-shell ionization of atomic carbon are identical to those produced by 1s-2p photoexcitation of a C^+ ion. The vacancy states stabilize by emission of an electron to produce C^{2+}.

Results are shown in Fig. 6 for photoexcitation of an admixture of ground-state and metastable C^+ ions. The metastable fraction was determined by use of theoretical cross sections calculated employing the R-matrix method. Lifetime linewidths were determined by fitting a Voigt profile (a convolution of a Gaussian and a Lorentzian profile) to experimental results.

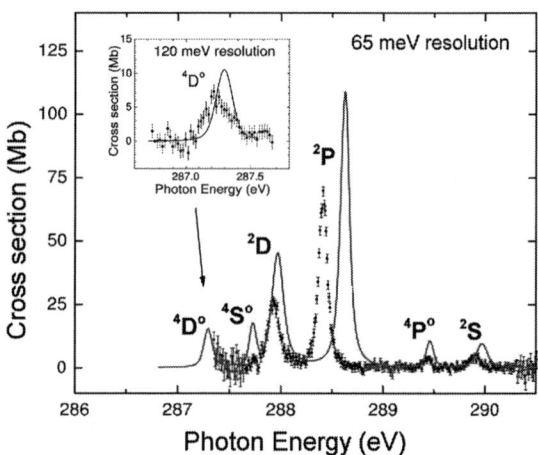

FIGURE 6. Experimental and theoretical cross sections for 1s-2p photoexcitation of C^+ ions for an admixture of 80% ground-state and 20% metastable-state ions [10].

Lifetimes are obtained from measured linewidths by use of the uncertainty principle. Lifetimes of K-shell vacancy states are found to be of the order of 10 fs, considerably longer than lifetimes observed for K-vacancy states in carbon-containing molecules, as there are additional L-shell electrons in molecules which are available to fill the K-shell vacancy.

TRUNCATION OF A 'GIANT' DIPOLE RESONANCE IN PHOTOIONIZATION OF Ti^{3+} IONS

Photoionization of triply charged titanium ions was studied using the merged-beams technique [11]. Cross sections are shown in Fig. 7. The energy range 42.6 to 49.4 eV encompasses the threshold for photoionization of the Ti^{3+} ground state. A giant dipole resonance associated with 3p-3d excitation in potassium-like ions decays by an extremely fast super-Coster-Cronig transition, and has a corresponding width of 1.5 eV. Since the central energy of this resonance in Ti^{3+} is located only 0.2 eV above the ionization threshold, the low-energy part of the resonance is missing in the photoion yield spectrum, producing an unusual truncation of its lineshape.

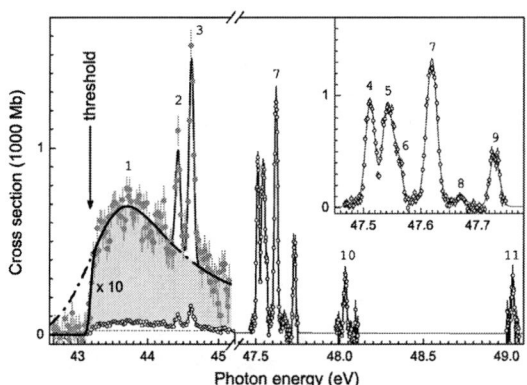

FIGURE 7. Cross section for photoionization of Ti^{3+} ions [11].

THEORETICAL APPROACH

R-matrix calculations have served a valuable role in guiding experiments at ALS, especially in locating resonances and regions of interest for experimental study. It is generally not possible to perform measurements over a large energy range with high spectral resolution to locate narrow resonances. Theoretical calculations were an invaluable guide in locating energy regions for measurement in the cases of K-shell photoionization of C^+ [10] and C^{2+} ions [12]. Theory has also been valuable in determining the metastable content of an ion beam from the ECR ion source, and in determining which resonances are attributable to ground-state ions and which to metastable ions.

CONCLUSION

The mating of an ECR ion source to the photon-ion end station at ALS makes possible systematic studies of photoionization of ions along isoelectronic and isonuclear sequences. Examples of recent investigations at ALS are the potassium and nitrogen isoelectronic sequences, and the iron, xenon, and krypton isonuclear sequences. Photoionization measurements have also been initiated with singly and multiply charged ion beams of C_{60} and C_{70} produced in the ECR ion source [13].

REFERENCES

1. Covington, A. M et al, Phys. Rev. A **66**, 062710 (2002) and references therein. Ph.D. Thesis, Alejandro Aguilar-Mendoza, University of Nevada Reno (2003).
2. Covington, A. M. et al, Phys. Rev. Lett. **87**, 243002 (2001).
3. Lyon I. C. et al, J. Phys. B **19**, 4137 (1986).
4. Kjeldsen et al, J. Phys B **32**, 4457 (1999).
5. Bizau, J.-M et al, Phys. Rev. Lett. **87**, 273002 (2001); Itoh, Y. et al, J. Phys. B **28**, 4733 (1995); Oura, M. et al, Nucl. Instrum. Meth. B **86**, 190 (1994).
6. Trassl et al, Phys. Scr. T**80**, 504 (1999). Schlapp et al, Nucl. Instrum. Meth. B **98**, 525 (1995), Trassl et al, Physica Scripta T**73**, 380 (1997).
7. Aguilar, A. et al, Astrophysical Journal Supplement Series **146**, 129 (2003).
8. Schippers, S. et al, Phys. Rev A **67**, 032702 (2003); Schippers, S. et al, Phys. Rev. Lett. **89**, 193002 (2002).
9. Müller, A. et al, J. Phys. B **35**, L137 (2002).
10. Schlachter, A. S. et al, J. Phys. B **37**, L103 (2004).
11. Schippers, S.. et al, J.Phys. B **37**, L209 (2004).
12. Scully, S. et al, article in preparation.
13. Scully, S. et al, article submitted for publication.

FORMATION OF ELECTRON DISTRIBUTION FUNCTION IN ECR DISCHARGE SUSTAINED BY STRONG MICROWAVE EMISSION IN AN OPEN TRAP

V.L.Erukhimov* and V.E.Semenov*

*Institute of Applied Physics, RAS,
Russia, 603950, Nizhny Novgorod, Ulyanova 46

Abstract. We consider a formation of Electron Distribution Function (EDF) in the Electron Cyclotron Resonance (ECR) discharge in an open trap. The ECR heating by strong microwaves, ionization, collisions and ambipolar losses are considered. The model is based on a system of two-dimensional Fokker-Plank equation for EDF. The stationary solution for EDF is investigated analytically. It consists of three groups of electrons: hot electrons with highly anisotropic velocity distribution that are heated in the ECR region, cold electrons with isotropic distribution that define the losses from the trap and warm electrons with considerably anisotropic distribution that are concentrated in the center of the trap and do not reach the ECR region. We build a qualitative model for the electron distribution function such that the original differential equation for EDF is transformed into two algebraic equations with two unknown parameters: neutral density and main plasma density. The latter can be solved analytically. The applicability of these results to a self-consistent model for ECR ion source is discussed. We show that the solution contradicts experimental results so that important effect is not taken into account in the model.

INTRODUCTION

The profile of the electron distribution function (EDF) plays an important role in the performance of ECR ion sources. The electrons with high energy provide the ionization of multicharged ions. Low energy electrons determine plasma confinement time in a magnetic trap. There have been quite a few attempts to develop a model for EDF. Many of them (see [1], [2]) assume a fixed EDF profile and calculate parameters such as electrons lifetime and density. Others (such as [4], [5], [6]) build a self-consistent model for EDF. The most of self-consistent models restrict their consideration to ECR heating, collisions, ionization and ambipolar losses. Even this limited set of processes makes the kinetic equation too difficult to solve analytically in general case. There are a number of numerical codes that integrate the kinetic equation (see [7] as one example) and few successful attempts to solve a kinetic equation analytically. The case of low microwave heating power is analyzed in [8], authors of [9] integrated a kinetic equation in case of small plasma density. Within this paper we show that EDF can be found analytically under conditions of high microwave power that are met for a wide range of experiments.

THE MODEL

We solve a system of bounce-averaged kinetic equation for an axial-symmetric EDF $f(v_x, v_y, v_z) = f(v_\parallel, v_\perp)$ and neutrals density balance N_0

$$\begin{cases} \frac{\partial f}{\partial t} = \{\hat{L}_i + \hat{L}_h + \hat{L}_c + \hat{L}_l\} f \\ \frac{dN_0}{dt} = F - \hat{L}_i f \end{cases}, \quad (1)$$

where $v_\parallel$ and $v_\perp$ are longitudinal and transversal components of electron velocity wrt the magnetic field of the axial-symmetric trap when the particles are located in the magnetic field minimum, $\hat{L}_i$ is an ionization operator, $\hat{L}_h$ corresponds to ECR heating, $\hat{L}_c$ is a collisions operator, $\hat{L}_l$ models electrons and ions losses from the trap and F is neutral gas input flux. For the sake of simplicity we consider a single fraction of ions with charge $-e$ and ionization energy ε_i.

ECR heating is introduced by a diffusion operator

$$\hat{L}_h f = v_\perp v_\parallel \frac{\partial}{\partial \varepsilon_\perp^r} D_h\left(\varepsilon_\perp^r, \varepsilon_\parallel^r\right) \varepsilon_\perp^r \frac{\partial}{\partial \varepsilon_\perp^r} \left(\frac{\omega_b f(v_\parallel, v_\perp)}{v_\parallel}\right), \quad (2)$$

where $\varepsilon_\perp^r$ and $\varepsilon_\parallel^r$ are electron longitudinal and transversal kinetic energies in the ECR section, i.e. where the electron cyclotron frequency ω_c equals the heating wave frequency ω, ω_b is the electron bounce oscillations frequency. We consider a parabolic trap where $\omega_b = v_\perp/L$, L is the characteristic longitudinal scale.

We consider the case of longitudinal propagation of the rf wave with respect to magnetic field[1]. Our model for D_h takes into account the following properties [10], [12]:

1. rf heating influences only electrons that reach ECR section;
2. electrons with sufficiently low energy $\varepsilon < \varepsilon_{max}$ are heated in a stochastic manner, i.e. their cyclotron phase is randomized during one bounce oscillation. On average an electron will increase its energy during one pass through ECR section, but depending on a cyclotron phase it can also be cooled. Electrons with $\varepsilon > \varepsilon_{max}$ are in a superadiabatic regime where they cannot significantly change their energy. ε_{max} depends on the parameters of the trap and the heating wave. For a typical trap it is subrelativistic or relativistic and has a simple dependence on the amplitude of the electric field E of rf wave: $\varepsilon_{max} \propto E^{3/4}$ (see [10] for details on ε_{max}). Following [9], we will take into account superadiabatic effect by imposing a condition on the diffusion coefficient: $D_h(\varepsilon \geq \varepsilon_{max}) = 0$.
3. The heating operator conserves the electron longitudinal energy in the ECR section:

$$\varepsilon_\parallel^r = \varepsilon - \frac{\omega}{\omega_0}\varepsilon_\perp, \qquad (3)$$

where ω_0 and $\varepsilon_\perp$ are the electron cyclotron frequency and transversal energy in the center of the trap. This implies that on the phase space shown in Fig.1 electrons interacting with rf wave move along hyperbolas such as lines 3 and 4. Line 1 corresponds to $\varepsilon_\parallel^r = 0$, i.e. to the particles that stop in the ECR section.

We model collisions with tau-approximation

$$\hat{L}_c f = \nu_c(\varepsilon)(f_M(n,T) - f), \qquad (4)$$

where f_M corresponds to the Maxwellian distribution function and ν_c is the Coulomb collision frequency. Density $n[f]$ and temperature $T[f]$ are functionals of f such that the collision operator does not change the total number of electrons and their total energy.

The losses of electrons and ions from the trap are modeled by the loss cone and the effective value φ of ambipolar potential in the plug section. The 0D model for φ from [1] is used. The electrons loss operator is represented in the following form: $\hat{L}_l f = -\nu_l(v_\parallel, v_\perp) f$,

---
[1] High density plasma in the region of ECR tends to turn wave vector of rf rays towards the direction of the magnetic field.

where the distribution of particles lifetime $\tau_l = 1/\nu_l$ depends on the ambipolar potential. The details of this dependence will be provided in the next section.

Ionization is given by its cross section σ_i: $\hat{L}_i f = \int \sigma_i v N f dv_\perp dv_\parallel$, where v is the velocity of electrons and N is the density of ions.

The boundary conditions for $v_\perp = 0$ and $v_\parallel = 0$ are usually obtained from the mathematical model. Since $\lim_{v_\perp \to 0} f(v_\perp, v_\parallel)/v_\perp = const$ (if there are no singularities in the distribution function in 3D cartesian coordinates) and $f(v_\perp, v_\parallel) = f(v_\perp, -v_\parallel)$, we can derive the following boundary conditions [5]:

$$\begin{cases} \left.\frac{\partial f}{\partial v_\parallel}\right|_{v_\parallel=0} = 0 \\ \left.f\right|_{v_\perp=0} = 0 \end{cases}.$$

The upper boundary conditions correspond to the zero flux of the particles through the boundary $\varepsilon = \varepsilon_{max}$.

STATIONARY SOLUTION

We start by noting that an electron in a typical experimental setup with a gyrotron power about $10 - 100kW$ can be heated up to 1-10keV during one pass through the ECR zone. In this range of energies collision processes are sufficiently weak compared to ECR diffusion. This implies that EDF contains a low energy isotropic fraction defined by collision operator and high energy anisotropic fraction formed by ECR diffusion. The region $\varepsilon < \varepsilon_2$ where collision operator prevails over ECR is given by the following expression:

$$\nu_c(\varepsilon_2)\tau_h(\varepsilon_2) \sim 1, \qquad (5)$$

where τ_h is the time from the first pass of the particle through ECR till it's energy reaches the superadiabatic limit ε_{max}:

$$\tau_h \sim \frac{\varepsilon_{max}}{A(\varepsilon)}\frac{2\pi}{\omega_b},$$

where

$$A = \frac{d}{d\varepsilon_\perp^r}(D_h \varepsilon_\perp^r)$$

is the mean energy that an electron gets from one pass through ECR section.

Let us denote by ε_c the energy such that the region $\varepsilon < \varepsilon_c$ contains the low energy isotropic fraction of the distribution function and $\varepsilon > \varepsilon_c$ contains the high energy anisotropic fraction. If $\varepsilon_c > e\varphi$ then $\varepsilon_c = \varepsilon_2$. In the opposite case ECR diffusion and collisions together form an isotropic distribution in the region $\varepsilon < e\varphi$ so the boundary energy will be equal to $\varepsilon_c = e\varphi$. The precise calculation of ε_c lies out of the scope of this paper (for more details see [1], [2]), for the rest of it we will assume $\varepsilon_c = 10^2 \div 10^3 eV$ which is true for a realistic experimental setup.

The electrons with $\varepsilon > \varepsilon_c$ can be divided into those that reach ECR point and those that don't. Following the terminology of [11] we will reference to the former as to hot electron fraction and to the latter as to the warm electron fraction. We will denote the distribution functions of these fractions by f_h and f_w correspondingly. The cold electrons fraction – below ε_c – with distribution function f_c defines the particles losses from the trap. This partition is illustrated in the phase space on Figure 1. Warm particles are located above lines 1 and 2, hot particles reside between lines 1 and 4 and above line 2, cold particles are bounded above by line 2. When the plasma mean energy $T \ll \varepsilon_c$ the distribution function is almost isotropic. This extreme case is well studied in [8]. Within this paper we will concentrate on the opposite case when $T \gg \varepsilon_2$ so that EDF is highly anisotropic. This is a typical inequality for modern ECR discharge setups. For instance, the estimation from (5) in the case of a trap with length $20cm$, plasma density $10^{13} cm^{-3}$ and rf wave intensity $100 kW/cm^2$ gives $\varepsilon_2 \sim 80 eV$ while temperatures observed in experiments are usually at least several times higher.

Let us consider a region where $\varepsilon \gg \varepsilon_c$. Since ECR heating prevails over other effects under these conditions, f_h corresponding to the stationary regime satisfies the equation $\hat{L}_h f_h = 0$. This equation coupled with boundary conditions gives

$$f_h(\varepsilon_\perp, \varepsilon_\parallel) = \frac{v_\parallel}{v_\perp} \tilde{f}_h(\varepsilon_\parallel^r). \qquad (6)$$

Due to the conservation of $\varepsilon_\parallel^r$ in the heating process the longitudinal energy of particles in the ECR section does not change much from its initial value. The only process that can change $\varepsilon_\parallel^r$ is collisions but the particles that have sufficiently large $\varepsilon_\parallel^r$ have a short lifetime because ECR heating makes them travel down to the loss cone. Hence $\tilde{f}_h(\varepsilon_\parallel^r)$ has a characteristic scale of order ε_c or less and drops with growth of $\varepsilon_\parallel^r$. Also one can note that the fraction in the right hand side of (6) is a constant when $\varepsilon \gg \varepsilon_\parallel^r$.

f_w is defined by $\hat{L}_c f_w = 0$ and boundary condition $f_w = f_h$ on the boundary $v_\parallel = v_\perp \sqrt{\frac{\omega}{\omega_0} - 1}$ (line 1 on Fig.1). If, for the sake of simplicity, we disregard the energy exchange between electrons (one can always check that taking it into account will bring nothing but an additional numerical factor into our estimations) the warm electrons EDF can be written as

$$f_w(v_\perp, v_\parallel) = f_h\left(\frac{\omega_0}{\omega} v, \left(1 - \frac{\omega_0}{\omega}\right) v\right). \qquad (7)$$

f_c is formed by collisions, ionization, losses and ECR heating. For the rest of the paper we will assume its characteristic scale to be close to ε_c and approximate it with a constant inside the region $\varepsilon < \varepsilon_c$. It is given by the following boundary condition: $f_c \sim \tilde{f}_h(\varepsilon_c)$.

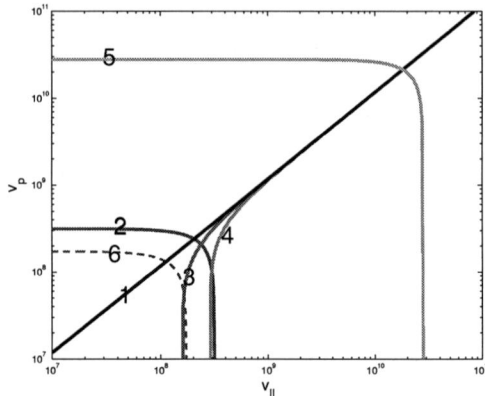

FIGURE 1. Phase space. Curve 1 corresponds to the particles that stop at the ECR point, curve 2 bounds the region of isotropic EDF, 3 and 4 correspond to electrons with constant $\varepsilon_\parallel^r$, curve 5 defines particles with energy $\varepsilon = \varepsilon_{max}$ and curve 6 corresponds to $\varepsilon = \varepsilon_i$.

We can immediately derive the relations between the numbers of particles in each of the fractions:

$$\begin{aligned} N_h &\sim N_c \left(\frac{\varepsilon_{max}}{\varepsilon_c}\right)^{1/2} \frac{\omega_0}{2\omega} \\ N_w &\sim N_c \left(\frac{\varepsilon_{max}}{\varepsilon_c}\right)^{3/2} \left(1 - \frac{\omega_0}{\omega}\right)^{1/2} \end{aligned}. \qquad (8)$$

Here $N_{(h,w,c)} = \int f_{(h,w,c)} dv_\perp dv_\parallel$ denotes the number of electrons in each fraction. For typical parameters ε_{max} is subrelativistic so that $\varepsilon_{max} \gg \varepsilon_c$. This implies that $N_w \gg N_h \gg N_c$. The main plasma is concentrated in the warm particles region.

Equation (1) can be rewritten as a balance between rates of electrons birth and loss as well as the input gas flux:

$$\int \hat{L}_i f_w dv_\perp dv_\parallel = -\hat{L}_l f_c dv_\perp dv_\parallel = F. \qquad (9)$$

The first equation defines the stationary value of neutrals density. We will do the calculations for the ionization cross-section corresponding to hydrogen using the approximation from [14]:

$$\sigma_i(\varepsilon) = \begin{cases} 1.4 \cdot 10^{-15} (cm^2)(\varepsilon_i/\varepsilon) ln(0.2\varepsilon/\varepsilon_i), \\ \qquad \varepsilon_i < \varepsilon < 10\varepsilon_i \\ 2 \cdot 10^{-15} (cm^2)\varepsilon_i(\varepsilon - \varepsilon_i)\varepsilon^{-1}(\varepsilon + 8\varepsilon_i)^{-1}, \\ \qquad \varepsilon > 10\varepsilon_i \end{cases} \qquad (10)$$

The stationary neutrals density is given by

$$N_0 \sim 2 \cdot 10^{15} \frac{\varepsilon_c^2}{\varepsilon_{max} \varepsilon_i} \frac{1}{L} \sqrt{\frac{m_e}{m_i}} \left(1 - \frac{\omega_0}{\omega}\right)^{-1/2} \frac{\omega_0}{\omega_m}, \qquad (11)$$

where ω_m is the cyclotron frequency in the trap bottleneck, m_i is ion mass.

Stationary plasma density can be calculated from the second part of (9). One can see that $N \propto F$. However it is usually hard to measure F with sufficient precision due to inohmogeneous distribution of neutral gas in the trap and strong interaction with vacuum chamber walls.

DISCUSSION

The stationary solution constructed in the previous section possesses several important features that contradict experimental results. One can note that the distribution function of warm particles over electron energy grows since f_h is constant and phase volume grows with energy. Energy exchange between electrons due to collisions does not play an important role because of (5). Another important observation is that the number and density of warm particles is much higher than of cold particles. This does not fit well into bremmstrahlung observations [15] indicating an almost flat distribution in the region of 5-50 keV and a cold (100eV-5keV) quasi-maxwellian fraction that is larger by several orders of magnitude. However the discharge in [15] had lifetime of about $t \sim 1ms$ which is much less than the time needed to reach the stationary EDF which is about $1s$. One can easily generalize the theory for this case by noting that for $\varepsilon > \varepsilon_c(v_c(\varepsilon_c)t)^{2/3}$ instead of (7) we will have

$$f_w(v_\perp, v_\parallel) = v_c(\varepsilon) f_h\left(\frac{\omega_0}{\omega}v, \left(1 - \frac{\omega_0}{\omega}\right)v\right). \quad (12)$$

Hence for the distribution over energies we will get $f_w \propto 1/\varepsilon$. This profile fits the results of [15] on a qualitative level. However, the warm fraction is still much larger than the cold one. This is a serious contradiction to [15] that cannot be explained within currently developed theory.

The inversion of the main plasma EDF directly follows from the model. It can be removed by several ways. One can speculate that if only energetic particles are heated, i.e. the heating is non-uniform in the phase space, the fraction of high energy electrons would be smaller. Also one can introduce a loss operator in high energy region. It can correspond to either losses of high energy particles from the trap or a process that significantly modifies the high energy distribution. The former can be induced by cyclotron instability that has been observed experimentally [16] as simultaneous bursts of high energy electrons ($> 7keV$) and microwave emission out of the trap. However the energies emitted with electrons are typically too low to assume that this process significantly influences for high energy EDF. The pitch-angle diffusion can be induced by the process that changes the invariant (3). This can take place due to a nonzero width of the heating wave spectrum. The cooling of electrons due to ionization of multicharged ions may play a significant role in the formation of high energy EDF. In addition there is no clear evidence on how dense plasma influences the heating wave so the contradictions above can be solved by introducing a different heating mechanism.

ACKNOWLEDGMENTS

We would like to thank A.Girard, M.D.Tokman, A.V.Vodopyanov, V.G.Zorin, and S.V.Golubev for fruitful discussions. The work has been supported by INTAS grant N 01-0373.

REFERENCES

1. A.V.TURLAPOV, V.E.SEMENOV, *Phys.Rev.E*, **57**, 5937 (1998).
2. V.E.SEMENOV, A.N.SMIRNOV, A.V.TURLAPOV, *Transactions of Fusion Technology*, **35**, N. 1T, 398 (1999).
3. G.SHIRKOV, V.ALEXANDROV, V.PREISENDORF et al, *Proceedings of 15th International Workshop on ECR Ion Sources* (2002).
4. A.GIRARD, C.PERNOT, G.MELIN, *Phys.Rev.E*, **62**, 1, 1182 (2000).
5. A.V.KASHEEV, N.V.SUETIN, *IEEE Trans. Plasma Science*, **23**, 4, 591 (1995).
6. V.L.ERUKHIMOV, V.E.SEMENOV, Proc. International Conference on Ion Sources, 2003.
7. A.GIRARD, C.LECOT, K.SEREBRENNIKOV, Journal of Comp.Physics, V.191, 1, pp.228-248, 2003.
8. S.V.GOLUBEV, V.E.SEMENOV, E.V.SUVOROV, M.D.TOKMAN, Proc. *Strong Microwaves in Plasmas* (1993).
9. E.V.SUVOROV, M.D.TOKMAN, *Plasma Physics Reports*, **15**, N8, pp. 934-943 (1989) [in russian].
10. A.V.TIMOFEEV, *Cyclotron oscilations of equilibrium plasma. Questions of plasma theory.*, (Energoatomizdat, Moscow, 1985), V.14 [in russian].
11. V.A.ZHILTSOV, A.A.SKOVORODA, A.V.TIMOFEEV et al, *Plasma Phys. Reports*, **20**, No.3, pp.242-251, 1994.
12. V.L.ERUHIMOV, V.E.SEMENOV, *Plasma Physics Reports*, **27**, No. 11, 932 (2001).
13. A.F.KUCKES, *Plasma Physics*, **10**, 367 (1968).
14. V.GOLANT, A.ZHILINSKY, I.SAKHAROV, *Fundamantals of plasma physics*, John Wiley & Sons, 1980.
15. S.V.Golubev et al, *Trans. Fusion Technology*, Vol.35, 288, (1999).
16. A.G.DEMEKHOV et al, *Proc. 15th workshop on ECRIS* (2002).

V TECHNIQUES AND GYROTRONS

Analysis of the SERSE Ion Output by Using Klystron-based or TWT-based Microwave generators

L. Celona, S. Gammino, G. Ciavola, F. Consoli, A. Galatà

Istituto Nazionale di Fisica Nucleare – Laboratori Nazionali del Sud
Via S. Sofia 64, 95123 Catania, ITALY

Abstract. A set of measurements has been carried out in order to confirm the previously observed enhancement of the SERSE source performances, when microwaves are fed by means of a TWT generator instead of a Klystron generator, which has been commonly used. An increase in the extracted currents takes place for the highest charge states by using a TWT at 14 GHz, while higher extracted currents were obtained for every charge state by using a TWT at 18 GHz. A description of the experimental set-up and of the obtained results is given in the following. The data will be analysed and a qualitative explanation in terms of ECRIS plasma model will be proposed.

INTRODUCTION

The SERSE source at LNS of Catania [1] (fig. 1) is one of the most effective ion sources for highly charged ions among those currently working in various laboratories spread over the world, either in terms of the extracted currents and in terms of the charge states distributions obtained. Many relevant studies on the physics and technology of electron cyclotron resonance (ECR) ion sources have been made possible by the versatility of such a source and in particular the studies upon the role of confining magnetic field and of the microwave frequency used to heat the plasma [1,2]. As in the recent past an experiment with the other ECR source working at LNS, the room temperature source CAESAR, has put in evidence the different behaviour of the ECR ion sources (ECRIS) when they are fed by klystron or TWT generator [3], a systematic series of measurements was carried out in the past months to have a precise comparison between the performance obtained in the two cases. The comparison regarded both the working frequencies commonly used, 14 and 18 GHz.

At LNS the microwaves needed to heat the plasma electrons can be produced by five different generators and can be injected into SERSE by means of two ports in the injection flange, as shown in fig.1. The list of the generators follows:

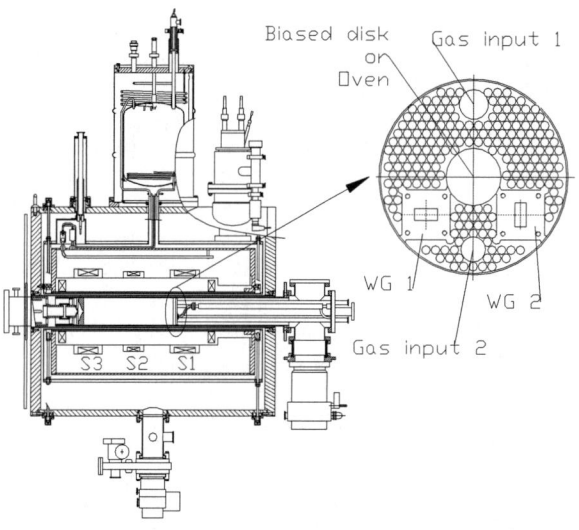

FIGURE 1. A sketch of the SERSE source with a scheme of the injection flange: WG1 and WG2 stay for microwaves ports.

1) One microwave generator (klystrons-based) able to reach a power up to 2 kW at 14 GHz.

2) Two microwave generators (klystron-based) able to reach a power up to 2 kW at 18 GHz.

3) A microwave generator (travelling wave tube based), called TWT1 from now on, able to reach a power up to 300 W with the possibility to vary continuously the emission frequency from 8 to 18 GHz.

4) A microwave generator (travelling wave tube based), called TWT2 from now on, able to reach a power up to 600 W with the possibility to change continuously the emission frequency from 13,75 to 14,5 GHz.

Some preliminary measurements have been carried out for oxygen by using the klystron generators: the charge state distributions have been collected at different power levels for both the working frequencies, by using only the biased disk method and $^{18}O_2$ as working gas in order to measure the fully stripped oxygen beams without contaminant. The gas mixing was avoided in order to simplify the tests; the working pressure was about 1×10^{-7} mbar while the ion species for which all the parameters were optimised was O^{7+}. Table 1 shows the magnetic field value (Tesla) used at both the working frequencies together with the fields at which ECR resonance occurs; it can be noted that the source works in the so called "High-B mode" [4] at both the working frequencies and that the field roughly scales as f_2/f_1, according to [5].

By reproducing exactly the same experimental conditions (magnetic field values, gas pressure, position and voltage of the biased disk), the spectra at different power levels have been recorded with the TWT1, in order to have a precise comparison with the Klystron at both the working frequencies. After these comparative measurements, preliminary tests have been carried out with the TWT2 at 14 GHz.

TABLE 1. Magnetic field values used.

Freq	B_{ECR} (T)	B_{hex} (T)	B_{max} (T)	B_{min} (T)	B_{extr} (T)
14 GHz	0.50	1.14	2.05	0.35	1.14
18 GHz	0.64	1.41	2.6	0.45	1.52

COMPARISON BETWEEN KLYSTRON AND TWT

The comparative measurements with klystron (KLY in the figures) and TWT1 at 14 GHz are presented in figure 2. The charge states distributions (CSD) were obtained, respectively, with the klystron at 250 W and with the TWT1 at 240 W. It can be noted from the figure 2 that the maximum values for the two distributions are obtained for different charge states (5^+ and 6^+ respectively); moreover, it can be also observed an increase in the extracted currents for the highest charge states. In fact for O^{6+} the extracted current is 1,5 times higher than the one obtained with the klystron, for O^{7+} it is 2 times higher and for O^{8+} it was increased 8 times or more (6 µA with TWT1, 0,7 µA with klystron).

Since SERSE has been designed to obtain high currents of high charge states, in order to have a measure of the effective gain in power connected with the use of a TWT, we concentrated our investigation in the comparison between the trends of higher charge states present in the distribution (O^{6+}, O^{7+}, O^{8+}). The results show a considerable gain in power by using the TWTs: the highest values of extracted currents for O^{6+} and O^{7+} (fig.3), obtained at 260 W for the TWT1, are obtained with the klystron at about 450 W with a net gain in power of a factor 1,5. The situation is even better for O^{8+}: 4 µA, already extracted at 240 W with the TWT1, are usually obtained at more than 500 W with the klystron with a gain in power greater than a factor two.

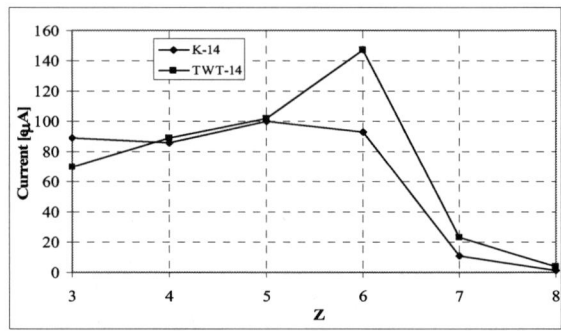

FIGURE 2. Comparison between the CSD of ^{18}O at 14 GHz with Klystron at 250W and TWT1 at 240W.

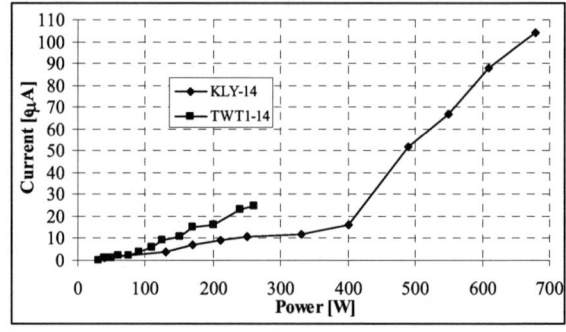

FIGURE 3. Comparison between trends of O^{7+} at 14 GHz for Klystron and TWT1.

In general the best results came out from the comparison between klystron and TWT1 at 18 GHz: in figure 4 the CSD obtained with the klystron at 190 W and with the TWT1 at 195W are shown. There is a different behaviour with respect to the results observed at 14 GHz: an increase in the extracted current is observed for every ion species present in the distribution, and the increase was greater as higher was the charge state. The extracted current is 1,1 times the one obtained with klystron for O^{3+}, 1,3 times for O^{4+}, 1,7 times for O^{5+}, 2,2 times for O^{6+}, 5 times for O^{7+}, concluding with 5,6 µA extracted current of O^{8+}, which was absent in the charge state distribution (CSD) obtained with the klystron.

In figure 5 the current of O^{7+} is shown for both the cases. We note that 42 µA of O^{7+} extracted with TWT at 235 W are obtained with klystron at about 700 W, i.e., at 3 times higher power. For O^{8+} the gain is even more than 3 times: in fact, 9 µA were obtained with TWT at 248 W and with klystron at a power close to 800 W. This result is particularly appealing for ECRIS operations, as a higher RF power entails a higher plasma chamber heating, leading to a higher pressure, that is detrimental for the build up of the highest charge states. Moreover, some technological aspects are eased by the lower RF power, i. e. the DC break and pressure window lifetime would be increased.

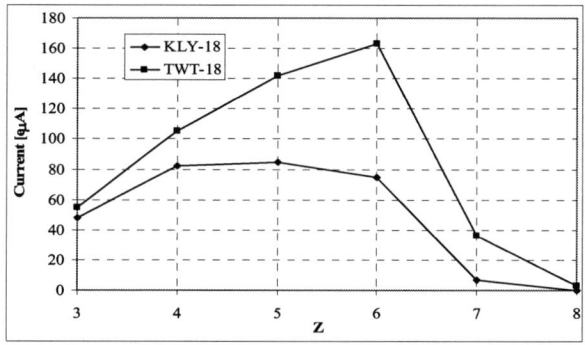

FIGURE 4. Comparison between CSD of ^{18}O at 18 GHz with Klystron at 190W and TWT1 at 195 W.

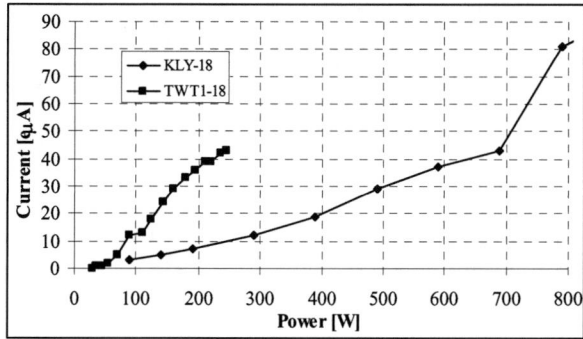

FIGURE 5. Comparison between the of O^{7+} current at 18 GHz for klystron (up to800 W) and TWT1.

DISCUSSION

In order to summarize the results obtained by using klystron and TWT, it is useful to focus on the highest charge states, i. e. O^{7+} and O^{8+}, as they need a higher electron energy and they are a clear signature of the plasma energetic content. It clearly comes out that the TWT generator permits to get at lower power the performance obtained with klystron at higher power levels. The gain in power, changing with the charge state and the working frequency, oscillates between 1,5 and more than 3 times (for O^{8+} at 18 GHz) and this result was experienced in various experiments and it was reproducible. A possible explanation is connected to the increase in electrons' temperature and density which is evident from the spectra analysis: since the only difference between the sets of measurements here compared is the generator used to fed microwaves into the plasma chamber, this increase can be attributed to the larger electromagnetic wave bandwidth provided by the TWT with respect to the klystron.

In order to confirm such a hypothesis, preliminary investigation of the –3dB width of the output signal and the presence of harmonics or spurious with a spectrum analyser have been carried out. All the generators are driven by a pure source signal: the klystron-based generators show a very clean and narrow frequency spectrum without any spurious or harmonics, while the Travelling-Wave-Tube based generators show a broader frequency spectrum with some spurious about –40 and –50 dB under the carrier in a frequency span of 200 MHz.

From the study of the propagation of a monochromatic plane wave inside a magnetized plasma, in a direction parallel to the uniform magnetic field and by taking into account the collisions of any kind undergone by the electrons, it can be shown that this plane wave will consist of two circularly polarized waves. Only the right-handed part has a resonant interaction with the cyclotron motion of the electrons [6].

For a quasi collision-less plasma the resonance peak is very narrow: consequently, the condition for which electron cyclotron resonance occurs, without taking into account Doppler and relativistic effects or the non-uniformity of the magnetic field, is exactly $2\pi\upsilon_{EM} = \omega_{EM} = \omega_g$ with υ_{EM} the frequency of the monochromatic electromagnetic wave and ω_g the electron cyclotron pulsation. Spatially speaking, this condition will be exactly achieved when electrons pass through the surface (egg-shaped) characterized by a static magnetic field equal to:

$$B = B_{ECR} = \frac{2\pi m_e}{|q_e|} \upsilon_{EM}. \qquad (1)$$

where: m_e and q_e are respectively the electron mass and charge and υ_{EM} is the frequency of the monochromatic electromagnetic wave. The gain in energy is depending on the electric field's amplitude. If microwaves with a given $2\Delta\upsilon$ bandwidth are provided to the cavity in presence of non uniform magnetic field, the condition for a resonant interaction will be not only given by (1) but it will be obtained in any point between the two closed surfaces:

$$\begin{cases} B_1 = \dfrac{2\pi m_e}{|q_e|}(\upsilon_{EM} - \Delta\upsilon) \\ B_2 = \dfrac{2\pi m_e}{|q_e|}(\upsilon_{EM} + \Delta\upsilon) \end{cases} \qquad (2)$$

Therefore because the electromagnetic wave injected in the chamber is not monochromatic but it has a certain frequency bandwidth, the electron resonance phenomena will take place in a volume, instead of a simple surface; more precisely inside the layer enclosed between the two surfaces B_1 and B_2. Consequently more electrons can be involved in a resonant interaction with the electromagnetic wave and they can also reach higher energy because, for a given power level, they are subjected to the interaction during the whole time spent by passing through this resonance zone. It can be easily understood that this effect positively influences the performance of the source making the ionisation process more probable since the number of energetic electrons is larger. Furthermore, the presence of energetic and better-confined electrons in the central part of the chamber creates a negative potential well $\Delta\varphi$ with respect to the overall positive plasma potential φ_{plasma}. The higher the number of energetic electrons, the deeper will be the negative well when the TWT is used (fig. 6). Being the confinement time τ_i of a given charge state z:

$$\tau_i \propto \exp\left(\frac{z|q_e|\Delta\varphi}{KT_i}\right). \qquad (3)$$

with kT_i equal for every ion species [6], an increase in the negative potential well produces a greater increase in τ_i as higher is the value of z. It was also qualitatively observed that the improvement obtained with a TWT is lost if the plasma become noisy, as the ion temperature increases and the potential well depth at the centre of the chamber decreases.

Of course, additional experiments and simulations will be carried out to get more insight; then simulations should consider the enlargement of the resonance zone caused by the generator together with the one caused by the gradient in the magnetic field and, eventually, by relativistic and Doppler effects which arises at high electrons' energy.

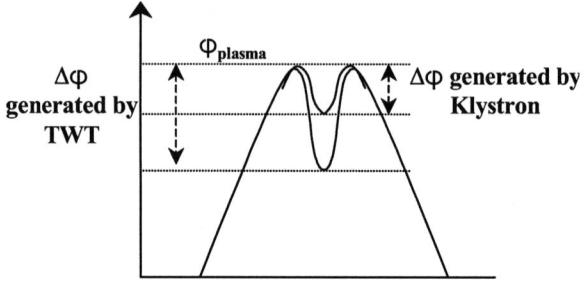

FIGURE 6. Comparison between the of O^{7+} current at 18 GHz for klystron (up to 800 W) and TWT1.

ACKNOWLEDGMENTS

The support of INFN Fifth Committee to the EDIPO experiment is gratefully acknowledged. The authors would like also to thank to A. Caruso, M. Castro and S. Marletta for their support.

REFERENCES

1. S. Gammino, G.Ciavola 2000, Rev. Sci. Instr. **71** (2), 631.
2. S. Gammino *et al* 2001, Rev. Sci. Instr., **72** (11), 4090.
3. S. Gammino *et al* 2002, Nucl. Instr. & Meth **A491** (4), 342.
4. G. Ciavola, S. Gammino 1992, Rev. Sci. Instr. **63**, 2881.
5. S. Gammino et al 1999, Rev. Sci. Instr. **70** (9), 3577.
6. R. Geller 1996, Electron Cyclotron Resonance Ion Sources and ECR plasma (Institute of Physics Publishing Bristol and Philadelphia).

A New Method for Enhancing the Performances of Conventional B-Geometry ECR Ion Sources

G. D. Alton, Y. Kawai, Y. Liu and H. Bilheux

*Physics Division, Oak Ridge National Laboratory**
Oak Ridge, TN 37831-6372

Abstract. The viability of the "volume"-effect for enhancing the high-charge-state populations and intensities of beams extracted from ECR ion sources has been clearly demonstrated at several laboratories. Enlarged ECR zones have been achieved by engineering the central magnetic field so that it is flat and in resonance with single-frequency *rf* power or alternatively, by using multiple frequency or broadband techniques to enlarge the ECR volumes within these sources. For example, the performances of conventional *B*-geometry sources have been ameliorated at several laboratories through the use of multiple frequency *rf* power sources. Although the multiple-discrete frequency method is a very effective means for enhancing the performances of traditional-*B* geometry sources, the practical application of the technique is very costly, requiring multiple independent single-frequency *rf* power supplies and serious modification to the *rf* injection systems of these sources. Broadband sources of *rf* power offer an low-cost and more effective alternative for increasing the physical sizes of the ECR volumes within these sources. Although, previously suggested, here-to-fore, broadband microwave power sources have not been available for use to the ECR ion source community. A special programmable additive "white" Gaussian noise generator (AWGNG) system for injecting broadband *rf* power into these sources has been conceived and developed in conjunction with a commercial firm for such applications. The noise generator, in combination with an external local oscillator, can be used to generate broadband microwave radiation for amplification with a TWT without having to modify the injection systems of these sources. The AWGNG and its use for enhancing the performances of conventional *B*-geometry ECR ion sources will be described in this document.

Keywords: Electron-cyclotron-resonance; ECR ion source; Broadband frequency; "White-noise"; Plasma heating
PACS: 07.77.Ka.; 29.25.Ni.; 41.75.-i.; 41.75.Ak.; 52.25.Jm.; 52.25.Xz.; 52.50.-b.; 52.50.Qt.

INTRODUCTION

Despite the steady advance in ECR ion source technology in recent years, ECR plasma heating has not yet reached its full potential in terms of charge-state and intensity within a particular charge-state, in part, because of the narrow bandwidth, single-frequency microwave radiation commonly used to heat plasma electrons in conventional *B*-geometry ECR ion sources. Since plasma confinement is effected by the vector sums of continually varying solenoid and multi-pole fields in these sources, the shapes, physical sizes, and locations of the ECR zones are determined by the frequency and bandwidth of the microwave radiation at points where the magnitude of the magnetic field is identically equal to the angular frequency of the narrow bandwidth microwave radiation. The ECR zones in these sources are often represented schematically as smooth ellipsoidal surfaces as illustrated in Fig. 1 (This source is a schematic representation of the 10 GHz Caprice ECR ion source, described in Ref. 1.) In reality, the ECR surfaces are usually thin, fluted, irregularly-shaped "surfaces",

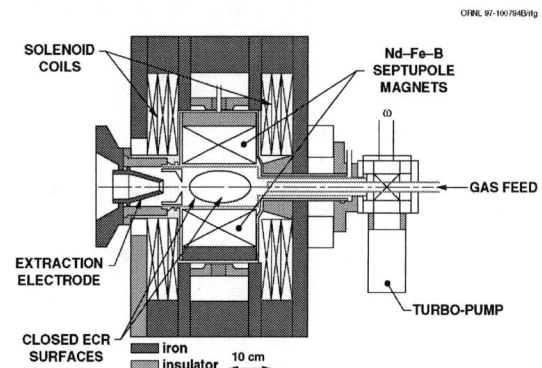

Figure 1. Schematic drawing of a single-frequency, conventional-*B* magnetic field geometry ECR ion.

rather than "volumes" that surround and intersect the axis of symmetry of the source at the injection and extraction ends of the source. The ECR zones in these sources are small in relation to the physical size of the plasma volume, and therefore, microwave absorption is determined not by the physical size of the plasma volume but by the size of the embedded ECR zone(s). Electrons can only be accelerated when they arrive in this zone in phase with the electric field vector of the electromagnetic wave. Hence the acceleration process is stochastic. Electrons that scatter from the zone have a reduced probability for returning in phase with the electric field of the wave, and, therefore, the probability for further stochastic acceleration is reduced. Thus, traditional sources suffer due to the fact that the ECR zones are too small to provide enough electrons with energies sufficiently high to optimize the ionization rate in the plasma volume of the source. Several methods have been proposed for ameliorating this problem [2-5]. The present article describes a practical means for implementing the broadband technique, described in Refs. 3-5, for enlarging the ECR zones in conventional-B geometry ECR ion sources.

EVIDENCES OF ECR "VOLUME" EFFECTS

In recent years, incontrovertible evidence has been provided of the viability of spatial domain technique (see, e.g., Refs. 6-11) and frequency domain, multiple-discrete frequency techniques (see, e.g., Refs. 12-14) for improving the charge-state distributions from these sources. Through these measurements, ECR ion sources equipped with enlarged ECR zones have been clearly shown to outperform their conventional-B single frequency counterparts in terms of high-charge-state capabilities. Although the multiple-discrete frequency method is a very effective means for enhancing the performances of traditional-B geometry sources, the practical application of the technique is very costly, requiring an inventory of independent single-frequency *rf* power supplies and complicated *rf* injection systems. Broadband sources of *rf* power offer an low-cost and more effective alternative method for increasing the physical sizes of the ECR volumes within these sources.

The "White-Noise" Broadband Frequency Generator

Fast scan rate, variable frequency and broadband frequency domain techniques have been previously proposed for increasing the resonant volumes in conventional B-geometry ECR sources [3-5]. The technology for generating broadband microwave power, developed originally for military applications and more recently for satellite communications applications, is now available for generation of broadband microwave power for ECR ion source applications [15]. Figure 2 schematically represents the 10-GHz Caprice source described in Ref. 1 when injected with broadband microwave radiation, as proposed in the present article.

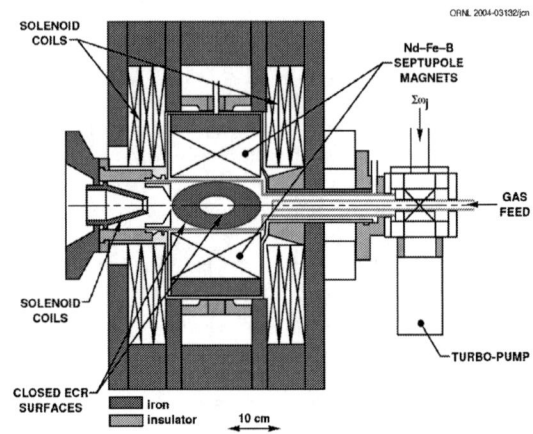

Figure 2. Schematic drawing of a broadband frequency ECR ion source.

Noise with power equally distributed over all *rf* frequencies is called "white-noise". Since the spectral density is evenly distributed over frequency, the output power is proportional to the bandwidth chosen for the particular application. The waveform of the signal generator for the TWT amplifier system, required for increasing the sizes of the ECR zones can be effected by introducing a white-noise generator between the local oscillator (signal generator) and the TWTA. A block diagram of a white noise generator/TWT based microwave power supply that can be used to effect increases in the resonant plasma volumes of any conventional B-geometry ECR ion source is shown in Fig. 3. The white noise generator is represented in the block diagram shown in Fig. 4. An example output frequency distribution of a signal from a white-noise generator, operating at a central frequency of 7.5 GHz with bandwidth of ~250 MHz, is displayed in Fig. 5. The bandwidth of the output signal from the white noise-generator is selected according to

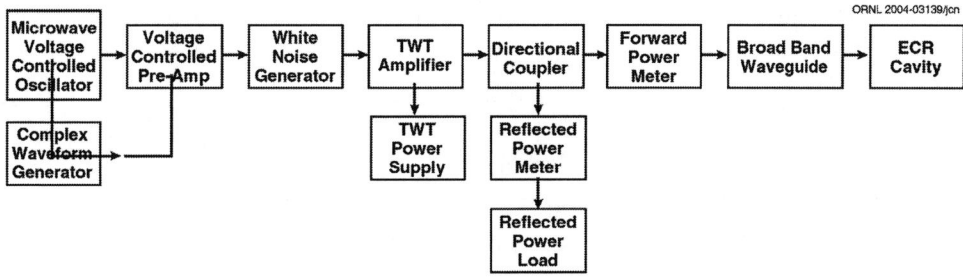

Figure 3. Schematic diagram of a "white-noise", TWTA microwave power supply system for creating large resonant plasma zones in conventional-B geometry ECR ion sources.

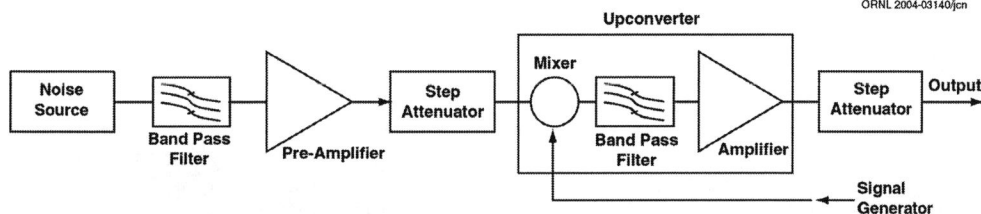

Figure 4. Schematic diagram of a "white-noise" generator for generating broadband width frequency distributions for creating large resonant plasma zones in conventional-B geometry ECR ion sources.

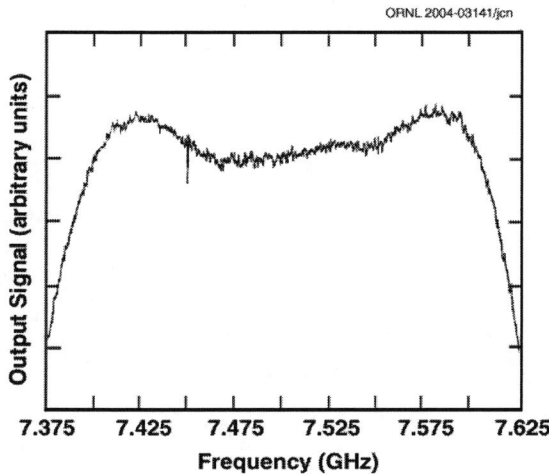

Figure 5. A frequency spectrum from a "white-noise" generator.

the band-width of the TWTA. For ECR ion source applications, the white-noise signals must be filtered to form the desired radiation bandwidth and amplified appropriately prior to injection into the TWTA.

Retrofitting existing ECR ion sources can be easily accomplished by replacing the single-frequency microwave power supply with a "white-noise" generator/TWTA. As illustrated in Fig 3, the signal from the local oscillator is fed into the white-noise generator that mixes a noise frequency distribution, divided equally about the frequency of the local oscillator (LO). Control of the bandwidth position is by moving the local oscillator signal within the bandwidth of the TWTA. The specification for a white-noise generator, on order for use in powering the conventional B-geometry configuration of the 6 GHz all-permanent magnet source described in Refs. 6–8, is displayed in Table 1. The white-noise generator system will be equipped with two filter systems for selecting either 200-MHz or 800-MHz bandwidth microwave radiation. The 200-MHz distribution can be positioned within the band-pass of the TWTA (800 MHz) either manually or by computer control by changing the frequency of the local oscillator (LO) signal. The 800-MHz bandwidth is fixed at the band-pass of the TWTA.

"White-noise" generators can also be procured for driving ECR ion sources that operate over a frequency range of 2 GHz to 19 GHz [15]. In choosing the bandwidth of the white-noise generator, care must be taken to ensure that the frequency distribution output from the TWTA is compatible with both the resonant frequency

Table I. Specification for a "white-noise" generator for the conventional *B*-geometry form of the ORNL 6 GHz all-permanent ECR ion source.

Noise Bandwidth (MHz)	Output Power	Attenuation
*200	0 dBm minimum at 0 dB attenuation setting	0 to 79 dB in 0.1 steps
800	0 dBm minimum at 0 dB attenuation setting	0 to 79 dB in 0.1 steps

- Tunable over the band-pass of the TWTA (5.85 GHz to 6.65 GHz).

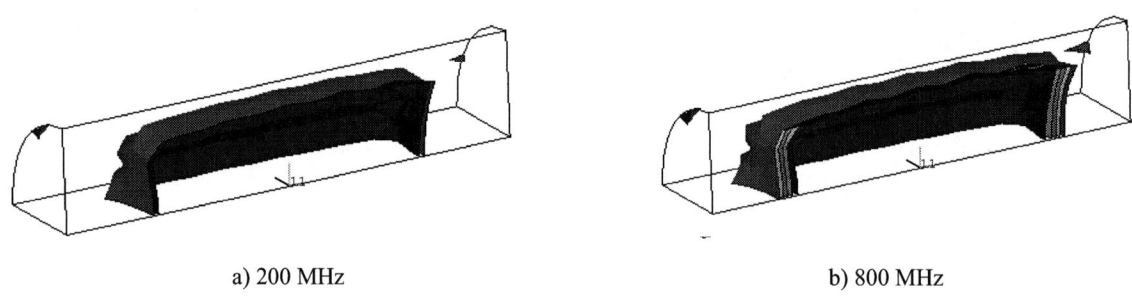

a) 200 MHz   b) 800 MHz

Figure 6. Comparison of the sizes of ECR zones magnet 6 GHz ECR induced in the conventional form of the all-permanent source by broadband white-noise. a); bandwidth: 200 MHz; b) bandwidth: 800 MHz.

of the confining magnetic field and the bandwidth of the TWTA used to inject microwave radiation into the source. Since the output power of the white-noise generator is proportional to the selected bandwidth, care must also be taken in specifying the gain of the particular device. The size of the ECR "volume" depends on the bandwidth of the variable-frequency or broadband microwave power supply in relation to the magnetic field distribution within the plasma volume. Fig. 6 displays ANSYS [16] simulation of ECR zones within the plasma volume of the conventional *B*-geometry form of the all-permanent 6-GHz ECR source, described in Refs. 6-8, when excited either with 200-MHz bandwidth (Fig. 6a) or 800-MHz bandwidth (Fig. 6b) microwave radiation. As noted, the ECR zone volume for heating the plasma with 800 MHz bandwidth radiation is ~4 times the physical size of the zone created with the 200 MHz bandwidth. Thus, the probability for acceleration of larger populations of electrons to higher energies should be greater when the broader bandwidth is utilized. However, one of the objectives of the present experiments will be to determine whether or not there is an optimum bandwidth size that maximizes the high-charge-state performances of conventional-*B* geometry ECR ion sources.

Discussion

The methods for increasing the physical sizes of the ECR zones described in this article are frequency domain complements of the single-frequency, spatial domain technique described in Refs. 2-5. The white-noise generator will be used to compare performances of the conventional-*B* geometry version of the ORNL ECR ion source [6-8] when operated *with* and *without* a white-noise microwave system using each of the two available bandwidths. In these studies, the charge-state distributions will be measured when the source is injected with ~2-MHz (klystron), 200-MHz (white-noise) and 800-MHz (white-noise) bandwidths of microwave radiation as functions of *RF* power. Since the 200-MHz injection system can be positioned either manually or by computer control within the bandwidth of the TWTA, this capability will permit examination of different regions of the magnetic field in search of the region most effective for producing high-charge-state ions.

These experiments are designed to provide information on the effect of bandwidth on source performance as well as insight as to the bandwidth that is most appropriate for operation of the conventional form of the source described in Refs. 6–8. This development will provide yet another method for ameliorating the performances of generic conventional-*B* geometry ECR ion sources.

Analogous experiments will be repeated using the spatial domain complement of the conventional-*B* form of the source described in Refs. 6-8. From previous experiments made with the spatial domain version of the conventional-*B* source, it is clear that the electrons get too hot. It is reasonable to think that there may be an optimum bandwidth and temporal microwave power combination in which the electron population and energy distribution can be more optimally controlled for producing multiply-charge ions. Therefore, another objective will be to examine results derived by temporally varying the microwave power at different bandwidths, during operation of the flat-*B* form of the source.

From these experiments, we will also further demonstrate the viability of the "volume" effect and in so doing, conclusively show the important role that zone size has on the performances of ECR ion sources, as predicted several years ago [2-5].

ACKNOWLEDGMENTS

The authors are indebted to present and past members of the research and development staff of the HRIBF, who through their diligent efforts, have contributed to the content of this paper.

REFERENCES

* Research sponsored by the U. S. Department of Energy under Contract No. DE-AC05-OR22725, managed by UT-Battelle, LLC.
1. Jaquot, B., and Pontonnier, M., *Nucl. Instrum. Meth. Phys. Res. A* **287**, 341 (1990).
2. Alton, G. D., and Smithe, D. N., *Rev. Sci. Instrum.* **65**, 775 (1994).
3. Alton, G. D., *Nucl. Instrum. Meth. Phys. Res. A* **382**, 276 (1996).
4. Alton, G. D., *Proceedings of the 14th Int. Conf. of Cyclotrons and Their Applications*, CapeTown, South Africa (1995) p. 362.
5. Alton, G. D., *Physica Scripta* **T71**, 656 (1997).
6. *"Design and Comparative Evaluation Studies of Conventional 'Surface' – and New Concept 'Volume'- Type, All Permanent Magnet Electron Cyclotron Resonance (ECR) Ion Sources,"* H. Bilheux, Ph.D Thesis, Université de Versailles Saint Quentin-en-Yvelines, France (2003).
7. Liu, Y., Alton, G. D., Bilheux, H., Cole, J. M., and Meyer, F. W., *Rev. Sci. Instrum.* (2004) in press.
8. Liu, Y., Alton, G. D., Bilheux, H., Cole, J. M., and Meyer, F. W., these proceedings.
9. Heinen, A., Rüther, M., Ortjohann, H. W., Vitt, Ch., Rhode, S., and Andrä, *Proceedings of the 14th Int. Workshop on ECR Ion Sources*, CERN, Geneva (1999) p. 224.
10. Heinen, A., Rüther, M., Ducrée, J., Leuker, J., Mrogenda, J., Ortjohann, H. W., Reckels, E., Vitt, Ch., and Andrä, H. J., *Rev. Sci. Instrum.* **69**, 729 (1998).
11. Müller, L., Albers, B., Heinen, A., Kahnt, M., Nowack, L., Ortjohann, H. W., Täschner, A., Vitt, Ch., Wolosin, S., and Andrä, H. J., *Proceedings of the 15th Int. Workshop on ECR Ion Sources*, University of Jyväskylä, Jyväskylä, Findland (2005) p. 35.
12. Xie, Z. Q., and Lyneis, C. M., *Rev. Sci. Instrum.* **66**, 4218 (1995).
13. Alton, G. D., Meyer, F. W., Liu, Y., Beene, J. R., and Tucker, D., *Rev. Sci. Instrum.* **69**, 2305 (1998).
14. Vondrasek, R. C., and Pardo, R. C., *Proceedings of the 15th Int. Workshop on ECR Ion Sources*, University of Jyväskylä, Jyväskylä, Findland (2005) p. 63.
15. Noise/Com, Parsippani, NJ 070545, USA (www.noisecom.com).
16. ANSYS is a finite element computer code designed to simulate thermal transport, mechanical stress, and electromagnetic problems; the code is marketed b ANSYS Inc., Houston, PA.

Ghost Signals In Allison Emittance Scanners

Martin P. Stockli[1,2], M. Leitner[3], D. P. Moehs[4], R. Keller[3], and R. F. Welton[1]

1) SNS, Oak Ridge National Laboratory, P.O. Box 2008, Oak Ridge, TN 37831, USA
2) Department of Physics, University of Tennessee, Knoxville, TN 37996, USA
3) SNS, Lawrence Berkeley National Laboratory, 1 Cyclotron Rd., Berkeley, CA, 94720, USA
4) Fermi National Accelerator Laboratory, P.O. Box 500, Batavia, IL 60510, USA

Abstract. For over 20 years, Allison scanners have been used to measure emittances of low-energy ion beams. We show that scanning large trajectory angles produces ghost signals caused by the sampled beamlet impacting on an electric deflection plate. The ghost signal strength is proportional to the amount of beam entering the scanner. Depending on the ions, and their velocity, the ghost signals can have the opposite or the same polarity as the main beam signals. The ghost signals cause significant errors in the emittance estimates because they appear at large trajectory angles. These ghost signals often go undetected because they partly overlap with the real signals, are mostly below the 1% level, and often hide in the noise. A simple deflection plate modification is shown to reduce the ghost signal strength by over 99%.

INTRODUCTION

The emittance of a particle beam is defined as the six-dimensional distribution of all position coordinates along the three configuration space directions and their associated velocity coordinates. It is normally projected into two-dimensional planes, {x-x'}, {y-y'}, and {z-z'}, reducing the emittance into three subsets.

The transverse subsets are measured with a slit that selects a narrow band from the beam at equidistant position coordinates, either x or y. Downstream, a second slit or a wire harp typically samples the evolved particle distribution to determine the distribution of the trajectory angles x' or y', respectively, with which the ions passed the upstream slit. Wire harps can measure the distribution in single shots but are subject to sagging, variation in the wire size, variations in the surface condition that affects the secondary electron emission coefficient, and variations in the gain and bias of the different amplifiers. A single secondary slit combined with some kind of scanning mechanism guarantees a uniform response, promising more reliable data for the trajectory angle distribution if the beam is stable during the time-consuming scan.

Most of the particle beam is intercepted on the first slit, where it generates a variety of scattered primary and secondary particles. When some of the charged particles reach the current collection device, they create ghost signals, i.e., signals that are not truly representative of the x/x' distribution being probed. A well-known example is slit scattering that alters the trajectory angle and possibly the charge of primary particles and can produce secondary charged particles. Ghost signals are normally small, but their appearance at extreme coordinate values can significantly alter the measured emittances. Accordingly, it is important to minimize all ghost signals. This paper discusses ghost signals that are characteristic for electrical sweep scanners and demonstrates their mitigation in an Allison scanner.

ALLISON EMITTANCE SCANNERS

Over the last 20 years, Allison scanners [1] have been implemented in many laboratories [2] to measure the emittance of low-energy ion beams. Allison scanners feature a base that supports two sets of slits, which can be aligned within tight tolerances. Electric deflection plates are located in the space between the slits, as shown in Fig. 1. A shielded Faraday cup with secondary electron suppression allows for reliable measurements of the small beam currents [3] that pass

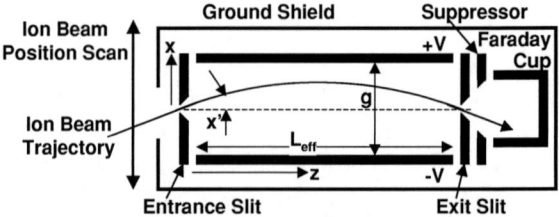

FIGURE 1. Schematic of an Allison emittance scanner.

through both slits. A grounded shield that surrounds the entire assembly minimizes ghost signals from the beam stopped on the entrance slits. A stepper motor moves the entire assembly to scan the position distribution of the beam. At each position both deflection plates are charged with ramped voltages V of opposite polarity to determine the trajectory angle distribution of the beamlet passing through the entrance slit.

Figure 2 shows the emittance distribution of a broad, slightly converging H⁻ beam from the SNS ion source [4] that was measured with our Allison scanner. The beam is measured as it emerges from an electrostatic lens [4], where aberrations cause tails at both ends of the position distribution, one of which is visible in Fig. 2.

The deflection voltage-to-angle conversion depends mainly on the deflection plate length L and the gap g between them. A fringing field correction [5] yields the more accurate effective length L_{eff}, although the difference is normally small because L >> g.

Integrating the transverse acceleration of the ions with charge q and energy q·U yields the equation of motion $x = \int v_x \cdot dt = x' \cdot z - V \cdot z^2/(2 \cdot g \cdot U)$, where x is the transverse position coordinate, z is the axial distance from the entrance slit, and x' is the initial angle (in radians) of the ions passing through the entrance slit (x=0=z). To pass through the exit slit ($x(z=L_{eff})=0$), ions must enter with an angle $x' = V \cdot L_{eff}/(2 \cdot g \cdot U)$, or, in other words, voltages of $V = 2 \cdot U \cdot x' \cdot (g/L_{eff})$ are required to measure ions entering with an angle x'.

The useful angular range is limited to $x'_{max} = 2 \cdot g/L$ because the transverse displacement cannot exceed g/2. This angle also determines the minimum taper that should be applied to the downstream side of both sets of slits to prevent scattering off the slit surface. Our Allison scanner features L = 115 mm, L_{eff} = 121 mm, g = 7 mm, and hence x'_{max} = 0.115 rad.

ALLISON SCANNER GHOST SIGNALS

When the emittance scanner is aligned with the scanned beamlet, its center (x'(z=0)=0) impacts on the second slit as long as the applied voltage is less than $V_0 = U \cdot (g/L)^2$ or when sampling angles that are less than $x'_0 = g/(2 \cdot L) = x'_{max}/4$. When these values are exceeded, the beamlet center impacts on a deflection plate at a distance $z_i = (g \cdot L/(2 \cdot x'))^{1/2} = g \cdot (U/V)^{1/2}$. The angle of impact with respect to the deflection plate, α, is given by $\tan(\alpha) = (2 \cdot x' \cdot g/L)^{1/2} = (V/U)^{1/2}$. These impact angles are small because V<<U.

The number of backscattered primary ions, as well as the number of secondary particles, increases rapidly with decreasing impact angle [6]. Practically all secondary particles have low kinetic energies. Therefore

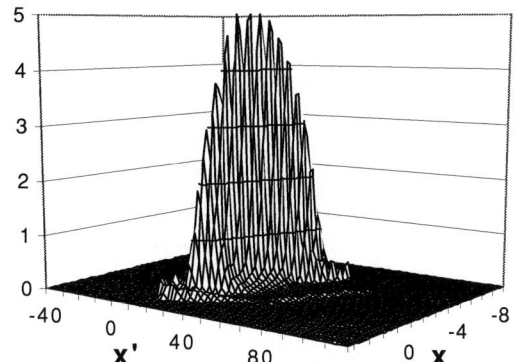

FIGURE 2. Measured beam current as a function of position x and trajectory angle x'.

charged secondary particles are quickly absorbed by the deflection plate with opposite polarity. Most scattered primary particles keep a large fraction of their forward momentum and therefore can reach the exit slit and generate ghost signals.

The emittance distribution of Fig. 2 is shown in the center of Fig. 3 as a density plot with gray tones that darken with increasing intensity. This central distribution is surrounded by an exclusively white zone, which indicates small signals that exceed the noise excursions. Further away, the white pixels are mixed with black pixels that indicate signals with a polarity opposite to the real signals. The zero of the scale was adjusted until the random-noise background in the lower left corner appeared as a random pattern with an equal mix of black and white. This method highlights ghost signals by revealing small deviations from random-noise background.

Figure 3 shows a large black area of non-random-noise background in the range between ~30 and ~80 mrad below the center of intense beamlets that pass through the entrance slit (-5mm < x < 2mm). A similar, but slightly smaller area is found above the center of the beamlets. The observed ~30-mrad gap between the beamlet centers and the ghost signals matches the

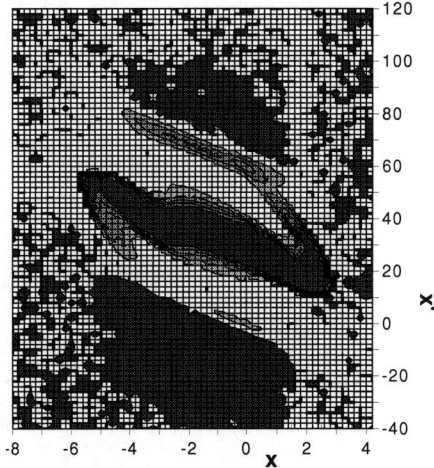

FIGURE 3. Density plot of the emittance data from Fig.2.

predicted 29-mrad deflection of the beamlet centers before they hit the deflection plates. An accurate comparison would require this threshold to be convoluted with the distribution of the beam trajectory angles and the acceptance of the exit slit, suppressor, and Faraday cup system. When scanning for ~30 mrad, the beamlet center impacts on the deflection plates a few millimeters in front of the exit slit with an impact angle of ~3°. When scanning for 80 mrad, the beamlet center impacts on the deflection plates ~50 mm in front of the exit slits with an impact angle of ~6°. The increasing impact angle and the increasing distance from the exit slit cause the ghost signals to fade away. The asymmetry of the ghost signal ranges is caused by the ~30-mrad misalignment between the axes of the Allison scanner and the ion beam, as one can see in Fig. 3.

The ghost signals have an inverted polarity because of most H⁻ ions being stripped during the backscattering process. Surface and near-surface scattering is often accompanied by a change of charge. Depending on the ions and their velocity, electron loss is likely for fast ions with low electron affinity. Electron gain is likely for highly charged ions. Accordingly, the net ghost signals have polarities that are equal or opposite to the real signals. Ghost signals that have the same polarity as the real signals are very difficult to detect unless emittance scanners are commissioned with highly collimated, narrow divergence beams.

GHOST SIGNAL MITIGATION

The small threshold of $x'_{max}/4$ causes the ghost signals to overlap with the real signals as one can see in the asymmetry of the two tails. A separation would require the gap-to-length ratio g/L to be increased. Maintaining the scanning range would require costly higher voltage supplies that are likely slower. Also likely, the deflection plate mount, the electrical connections, and vacuum feedthroughs require a redesign for significantly higher voltages. The gap has to remain smaller than the length ($g \ll L$) to avoid the significant shortening of the effective length.

A dramatic improvement can be achieved by machining a staircase into the deflection plate surface. As

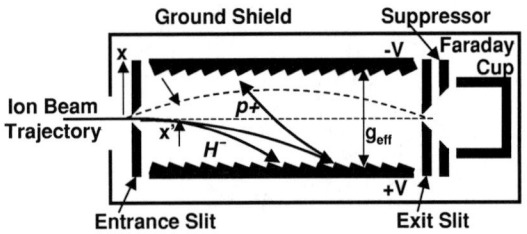

FIGURE 4. Staircase deflection plates bust ghost signals.

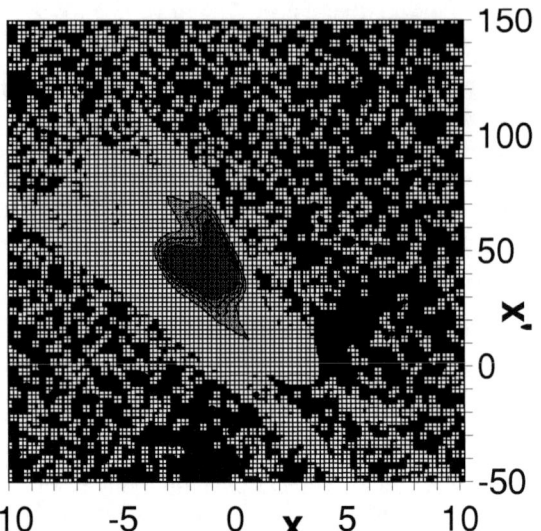

FIGURE 5. With stair-cased deflection plates, the emittance data exhibits random-noise background.

shown in Fig. 4, ions impact on the faces of the stairs with an almost normal impact angle. Backscattering is significantly reduced and results in trajectories that track backwards. The tilt angle must ensure that no ions impact on the flats of the stairs to avoid the increased backscatter probability at even shallower angles. The beamlet centers encounter the largest trajectory angle, α_{max}, when scanning for the maximum useful angle x', thus $\tan(\alpha_{max}) = 2 \cdot g/L$. Half of the beam's divergence and misalignment allowance need to be added. After adding another 50% safety margin, we selected 70° for the faces and 20° for the flats of the stairs. After the modification, the emittance data show a central distribution that is surrounded by random-noise background, as seen in Fig. 5.

The steps of our stairs are 1 mm high and 3 mm apart. After the steps were machined, a small, final cut was taken to obtain sharp edges that are ~25 µm wide. The edge width-to-separation ratio and the roughness of the edge surfaces suggest a ghost signal suppression in excess of 99%. We approximate the new effective deflection plate gap g_{eff} with the sum of the gap between the ridges and height of one step.

EMITTANCE ANALYSIS

Adding a miniscule 0.01% bias to the ghost-free data increases the rms emittance by 33% when calculated from the raw data. More reliable rms-emittance estimates require an analysis with SCUBEEx, the Self-Consistent, UnBiased Elliptical Exclusion method [7], part of a free emittance analysis code [8]. It uses ellipses to separate the real signals from areas that contain pure background. In this work the Twiss parameters of

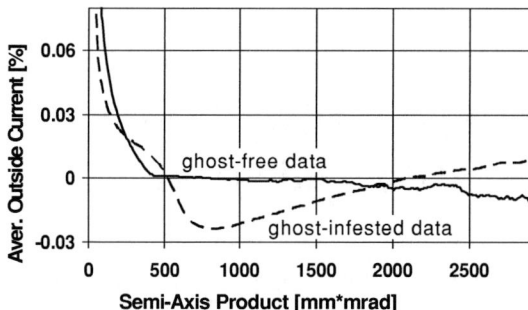

FIGURE 6: SCUBEEx bias estimates for both data sets

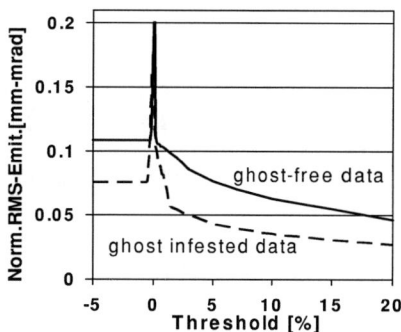

Figure 8. Rms-emittance estimates using thresholds

tightly fitting ellipses were determined from the data that exceeded a 10% threshold. Figure 6 shows the average current measured outside such an ellipse as a function of its semi-axis product. When all real signals of the ghost-free data are included at 430 mm·mrad, the average outside current no longer changes when increasing the ellipse size. However, statistical fluctuations become evident above 1000 mm·mrad. Therefore the values in the intermediate range represent self-consistent bias estimates. No such self-consistent bias estimates can be established for the ghost-infested data because the ghost signals significantly alter the average outside current.

Likewise, Fig. 7 shows the rms-emittance estimates as a function of the semi-axis product of the same ellipses, as in Fig. 6. These estimates are calculated from the data within the ellipse after subtracting the bias estimated from the average current found outside the ellipse. Again, when all real ghost-free signals are included above 430 mm·mrad, the rms-emittance estimate no longer changes when increasing the ellipse size. And once more, statistical fluctuations become evident above 1000 mm·mrad. Accordingly, the values in the intermediate range represent a self-consistent estimate of the semi-axis product of the normalized rms emittance, namely 0.115 ±0.002 mm·mrad.

As before, SCUBEEx cannot self-consistently estimate the rms emittance of the ghost-infested data. Small ellipses underestimate the rms emittance because the ghost signals reduce the signals in the tails. Large ellipses underestimate the rms emittance because the negative contributions from the ghost signals with large x' start to dominate until they exceed all positive contributions at 2620 mm·mrad where the rms emittance becomes imaginary. The two effects discussed previously could cause intermediate ellipses to underestimate the rms emittance, while an overestimation could be caused by the bias underestimation observed in Fig. 6. The two data sets have been measured under very different conditions. It is therefore purely coincidental that the rms-emittance peak value of the ghost-infested data matches the plateau value of the ghost-free data.

Figure 8 shows the rms emittance estimated from thresholded data. Applying a 1, 5, or 10% threshold underestimates the rms emittance of the ghost-free data by 11, 39, or 46%, respectively. The ghost infested data reveal much stronger threshold dependence because the ghost signals significantly reduce the signals in the tails. The lack of a self-consistent rms-emittance estimate prohibits the evaluation of the associated errors, but its threshold dependence suggests errors that could be twice as high.

REFERENCES

1. P. W. Allison, J. D. Sherman, and D. B. Holtkamp, IEEE Trans. on Nucl. Sci. NS-30, 2204-2206 (1983).
2. e.g. M. Dombsky et al., Rev. Sci. Instrum. 60, 1170-1172 (1998); D. Wutte, M. A. Leitner, and C. M. Lyneis, Physica Scripta T92, 247-249 (2001); Y. J. Kim et al, Rev. Sci. Instrum. 75, 1681-1683 (2004).
3. e.g. M. P. Stockli and S. Winecki, Physica Scripta T71, 164-174 (1997).
4. R. Keller et al., Rev. Sci. Instrum. 73, 914-916 (2002).
5. H. Wollnik and H. Edwald, Nucl. Instr. and Meth 36, 93-104 (1965) and its Ref. 2.
6. O. S. Oen and M. T. Robinson, Nucl. Instr. and Meth 132, 647-653 (1976).
7. M. P. Stockli, R. F. Welton, and R. Keller, Rev. Sci. Instrum. 75, 1646-1649 (2004); M. P. Stockli et al. in *Production and Neutralization of Negative Ions and Beams,* M. P. Stockli, edt., AIP Conference Proceedings 639, Melville, New York, 2002 pp. 135-159; R. F. Welton et al. in ibid, pp. 160-174.
8. https://www.sns.gov/APGroup/Codes/EAS/eas.htm.

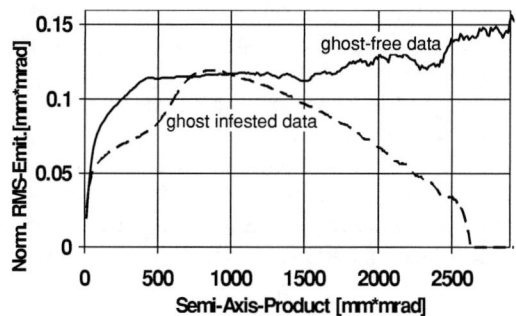

Figure 7. SCUBEEx rms-emittance estimates

Multicharged Ion Generation in Plasma Confined in a Cusp Magnetic Trap at Quasigasdynamic Regime

V. Skalyga, V. Zorin

Institute of Applied Physics, Nizhny Novgorod, Russia

Abctract. Modern way of ECR multicharged ion sources development is connected with increasing of microwave frequency up to tens of GHz. Millimeter wave gyrotrons are used now in several laboratories. Traditional mirror magnetic traps with min B configuration for suppressing of MHD instabilities became too expensive because very strong magnetic fields are required. So prospection of axisymmetric plasma trap with MHD stabilization is quite relevant subject of modern plasma physics. The simplest trap of such kind is cusp one. The zero level of magnetic field in the center of the trap doesn't allow to achieve a good confinement in classical regime. A gasdynamic regime of plasma confinement was studied in [1]. The same regime may be realized in the cusp geometry trap if plasma density is high enough.

In present work a theoretical model of quasi-gasdynamic plasma confinement in a cusp trap is developed. The possibilities of multicharged ion generation in a plasma confined in a cusp trap under the conditions of powerful ECR heating by pulsed radiation of a millimeter wavelength gyrotron were investigated numerically. Calculations were made on basis of [2]. Confinement time was estimated with using of approach developed in [3].

The results of experiments with simplest kind of cusp trap have demonstrated good correspondence with theoretical calculations and therefore the adequacy of the developed approach and possibility to build more effective source on this base. Two ways of possible evolution of ECR ion sources based on a cusp magnetic trap are proposed.

INTRODUCTION

The possibility to realize two different regimes of non-equilibrium plasma confinement under condition of powerful ECR heating by millimeter wave radiation was shown in [1]. In greater part of modern sources the classical confinement is realized. In this case electrons are confined by the magnetic field and leave the trap scattering into the loss cone in the velocity space. Ion confinement is determined by distribution of ambipolar potential in the trap. For the conditions of the mirror trap the electron scattering into the loss cone is related with either collisions with ions and among themselves or quasilinear diffusion in velocity space due to the intensive ECR heating. At the increasing plasma density the transition from classical regime of plasma confinement to the so-called quasigasdynamic one is observed experimentally [1]. It is connected with the increasing frequency of Coulomb collision scattering of electrons into the loss cone. The change of plasma confinement occurs at the values of plasma density when the rate of the loss cone filling by electrons is higher than rate of plasma losses out from trap. In this case the loss cone is filled, the electrons are confined by ambipolar potential along the magnetic field lines and plasma losses are determined by gasdynamic ion escape. Such regime of non-equilibrium plasma confinement with filled loss cone was called quasigasdynamic one. The transition to this regime of plasma confinement is inevitable while its density grows.

It should be mentioned that the realization of classical confinement in cusp trap is accompanied with a number of difficulties due to the peculiarities of magnetic field lines configuration. In central part of the trap, in the area of low magnetic field the approximation of adiabatic motion of electrons on Larmor orbits can not be valid, the occurrence of electron in saddle area can be equivalent to the electron collision with its transition into the loss cone and its lifetime is equal to the time of single flight along the trap. In the case of quasigasdynamic confinement regime such processes aren't important.

PLASMA LIFETIME AT QUASIGASDYNAMIC CONFINEMENT REGIME

The problem of determining plasma lifetime in the magnetic traps was considered in a number of studies of various authors. One of solutions of this problem for various given electron velocity distribution functions was proposed in [4]. However in the case of quasigasdynamic confinement regime the distribution function is isotropic due to the high collision frequency, so plasma lifetime can be found from more simple considerations. The approach developed for the investigation of gasdynamic traps [3] is convenient. It is supposed that plasma density N_e is approximately uniform along the magnetic field tube. Plasma flux S through the plugs of this tube can be written in the form:

$$S = N_e s V_s \quad (1)$$

where s is total square of cross-sections bounding the volume of the tube between magnetic plugs (see fig. 1), V_s is ion sound velocity. Then the longitudinal plasma lifetime in this tube τ_l (time of plasma escape along the magnetic field lines) is equal to ratio of total plasma amount in the tube to the outgoing flux

$$\tau_l = \frac{N_e \cdot V_p}{S} = \frac{V_p}{s \cdot V_s}, \quad (2)$$

here V_p is the volume of force tube limited by magnetic plugs. For the traps with long uniform part (length>>radius) the expression (2) gives simple well-known formula $\tau = \frac{R \cdot L}{2 \cdot V_s}$. We will call ratio V_p/s in (2) the effective trap length (the analogue of $RL/2$ value for mirror trap with long uniform part of the magnetic field, R is plug ratio). It can be found in the case of quasigasdynamic regime that plasma lifetime is determined only by ion sound velocity and trap geometry.

Due to rather complicated configuration of the magnetic field lines in a cusp trap it is necessary to use numerical methods for lifetime estimation. The ion beam is formed from plasma along the magnetic field lines by means of the extractor placed on the system axis, so from the viewpoint of ion beam forming the most important plasma area is axisymmetrical magnetic tube (placed evidently near the axis), abutting on the hole for ion extraction (see Fig. 1).

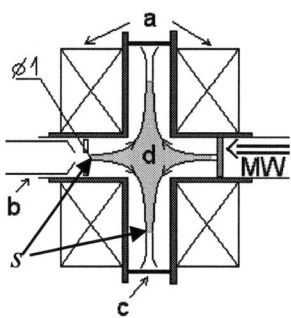

FIGURE 1. A sketch of the discharge vacuum chamber with drawing of magnetic field lines of the trap. a – magnetic coils, b – extractor, c – vacuum chamber, d – axisymmetrical force tube.

Present work contains the estimation of lifetime for the trap designed for the experiments on SMIS-37 setup [5]. For this trap the volume of considered axisymmetrical magnetic tube is approximately equal to 0.8 cm³ under condition that plug diameter on the longitudinal axis is equal to the diameter of hole for the ion extraction (1 mm for SMIS'37). In this case the effective trap length is equal to 13 cm, longitudinal plasma lifetime being equal to $\tau_l \approx 5$ µs.

BASIC EQUATION

Theoretical analysis of formation of multi-charged ions in plasma at quasigasdynamic confinement in cusp trap can be carried out using the equation set described in details in [2]. It is 0-dimensional non-stationary set of differential equations of ionization balance for ions of all charge states i:

$$\frac{dN_i}{dt} = (k_{i-1,i} N_{i-1} - k_{i,i+1} N_i) \cdot N_e - \frac{N_i}{\tau_i} \quad (3)$$

Here τ_i are lifetimes for ions with various charge states, $k_{i,i+1}$ are rate of electron-impact ionization. The equation for electrons:

$$\frac{dN_e}{dt} = N_e \cdot \sum_{i=0}^{6} k_{i,i+1} N_i - \frac{N_e}{\tau_e} \quad (4)$$

The equation for neutral gas density:

$$\frac{dN_0}{dt} = I(t) - k_{0,1} N_0 N_e \quad (5)$$

here $I(t)$ is the rate of neutral gas inlet into trap. Also we add the condition of plasma quasineutrality to the set:

$$\frac{1}{\tau_e} = \frac{1}{N_e} \sum_{i=1}^{7} \frac{i \cdot N_i}{\tau_i} \quad (6)$$

This set should be also enlarged with the equations for plasma lifetime. As it was already mentioned above, the lifetime of ions of kind i at quasigasdynamic confinement is determined from expression (2). The set of equations is written for the case when nitrogen is used as an actuating gas. All calculations were carried out in the approximation of constant electron temperature and neutral gas inlet rate to the trap.

THE RESULTS OF NUMERICAL SIMULATION

The ion charge state distribution in the extracted beam depends on electron temperature and density as well as on the trap length and distribution of magnetic field in the trap. It was shown in [1] that plasma density in the experiment is limited by critical value for the frequency of used microwave radiation. Hence maximum density for SMIS 37 setup is equal to $2 \cdot 10^{13}$ cm^{-3}. As it was shown in a number of experiments its real value is close to critical one [1].

The electron temperature can be estimated from the considerations of balance between energy of microwave radiation being input into the trap and energy being output from the trap by plasma. Energy flux density carried by plasma from the trap at quasigasdynamic confinement regime can be calculated by the following expression [2]:

$$S = N_e \cdot V_s \cdot kT_e \cdot \ln\sqrt{\frac{M}{<Z> \cdot m_e}} \quad (7)$$

The results of calculations of the energy flux density carried by particles from the trap versus the electron temperature at their density $Ne = 1 \cdot 10^{13}$ cm^{-3} are shown on Fig. 2.

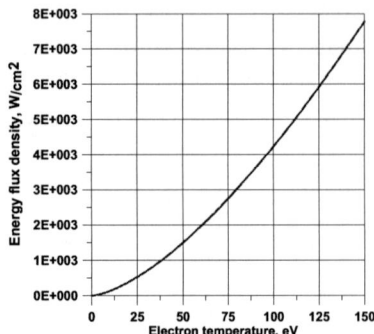

FIGURE 2. The energy flux density carried by particles from the trap versus the electron temperature at their density $Ne = 1*10^{13}$ cm^{-3}.

The power of gyrotron used at SMIS'37 setup is approximately equal to 100 kW. But the density of microwave energy flux is not higher then 2 kW/cm^2, so according to Fig. 2 electron temperature is not more than 70 eV. If we suppose that absorption factor for microwave power in experiments is 50% then maximum possible electron temperature will be 50 eV. The calculated ion charge state distribution for this electron temperature at electron density averaged over trap and equal to $N_e = 1 \cdot 10^{13}$ cm^{-3} is shown on Fig. 3.

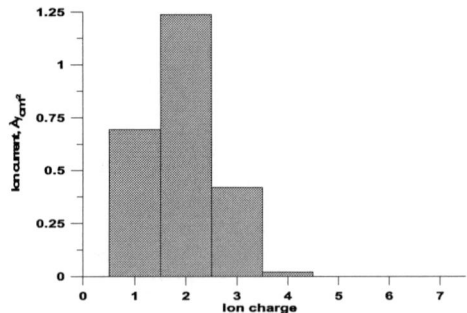

FIGURE 3. Theoretical charge state distribution of nitrogen ions for $Ne = 1 \cdot 10^{13}$ cm^{-3} and $Te = 50 eV$ for cusp trap under conditions of SMIS 37 setup.

In this case the calculated density of total ion current from the trap through its plug on longitudinal axis is about 2.5 A/cm^{-2}.

Thus the presented estimations and the results of numerical simulation allow to expect the successful demonstration of MCI generation in such system in the condition of SMIS 37 setup as well as the getting of high density of the extracted ion beam.

COMPARISON WITH THE EXPERIMENT

The detailed ion charge state distributions were measured in the experiments on SMIS'37 setup. It was shown that under optimal condition for generation of N^{3+} the ratios of currents for main ions have the following values: $I_N^{2+}/I_N^{+} \approx 1.8$, $I_N^{2+}/I_N^{3+} \approx 3.5$. This measurements are corresponding with the accuracy about 10% to the results of numerical simulations for the case of electron density $Ne = 1 \cdot 10^{13}$ cm^{-3} and their temperature $Te = 50$ eV. It should be mentioned that correspondence of theoretical and experimental values of total ion current density from the trap is rather good too.

The experiments have demonstrated the possibility to realize quasigasdynamic regime of plasma confinement in a cusp trap and to obtain the ion beam with high current density. The coincidence of the experimental results with calculations makes further successful optimization of multi-charged ion source based on a cusp trap really promising.

CONCLUSION

In this part of the paper two possible variants of ECR ion source based on cusp trap are discussed.

Formula (2) shows that plasma lifetime can be increased by using magnetic trap with bigger volume of axisymmetrical magnetic tube from those ions are extracted. Such modified discharge chamber and system of magnetic coils were designed. Using of new magnetic trap will give an opportunity to increase plasma lifetime up to 10 μs.

The first variant of the source is based on modern gyrotron with operating frequency 75 GHz and power 400 kW, which combined with modified discharge chamber. Such combination will provide us to create and sustain plasma with density $6 \cdot 10^{13}$ cm^{-3} and electron temperature about 100 eV. In this case one can expect significantly better ion charge state distribution shown on Fig. 4. The ion current density through the plug on the trap longitudinal axis will be about 10 A/cm².

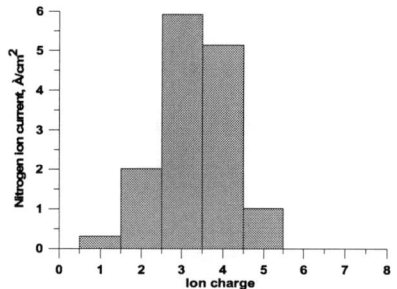

FIGURE 4. The charge state distribution of nitrogen ions. $Ne = 6 \cdot 10^{13}$ cm^{-3}, $Te = 100$ eV, lifetime 10 mks.

The second variant of ECR ion source based on cusp trap would be powered by cheaper and frequently used gyrotron with radiation frequency 28 GHz and power 15 kW with good microwave focusing system. In this case it would be possible to achieve a charge state distribution close to the one shown on Fig.3.

It also should be mentioned that cusp magnetic trap provides to apply several extracting systems along the ring plug of the trap. It can be useful for creation of many ion beams from one source on technological setups.

ACKNOWLEDGMENTS

This work was supported by INTAS grant # 01 0373.

REFERENCES

1. S.V. Golubev, S.V. Razin, A.V.Vodopyanov, V.G. Zorin, M.A. Shilov, *Tech. Phys. Letters*, 1999, v. 25, n. 7, p. 588 - 589.
2. V.E. Semenov, V.A. Skalyga, V.G. Zorin, Review of Scientific Instruments, v. 73, n2, Part II, p. 635 – 637, 2002.
3. Mirnov V.V., Ryutov D.D. *Pisma v Zhurnal Theknicheskoi Fiziki.*, 1979, v.5,p. 678.
4. A. V. Turlapov, V. E. Semenov, *Phys. Rev.* E, 57, 5937, 1998.
5. S.V. Golubev, S.V. Razin, V.G. Zorin, *Review of Scientific Instruments*, v. 69, 1998, N. 2, p. 634 – 636.

Multiple Ionization Of Metal Ions By ECR Heating Of Electrons In Vacuum Arc Plasmas

A.V. Vodopyanov[1], S.V. Golubev[1], D.A. Mansfeld[1], A.G. Nikolaev[2], E.M. Oks[2], S.V. Razin[1], K.P. Savkin[2]

[1] *Institute of Applied Physics (IAP RAS), Nizhny Novgorod, RUSSIA*
[2] *High Current Electronics Institute (HCEI RAS), Tomsk, RUSSIA*

Abstract. A joint research and development effort has been initiated, whose ultimate goal is the enhancement the mean ion charge states in vacuum arc metal plasmas by a combination of a vacuum arc discharge and an electron cyclotron resonance (ECR) heating. Metal plasma was generated by a special vacuum arc mini-gun and injected into mirror magnetic trap. Plasma was pumped by high frequency gyrotron-generated microwave radiation (frequency 37.5 GHz, max power 100 kW, pulse duration 1.5 ms). Using of powerful microwaves makes it possible to sustain sufficient temperature of electrons needed for multiple ionizations at high plasma density (more then 10^{13} cm^{-3}). Parameter of multiple ionization efficiency $N_e\tau_i$, where N_e is plasma density, τ_i is ion lifetime, in such a case could reach rather high value $\sim 10^9$ cm$^{-3}\cdot$s. In our situation $\tau_i = L_{trap}/V_i$, where L_{trap} is trap length, V_i is plasma gun flow velocity. The results have demonstrated substantial multiple ionization of metal ions (including metals with high melting temperature). For a metal (lead, platinum) plasma, ECR heating shifted the average ion charge up to 5+. Further increase of the ion charge states will be attained by increasing the vacuum arc plasma density and optimizing the ECR heating conditions.

INTRODUCTION

Novel fundamental matter research (for example, new element synthesis) requires acceleration of heavy particles to high energies due to high energy threshold of nuclear reaction. The energy of a particle at the exit of accelerator (linear or cyclotron) is more when charge of an ion injected is higher. Problem of fundamental matter research is associated with development of highly charged heavy ion sources.

Sources of multicharged ions based on low pressure electron-cyclotron resonance (ECR) discharge in a magnetic trap are able to produce ion beams with high charge states and currents. ECR sources can produce gaseous ion beams, but all the elements heavier then xenon are solids in normal conditions. It is possible to use an oven where not refractory solids can be melted and vapoured. ECR discharge in a vapour goes in usual way like in gas injection situation. The significant lack of this scheme is inability to operate with refractory solids (with melting temperature > 1000 degrees).

A vacuum arc is an attractive method to produce metal ions. Metal ions are generated by the intense ionization of metal atoms evaporated at the cathode spot [1]. Such sources are able to produce high current ion beams (up to several amperes) in pulse or DC operation mode. The cathode material determines ion species of the plasma. It can be made of any conductive material including high melting temperature metals. Mean charge states of vacuum arc plasma are close to 2+ for a wide range of cathode materials [1].

It appears reasonable to use vacuum arc plasma gun for refractory metal plasma injection into the ECR source to increase the charge of metallic ions of vacuum arc plasma in a magnetic trap with ECR heating by strong microwaves.

The ion confinement time τ_i of vacuum arc plasma injected into the magnetic trap is not long enough. The fact is that decelerating of the plasma jet or trapping ions are very complicated problems; τ_i is determined by the magnetic trap size L_{trap} and vacuum arc plasma jet velocity V_i, $\tau_i \approx L_{trap}/V_i$. The plasma flux velocity equals 10^6 cm/s in order of magnitude [1]. Thus, rather high plasma density is required for significant additional ion stripping (the confinement parameter $N_e\tau_i \sim 10^9$ cm^{-3}·s [2]). Plasma density N_e should be $\sim 3\cdot 10^{13}$ cm^{-3} for the L_{trap} = 30 cm. The use of microwave pumping with more power (upto 100 kW) and with higher frequency (37.5 GHz), in contrast to traditional ECR sources, helps us to carry out experiments at much more dense (more than an order of magnitude) plasma. It is important to note that keeping the electron temperature at the high level necessary for multiply ionization requires rather high microwave power. Estimations based on the methods described in [3] give values of power flux density about several tens of kW/cm^2. Such power is easily achievable when modern gyrotrons are used.

This work is continuation of the investigations initiated in [4].

EXPERIMENTAL SETUP

Sketch of the vacuum arc plasma gun is presented on Fig.1. Vacuum arc discharge between cathode (1) and hollow anode (2) is initiated by the auxiliary discharge between igniting electrode (3) and cathode (1) on the ceramic surface (4). It produces plasma of the cathode material. This plasma is filling hollow anode and propagates into the trap along the magnetic field lines. Diameter of orifice in the anode is 2 mm.

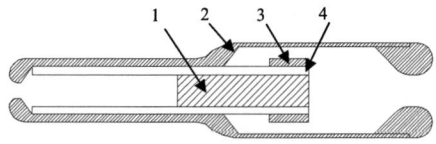

FIGURE 1. Sketch of the vacuum arc plasma gun. 1 - cathode, 2 - anode, 3 - igniting electrode, 4 - ceramic insulator.

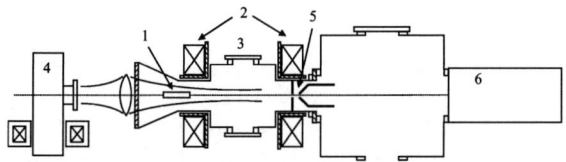

FIGURE 2. Sketch of the experimental setup. 1 - mini plasma gun MEVVA, 2 - magnetic trap, length 27 cm, mirror ratio 5, max field in the plug 3 T, 3 - discharge chamber, 4 - gyrotron, frequency 37.5 GHz, max power 100 kW, pulse duration 1.5 ms, 5 - extractor, hole 1 mm, voltage upto 10 kV, 6 - ion beam analyzer.

The outline of the experimental facility is presented on Fig. 2. The plasma gun (1) was placed on the magnetic axis outside the trap, close to the plug. The magnetic simple mirror trap was produced by two "warm" identical coils (2). The power supply of the plasma gun provides arc current from 50 A to 3 kA. Pulse duration ~100 μs. Plasma gun power supply was placed on the high voltage platform because anode connected with discharge chamber (3 on figure 2) was biased by the extracting voltage upto 10 kV. Microwave radiation of the gyrotron (4) was injected into the magnetic trap along the magnetic field lines through the teflon window of the discharge chamber. ECR heating was realized at the fundamental gyrofrequency and it was followed by the additional ion stripping by the electron impact. Two-electrode extractor (5) was installed in the second plug of the magnetic trap for ion acceleration. The hole of the plasma electrode was 1 mm. Charge state composition of the beam was analyzed in the beam line with the bending magnet (6).

CONFINEMENT PARAMETER ESTIMATION

The key parameter of our system is the vacuum arc plasma flux velocity. It limits the hot electron - metal ion interaction time when trap length is fixed and determines the efficiency of multicharged ion formation. Another important parameter of the system is density of the plasma generated by the plasma gun. Experiments were carried out for the parameters determination. Platinum cathode was used. It is already known fact that application of the magnetic field to the cathode area leads to vacuum arc voltage increase. Velocity of ions emitted by cathode spots increases also [5]. In our case measure of ion velocity was based on the analysis of extracted current oscillograms registered at different values of the magnetic field of the trap (See Fig. 3). It shows that ion's velocity is more when magnetic field is stronger. The dependence of the ion flux velocity from the magnetic field value gathered from these oscillograms is shown on Fig. 4. Significant increase of the vacuum arc voltage was observed when magnetic field is applied. It is in accordance with [5] and additionally proves that velocity of ions emitted by cathode spots is more when magnetic field is stronger. As it is goes from the measurements described above, influence of the magnetic field on the metal ion velocity saturates at magnetic field values more than 1.5 T. ECR heating at fundamental gyrofrequency needs to maintain magnetic field more than 1.5 T in the plug.

So, it is possible to put averaged ion velocity $V_{Pt} \approx 1.5 \cdot 10^6$ cm/s for estimations.

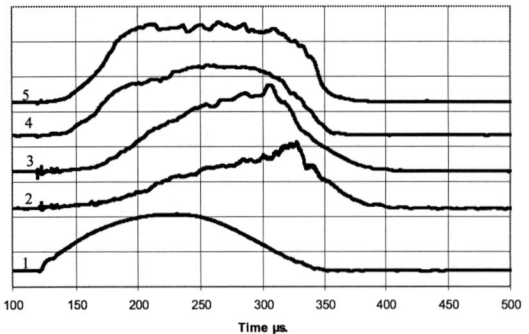

FIGURE 3. Oscillograms: 1 - vacuum arc current; 2, 3, 4, 5 - extraction current at magnetic field value 0, 1.5, 1.8, 2.6 T in the plug. Maximum vacuum arc current is 150 A. Platinum cathode.

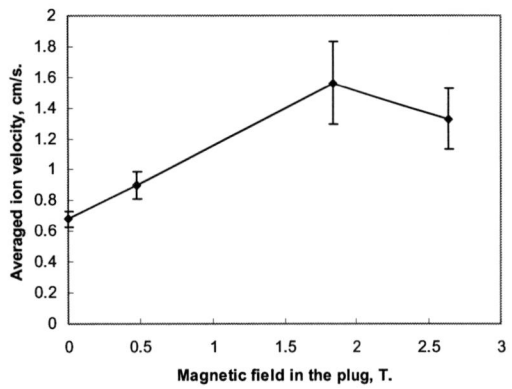

FIGURE 4. Averaged ion velocity versus magnetic field value in the plug. Arc current is 150 A. Platinum cathode

It is possible to estimate metal plasma density produced by the plasma gun by the extracted current measurements. $N_e \approx I_{extr}/(V_{Pt} \cdot e \cdot S_{extr})$, where I_{extr} is total extracted current, V_{Pt} - ion flux velocity, e is electron charge, S_{extr} - is extractor hole section. Thus, plasma gun fill the trap with metal plasma with $N_e \approx 4 \cdot 10^{12}$ cm^{-3} at vacuum arc current of 115 A.

So, hot electron - ion interaction time is $\tau_i = L_{trap}/V_{Pt} \approx 15$ µs, where $L_{trap} = 27$ cm is magnetic trap length. At the same time, density of the plasma filling the trap can vary. Roughly, it is proportional to the vacuum arc current. If cut-off density for pumping wave is proposed to be the optimal, $N_e \approx 2 \cdot 10^{13}$ cm^{-3}, one can realize confinement parameter $N_e \cdot \tau_i \approx 3 \cdot 10^8$ cm^{-3}·s. Such a value of confinement parameter let us hope to observe significant additional stripping of metal ions, if the electron temperature will be high enough.

MULTICHARGED METAL ION FORMATION

Investigations of additional stripping of metal ions of vacuum arc plasma injected into magnetic trap with ECR heating by gyrotron emission were performed for two elements: lead and platinum. Melting temperature of the lead is about 327 degrees, platinum is a refractory metal, and its melting temperature is 1770 degrees.

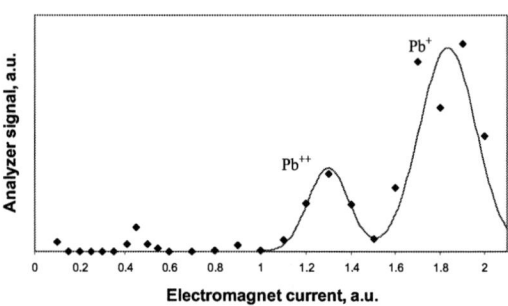

FIGURE 5. Ion charge state distribution of vacuum arc plasma. Lead cathode. Arc current 130 A. Magnetic field 1.9 T.

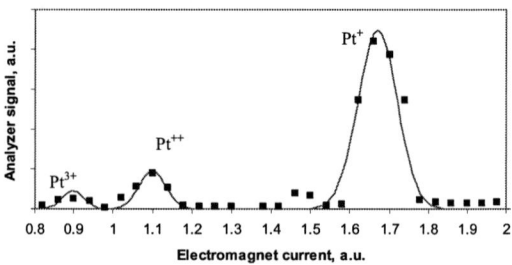

FIGURE 6. Ion charge state distribution of vacuum arc plasma. Platinum cathode. Arc current 140 A. Magnetic field 1.3 T.

Charge state distributions of metal ions of vacuum arc discharge propagated through the magnetic trap and accelerated in the extractor gap are shown at Fig. 5 and 6. The average ion charge of lead plasma is 1.2. The average ion charge of platinum plasma generated by the plasma gun is 1.3. Powerful microwave heating of the metal plasma under ECR conditions results in additional multiple ionization. Ion charge state distribution depends on three main parameters of the system. They are: vacuum arc current, magnetic field value, and microwave power. Figures 7 and 8 shows metal ion charge state distributions at optimal for additional stripping parameters. For lead plasma: vacuum arc current is 50 A, magnetic field value in the plug is 2.6 T, microwave power injected is about 60 kW. For platinum plasma: vacuum arc current is 80 A, magnetic field is 2.6 T, microwave power is about 60 kW.

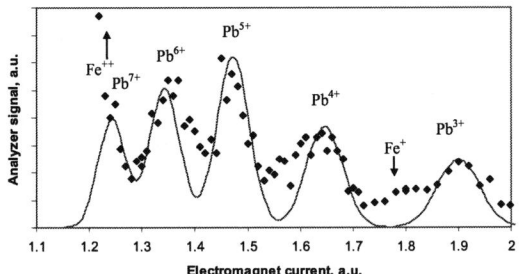

FIGURE 7. Ion charge state distribution at optimal parameters. Lead cathode. Arc current is 50 A. Magnetic field is 2.6 T. Microwave power is 63 kW.

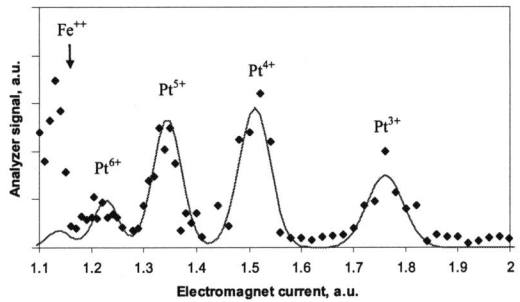

FIGURE 8. Ion charge state distribution at optimal parameters. Platinum cathode. Arc current is 80 A. Magnetic field is 2.6 T. Microwave power is 63 kW.

As it can be seen from the spectra, there are a lot of different impurities in the plasma. Experiments shows that it is mainly ions of elements composing stainless steel (Fe, Ni, Cr) and gaseous ions (H, O, N). Ions of Fe, Ni, and Cr go into the plasma mainly because of erosion of stainless anode of the plasma gun. Gaseous fractions are from residual gas adsorbed between discharge pulses due to small pulse repetition rate [6] (1 pulse per 20 s).

The confinement parameter realized in the experiment $N_e \tau_i = 3 \cdot 10^8$ cm^{-3}s approximately corresponds to maximum averaged ion charge obtained [2]. The ion current density which could be extracted from such plasma is $J = e \cdot N_e \cdot V_{Pt} \approx 4$ eA/cm^2.

Thus, source of multicharged metal ions is demonstrated. It can successfully operate with refractory metals.

REFERENCES

1. Mesyats G. A., and Barengol'ts S. A., *Phys. Usp.* **45**, 1001-1018 (2002).

2. Golovanivsky K. S., *Instruments and experimental techniques* **28**, 989 (1985).

3. Semenov V. E., Skalyga V. A., Smirnov A. N., and Zorin V.G., *Rev. Sci. Instrum.* **73**, 635 (2002).

4. Vodopyanov A. V., Golubev S. V., Zorin V. G.,. Razin S. V, Vizir A. V., Nikolaev A. G., Oks E. M., and Yushkov G. Yu., *Rev. Sci. Instrum.* **75**, 1888-1890 (2004).

5. Anders A., and Yushkov G.Yu., *Journal of Applied Physics* **91**, 4824-4832 (2002).

6. Yushkov G.Yu., Anders A., *IEEE Transactions on Plasma Science* **26**, 220-226 (1998).

VI NEW SOURCES

An All-Permanent Magnet ECR Ion Source for the ORNL MIRF Upgrade Project

D. Hitz[1], M. Delaunay[2], A. Girard[1], L. Guillemet[1], J.M. Mathonnet[1], J. Chartier[1], F.W. Meyer[3]

CEA-Grenoble, Département de Recherche Fondamentale sur la Matière Condensée, [1]Service des Basses Températures, [2]Service de Physique des Matériaux et Microstructures, 17 Rue des Martyrs, 38054 Grenoble Cedex9 , France
[3]Physics Division, Oak Ridge National Laboratory, Oak Ridge, TN 37831-6372, USA

Abstract. A new high voltage platform has been installed at the ORNL Multicharged Ion Research Facility (MIRF) to extend the energy range of multicharged ions available for experiments studying their collisional interactions with electrons, atoms, molecules, and solid surfaces. For the production of the multiply charged ions, a new all-permanent magnet ECRIS has been designed and fabricated at CEA/Grenoble. After a brief overview of the basic features of the new platform, and associated beam transporta detailed description of the new ion source design and performance is provided, together with some typical Ar, Xe, and O beam intensities obtained during source commissioning prior to shipment to ORNL.

INTRODUCTION

While developed initially as an injector for high-energy accelerators, in recent years the Electron Cyclotron Resonance (ECR) ion source has become an indispensable tool in the field of multicharged ion (MCI) collision physics. In fact, the availability of low energy, intense, high charge state, high duty factor, ion beams from ECR sources has opened many areas of MCI research that have previously been inaccessible. To extend the present energy range available at the ORNL – MIRF (Multicharged Ion Research Facility), a new high voltage platform has recently been built. For installation on this platform, a new ECRIS has been designed and built whose goal is to reach beam intensities equivalent to the Caprice ion source. General trends of this new ECRIS are presented along this article.

CONCEPTUAL DESIGN

Permanent magnet qualities are nowadays good enough to let the ion source designer build an all-permanent magnet ECRIS able to fulfill todays requirements needed by atomic physics programs like those performed at MIRF. However, the design of any ion source of this type must follow the same rules as for any other type of ECRIS. For example, our source design is based on previous studies of optimum magnetic confinement [1,2]. Then a compromise has been chosen to manage at the same time:
- plasma chamber size: the larger the diameter, the better the ion confinement
- resonance zone length
- magnets size which defines the cost of the source.

Another permanent magnet ECRIS based on the same principles is under construction [3,4]. This latter version is however much larger and consequently more expensive than our design and will certainly deliver larger beam intensities and higher charge states than ours.

Furthermore, the overall design of this source is based on GTS ECRIS which has given a large variety of beams either of low or high charge states [5, 6]. For example, the double wall plasma chamber is made of aluminum for better performances thanks to higher secondary electron emission and for a more efficient cooling.

Magnetic Field Configuration

The axial magnetic field shape follows the "High-B" mode principle leading in such a way to a very good confinement at the rear side of the plasma chamber. A 1.8 T magnetic field maximum is achieved by use of an iron plug inserted from the rear into the plasma chamber. The source also features an axial magnetic field maximum of 0.9 T on the extraction side and a minimum axial field of about 0.4 T. **Figure 1** presents the calculated axial field. To follow our scaling laws and get a minimum-B value at about 0.4 T, a magnet ring is added at the outer part of the hexapole (**figure 2**).

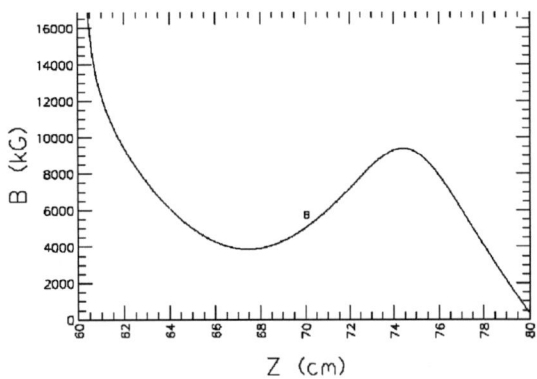

Figure 1: Calculated axial magnetic field from the Poisson code.

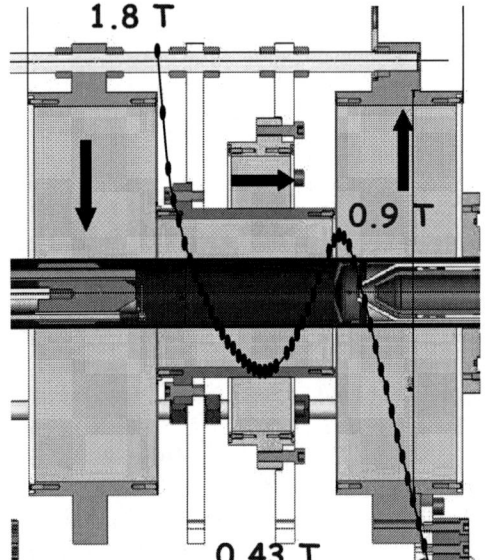

Figure 2 : Measured axial magnetic field when the source is equipped with an iron plug.

The radial magnetic field is created by a 24 pole Halbach type configuration. Typical magnetic field at chamber wall (Ø 50 mm) is 1.05 T.

Other components

For an all permanent magnet ECRIS, there is no way to tune the axial field as it is done for other ECRIS equipped with coils. Such a tuning is possible by moving different magnetic pancakes. Alternatively, one can change the frequency value at fixed magnetic field. For this reason, a variable frequency transmitter (12.5 GHz – 14.5 GHz) has been chosen for fine tuning of the source,. In addition, it has been noticed that, for a certain amount of rf power, a TWT is more efficient than a klystron, thanks to its larger bandwidth [7].

For the rf coupling to the plasma, a basic rectangular waveguide crossing the iron plug has been installed. Within such a configuration, the microwaves encounter 3 resonance points before entering the plasma chamber. Such a configuration needs some hours of plasma to be operational, since outgassing occurs in the waveguide. But, after this outgassing period, this rf coupling is by far more efficient than the coaxial one used in the Caprice type ECRIS.

Just as for the GTS, a biased disk is placed in front of the plug, the axial magnetic field at the disk being 1.8T. In addition, an oven is available for metallic ion production.

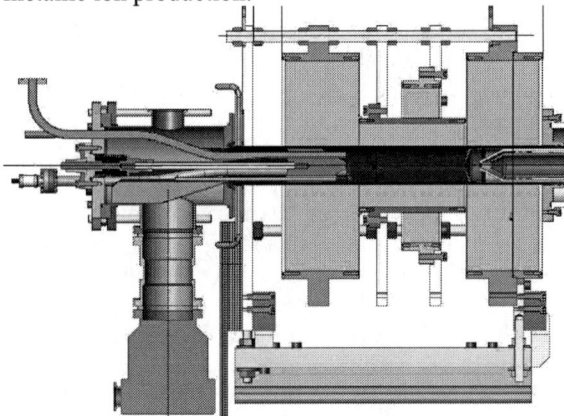

Figure 3: General aspect of the permanent magnet source showing the so called first stage with one waveguide, one gas inlet and one oven

Finally, the extraction, calculated with PBGun code, is made of 3 parts: 1 plasma electrode set at source potential, 1 negatively biased puller placed at about 30 mm from the plasma electrode and 1 grounded puller. Special care has to be taken to avoid any Paschen discharge in this extraction region: for example, base vacuum has to be better than 10^{-7} mbar. An artist view of this source is presented in **figure 3**.

TYPICAL BEAM INTENSITIES

After a short outgassing period, this new source has been tested in Grenoble over a period of several weeks before its shipping to Oak Ridge. During this commissioning period, several tests have been performed with permanent gases. Since very high charge states are not required by MIRF, the tests focused on medium charge states. **Figure 4** shows a typical argon charge state distribution when the source is tuned on 8+. This CSD is similar to that which could be obtained by an ECRIS equipped with coils. Getting ½ mA of Ar^{8+} is proof that magnetic scaling laws are also valid for such type of machine.

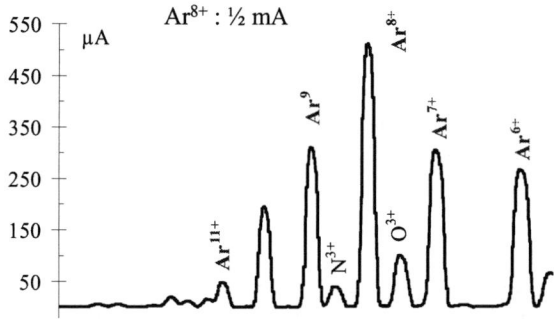

Figure 4: Argon CSD for a tuning on 8+.

The rf coupling is another key point of such a source: if the vacuum is not good enough in the waveguide, either the reverse rf power is too high or microwaves are not coupled to the main plasma but are absorbed at the parasitic resonance zones. In both cases, the extracted beam intensity would be very low. **Figure 5** shows the evolution of O^{6+}. It also shows the behaviour of the reverse power. Typically, such an rf coupling gives 1 µA of beam per injected watt which is reasonable. This value is comparable to GTS which delivered 1.5 mA of O^{6+} with 1.5 kW and 1.9 mA with 2 kW.

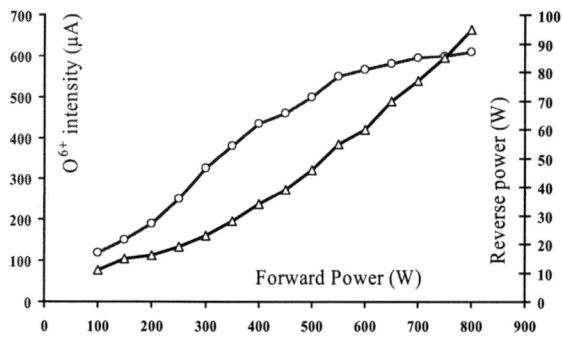

Figure 5: Evolution of O^{6+} (dots) with microwave power and evolution of reverse power (triangles)

On the other hand, such an ion source can deliver a wide range of charge states. **Figure 6** presents typical xenon CSD tuned on 10+, while **Figure 7** shows higher charge states.

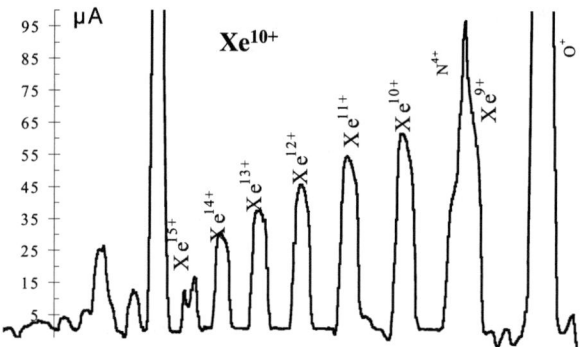

Figure 6: Xenon CSD for the production of Xe^{10+}.

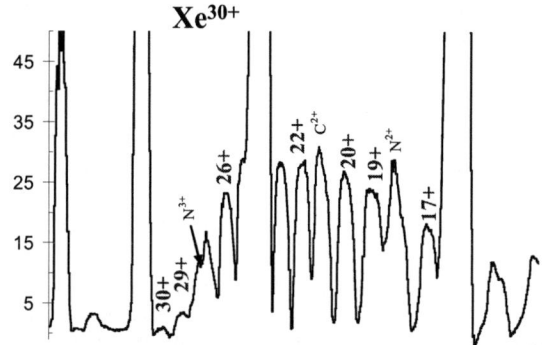

Figure 7: Xenon CSD for the production of Xe^{26+}

The CSD presented in **Figure 7** shows that source parameters are good enough to allow long ion lifetimes which are necessary to get Xe^{30+}, while **Figure 4** shows that intense beams are also obtained.

BEAM EXTRACTION AND TRANSPORT

Beam extraction is another sensitive phase of any ion source and especially with an ECRIS, since extracted ions are those which are lost by the confinement. On the other hand, beam extraction from a permanent magnet ECRIS needs special care because of the magnetic field shape.

Table 1 presents different oxygen beam intensities obtained with our system for different charge states. This table clearly shows that 700 µA is the upper limit, not of the ion source, but the present extraction system. Although our electrode design allows the source to reach 30 kV source potential, the design of both pullers shape is not yet optimized and another extraction system is under construction.

TABLE 1. Typical source parameters for various oxygen charge states

Mode	RF power (W)	Extraction Voltage (kV)	Drain (mA)	Intensity (μA)
O^+	10	20	7	700
O^{2+}	30	20	7	700
O^{3+}	130	20	7	700
O^{5+}	600	23	6	600
O^{6+}	650	27	6.5	650
O^{7+}	650	27	4.7	90

BEAM TRANSPORT SYSTEM

The design energy range of the high voltage platform is 20 – 270 x q keV. To achieve this wide energy range with maximum beam transmission a number of specific features were incorporated into the design of the beam transport system. All effective optic element apertures on the platform were designed for 100% beam transport for beams having an unnormalized emittance of 160 π·mm·mrad at 20 x q keV.

Figure 8 shows calculated beam envelopes up to the waist produced by the first 65 degree deflector for the two extremes of the beam energy range to illustrate the action of the tandem einzel lenses at low energies, and the stronger focusing action of the acceleration column at the higher energies.

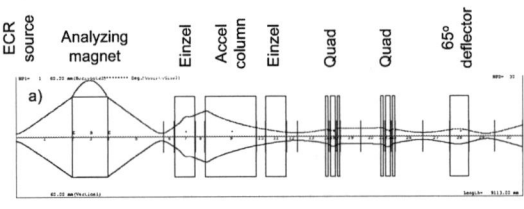

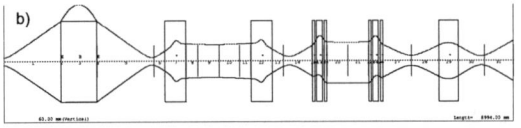

Figure 8: Horizontal (upper traces) and vertical (lower traces) beam envelopes from ECR source to the exit waist of the 65 degree deflector at the two energy extremes: a) for a 270 x q keV beam, and b) for a 20 x q keV beam. The transverse distance scales are 60 mm, total longitudinal distance is about 9 m.

All major fabrication and installation tasks of the HV platform part of the MIRF upgrade project have now been completed. The ECR source, as well as the platform- and ground- potential beamlines, are under vacuum of ~ 10^{-7} Torr. Initial testing of the various components is presently underway, and the first experiment is planned for October, 2004.

ACKNOWLEDGMENTS

Design and construction of this new source has been performed under contract CEA-ORNL #4000019233.

REFERENCES

1. S. Gammino et al, Rev. Sci. Inst. 72 (2001) 4090.
2. D. Hitz et al. Rev, Sci. Inst. 73 (2002) 509.
3. LT. Sun et al., Proc. 10th Int., Conf. Ion Sources, Dubna, 2003, Rev. Sci. Inst., 75 5 (2004) 1514.
4. H.W. Zhao, this workshop
5. D. Hitz, D. Cormier, J.M. Mathonnet, Proceedings of 8th European Particle Conference, Paris 3-7 June 2002.
6. D. Hitz et al., "Production of Highly Charged Ions with the Grenoble Test GTS Ion Source", Proc. 10th Int., Conf. Ion Sources, Dubna, 2003, Rev. Sci. Inst., 75 5 (2004) 1403.
7. L. Celona et al, this workshop

GTS-LHC: A New Source For The LHC Ion Injector Chain

C.E. Hill[1], D. Küchler[1], R. Scrivens[1], D. Hitz[2], L. Guillemet[2], R. Leroy[3] and J.Y. Pacquet[3]

[1] Department AB/ABP, CERN, 1211 Geneva, Switzerland
[2] CEA, DSM/DRFMC/SBT, rue des Martyrs, 38054 Grenoble, France
[3] GANIL, Boulevard Henri Becquerel, 14076 Caen, France

Abstract. The ion injector chain for the LHC has to be adapted and modified to reach the design beam parameters. Up to now an ECR4 delivered the ion beam for the SPS fixed target physics programme. This source will be replaced by a higher intensity source to produce the Pb^{27+} ion current required to fill the Low Energy Ion Ring (LEIR). The new ion source will be based on the Grenoble Test Source which was itself based on empirical scaling laws derived from the Framework 5 "Innovative ECRIS" collaboration. This paper will describe the design principle, the commissioning timetable and the present status of the source development.

INTRODUCTION

Following the success of the light ion fixed target programme at the SPS, an international collaboration was formed to build a heavy ion injector [1] for the CERN complex and most especially for the SPS. This installation became operational in 1994 [2] providing Lead ions to physics. This programme came to an end in 2003 with a final operational period using Indium ions instead of the habitual Lead. The experience gained with this beam will be described later in another paper at this workshop [3].

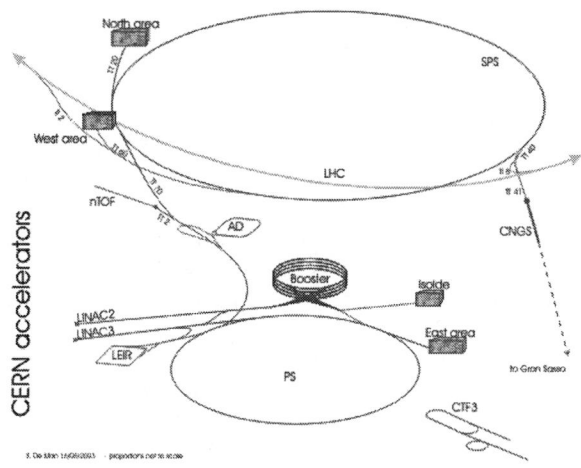

FIGURE 1. The CERN Accelerator Complex

Within the CERN accelerator complex (Fig. 1) the current major accelerator project is the Large Hadron Collider (LHC) which is being constructed in the border zones between France and Switzerland to collide 2 * 7 TeV proton beams deep underground [4]. Although the emphasis for the project has been on protons Lead ion beams at 2.76 TeV/u have been requested by the physics community. Although initially only one experiment wanted ion/ion collisions a number of the other experimental groups have expressed a desire to benefit from the presence of these ions in the rings. The changes needed to the Linac injector needed to meet the stringent requirements of the heavy ion beam will be presented together with some ideas schedule for construction and commissioning of the new source.

INJECTION OF LEAD BEAMS

As mentioned above, an option for heavy (and eventually medium and light) ions has been retained for the LHC machine. The present injection scheme used for fixed target physics is via the PS Booster (PSB) but using this route results in missing a factor of 30 in beam brightness. Naturally this could be overcome by increasing the source current by the same factor. Unfortunately, the LHC has a very tight emittance and beam quality budget to avoid unnecessary quenches of the superconducting magnets.

This emittance budget reflects back down the injector chain and it would be highly unlikely that the multiturn injection process used at the PSB with this higher source current meets this criterion. Even the emittances resulting from multiturn injection with the present beam could give rise to problems. A high instantaneous current short pulse laser ion source project was instituted to allow single turn injection but this was abandoned due to lack of satisfactory performance and stability of this type of source [5].

CERN has considerable experience in phase space cooling of antiprotons and had a mothballed antiproton decelerator and stretcher ring available (LEAR). Experiments to prove the feasibility of stacking and electron cooling of highly charged ions in this ring proved successful. However, it did demonstrate that although the lifetime of lead ions was interesting, the lifetime of Pb^{53+} was considerably shorter than that of Pb^{54+} under electron cooling. Fortunately stripping lead ions at 4.2 MeV/u gives virtually the same yield of both charge states.

Further experiments using phase space stacking in three planes in the storage ring together with electron cooling were carried out to determine the possibilities of these techniques. With the linac injecting at 2.5 Hz (instead of 0.8 Hz) it was shown that at least 25% of the required number of ions per PS pulse could be accumulated in this ring [6]. The limitation in the accumulation was the vacuum. Ion induced desorption caused pressure bumps which increased recombination losses until an equilibrium was reached with the injected beam.

Phase space cooling applied to the stacked beam increases its brightness and with the ions accelerated to a 77.2 MeV/u in this ring space charge blow up during injection, this time into the Proton Synchrotron, would be alleviated. Hence by doubling the repetition rate of the linac again, to 5 Hz, doubling the source current and improving the LEIR dynamic vacuum, the desired number of ions per injection should be achieved. This is the scenario that has been retained for the Ions for LHC project (I-LHC) [7]. Fortunately most of the ion linac was designed for 10 Hz operation and thus a maximum amount of the existing infrastructure and equipment can be retained. The storage ring is being rebuilt and renamed LEIR (Low Energy Ion Ring).

Lead ions at 4.2 MeV/u have been used in an extensive programme to understand ion induced desorption by heavy ion beams and the lessons learnt from these tests will be used to reduce pressure bumps due to beam loss in the already very good vacuum of the ring (10^{-12} mbar).

ION CURRENT INCREASE

Once the basic scenario had been defined, it was necessary to plan an upgrade of the present ECRIS. A European collaboration had been set up under the European Union Framework 5 research programme [8] to study scaling in ECRIS with the objective of trying to understand the parameters affecting source performance. A study for a high intensity source that could possibly be used for LHC or other laboratories was included. A large portion of this work was directed to understanding the scaling laws not only for an increase in frequency, but also the relationship between the longitudinal and radial magnetic fields so as to maximise a desired ion species. 28 GHz was chosen as the next frequency step because it is a standardised frequency and a power source was available.

It was quickly realised that an upgrade to this frequency at CERN would be extremely expensive. To combat space charge blow up the extraction energy from the source would have to be upgraded from 2.5 keV/u. This would require a new, longer, RFQ, a new Low Energy Beam Transport (LEBT) and spectrometer line and civil engineering works. Following the budget crisis of LHC in 2001 this type of money was not available. It was felt that by increasing the frequency to 18 GHz, by accelerating Pb^{25+}, by studying and eliminating possible bottlenecks in the linac and by tuning to peak performance on demand rather than stability (LHC filling will be programmed) the desired number of ions could be produced.

SOURCE UPGRADING

This initial source upgrade was pursued in spite of some doubts being expressed as to the adequacy of the radial confinement provided by the hexapole. In view of the fact that the longitudinal fields used in the 14 GHz ECR4 in optimised afterglow mode were weaker than those expected from CW operation, it was felt that a weaker radial confinement could be an advantage.

During 2003 information became available on an ECRIS that had demonstrated 200 eµA of Bi^{24+} in afterglow mode at 14.5 GHz with moderate RF power [9]. This source, the Grenoble Test Source (GTS) [10] [11], uses a traditional minimum-B configuration optimised to obtain a better compromise between plasma confinement and ion losses. The principles of this optimisation was one of the spin-offs of the Innovative ECRIS collaboration [12]. Apart from the improved magnetic field configuration and a larger plasma chamber this source could be adapted into the existing infrastructure in the ion linac building with a minimum of expenditure and modification.

The final design of the source, which is currently under construction after approval from the CERN project management, will include the possibility of an upgrade to 18 GHz or to 14 + 18 GHz with only very minor modifications. Two medium temperature ovens designed and constructed in collaboration with GSI will be installed in the source (another spin-off of "Innovative ECRIS").

CERN will provide the infrastructure and commission the source in collaboration with CEA, Grenoble and GANIL. Overall, the cost of this upgrade will fall within the budget envelope defined for the earlier upgrade ideas. Additionally the source could be commissioned within the current timetable of the LHC project, Fig. 2 is an drawing of the source.

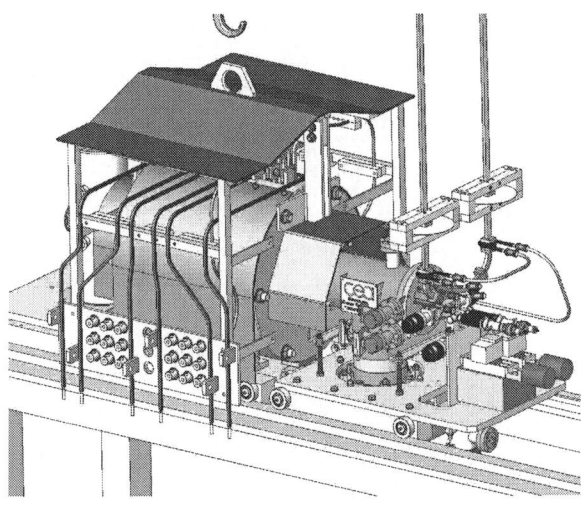

FIGURE 2. An impression of the GTS-LHC source with its lead shielding.

TIMETABLE

The planning of the source construction and commissioning is constrained by the need to have an ion beam ready for the start of testing of the injection line for LEIR and by the desire to utilise the lead beam to its maximum for equipment tests. One important test that need to be carried out is to observe the behavior of the stripper foil with a 5 Hz. beam It is known to have resisted destruction at 2.5 Hz.

The procurement procedure within CERN proved more onerous than anticipated but the source should arrive at CERN in November 2004, well in advance of the required test beam for the injection line is scheduled for the beginning of May 2005. Installation and the start of testing of the source will be controlled by the needs of the annual maintenance of the water and electrical services but is scheduled to start in mid January 2005.

Until the beam is required by the LEIR machine physicists, opportunity will be taken to acquire running and operational experience with the new source. In parallel with all this work will be the design and construction of the ovens so that they can be ready for this running-in phase.

FINAL COMMENTS

Because the LHC is a big machine it must not be considered that it needs a big ion current from the injectors. The minimum circulating intensity in LHC is defined by the beam observation system sensitivity and the maximum was defined by Electromagnetic Dissociation and probably by Electron Capture by Pair Production.. Collimation of the ion beam may also be a problem. The range of Luminosities available will be $5 * 10^{25}$ to $1 * 10^{27}$ cm^{-2}s^{-1} with $7 * 10^7$ ions per bunch which means that regardless of the final luminosity, the Linac will have always to supply the same ion current.

The presence of a circulating heavy ion beam at LHC energies could give rise to unexpected surprises that can not be anticipated from the proton commissioning period. Ion physics in LHC is scheduled to start in April 2008.

REFERENCES

1. D. Warner, Editor. "CERN Heavy-Ion Facility Design Report", CERN 93-01, 1993.
2. H. Charmot et al. "Operational Experience with the CERN Hadron Linacs", Proc. 18th Int. Linac Conf., Geneva, 1996, CERN 96-07, 360, 1996
3. C. Andresen et al., "Characterisation and Performance of the CERN ECR4 Ion Source", This Workshop.
4. O. Brüning et al .(Eds) "LHC Design Report, Vol 1, The Main Ring", CERN 2004-003, 2004.
5. A. Balabaev et al., "Laser Ion Source Based on a 100 J 1 Hz CO2 Laser System", Proc. 10th Int., Conf. Ion Sources, Dubna, 2003, Rev. Sci. Inst., 75, 1572,. 2004

6. J. Bosser et al. "Experimental Investigations of Electron Cooling and Stacking of Lead Ions in a Low Energy Accumulation Ring" Particle Accelerators, 63, 171, 1999
7. O. Brüning et al.. (Eds), "LHC Design report, Vol 3, The LHC Injector Chain", CERN 2004-003, In publication, 2004.
8. European Commission Framework 5 Contract HPRI-1999-50014 "New Technologies for the Next Generation ECRIS".
9. D. Hitz, Private Communication.
10. D. Hitz et al., "Grenoble Test Source (GTS): A multipurpose Room Temperature ECRIS", Proc. 15th Int. Workshop on ECR Sources, Jyväskylä, 2002, JYFL Research Report 4/2002, 2002.
11. D. Hitz et al., "Production of Highly Charged Ions with the Grenoble Test ECT Ion Source", Proc. 10th Int., Conf. Ion Sources, Dubna, 2003, Rev. Sci. Inst., 75,. 1403, 2004
12. D. Hitz et al., "Results and interpretation of High Frequency Experiments at 28 GHz in ECR Sources: Future Prospects", Proc. 9th Int., Conf. Ion Sources, Oakland, 2001, Rev. Sci.. Inst., 73, 509, 2002.

Design of SuSI – Superconducting Source for Ions at NSCL/MSU – I. The Magnet System

P. A. Zavodszky, B. Arend, D. Cole, J. DeKamp, G. Machicoane,
F. Marti, P. Miller, J. Moskalik, J. Ottarson, J. Vincent and A. Zeller

National Superconducting Cyclotron Laboratory, Michigan State University, East Lansing, MI 48824, USA

Abstract. An ECR ion source is being designed to initially serve as a test bench for development and later will replace the existing 6.4 GHz SC-ECRIS. This ECRIS will operate at 18+14.5 GHz microwave frequencies. The radial magnetic field will be produced by a superconducting hexapole coil, capable of 1.5 T at the aluminum plasma chamber wall (R=50 mm). The axial trapping will be produced with six superconducting solenoids enclosed in an iron yoke. We will present the Flexible Axial Magnetic Field Concept, introduced for the first time in this design, which will allow tuning the distance between the plasma electrode and resonant zone in the plasma. The distance between the two axial magnetic maxima will be also tunable in the range of 340 to 460 mm.

INTRODUCTION

The National Superconducting Cyclotron Laboratory at Michigan State University operates two cyclotrons in coupled mode in order to produce radioactive ion beams by projectile fragmentation [1]. The primary beam energy is up to 200 MeV/u, and since October 2000 many different primary beams were accelerated between ^{16}O and ^{209}Bi. The primary ions are produced by two ECR ion sources, one superconducting (SC-ECR) built in the early 90's [2], and the other (ARTEMIS) with room temperature magnets [3] built from a design based on the AECR-U at LBNL [4].

During the four years of experience since commissioning the coupled cyclotrons, it became evident that the emittance of these ion sources poorly matches the acceptance (about 75 π mm mrad) of the cyclotrons. In order to achieve good transmission from the ion sources to the inflector of the K500 cyclotron, the ion beam has to be collimated strongly, loosing a large fraction of the extracted ion beam from the ion sources. Besides a planned upgrade of the existing ECR ion sources extraction systems and further studies to improve the transport efficiency of the injection beamline, the other approach is to build an ECR ion source, which is more flexible than the existing ones in order to better match the emittance of the source with the acceptance of the accelerators. Because the coupled cyclotrons require intense medium charged ions, the emphasis is shifted from the production of very high charge states to medium charge states, but the increased intensities make the effect of the space charge more important than in the previous stand-alone operation mode. Thus, the new ECR ion source with the associated focusing and analyzing system has to be capable of handling total extracted currents of several milliamperes.

After studying several technical options, we decided that we would design and build in NSCL an ECR ion source capable of operating at 18+14.5 GHz, using fully superconducting magnets.

DESIGN PARAMETERS OF SUSI – SUPERCONDUCTING SOURCE FOR IONS

According to the currently accepted semi-empirical design criteria [5], an ECR ion source should have a magnetic confinement characterized by the following field values: an axial magnetic trap with $B_{inj} \approx 4\ B_{ECR}$ at the injection side, $B_{ext} \approx 2\ B_{ECR}$ at the extraction side, with a minimum magnetic field $B_{min} \approx 0.8\ B_{ECR}$. The radial confinement magnetic field value at the plasma chamber walls should be $B_{rad} \approx 2B_{ECR}$. The extraction magnetic field is also correlated to the radial field through the relationship: $B_{ext} \approx 0.9\ B_{rad}$. The resonant magnetic field value can be obtained from the following equation:

$$\omega_e = \frac{qB_{ECR}}{m} = \omega_{rf} \qquad (1)$$

where q is the charge of the electron, m is the mass of the electron, B_{ECR} is the resonant magnetic field value, ω_e is the gyrofrequency of the electron and ω_{rf} is the microwave frequency used to heat the electrons in the plasma.

In order to reach these magnetic field values for 18 GHz microwave frequency (B_{ECR} = 0.64 T) and a plasma chamber of 100 mm diameter, it is difficult to use room temperature solenoids and permanent magnet hexapole. It is more convenient to construct a fully superconducting magnet system. This has the advantage of a tunable radial magnetic field, lower electric power consumption for the axial solenoids and no risk of demagnetization of the permanent magnets used in a room temperature hexapole system for the radial confinement. With a superconducting solenoid magnet there is no need for an iron plug in the injection side, leaving more room for different devices necessary to produce metallic beams, for multiple waveguides, bias disk and good vacuum pumping.

Considering the existing 2 kW LHe plant at NSCL, it is not necessary to use cryocoolers and high-Tc superconductor current leads to minimize the LHe consumption, simplifying the design and lowering the initial capital investment.

The Flexible Magnetic Field Concept

The Nakagawa group in RIKEN reported [6] that the extracted beam intensity for a specific ion type depends on the position of the plasma electrode relative to the plasma. In fact, it seems that each charge state of a particular ion has an optimum plasma electrode position. Furthermore, as reported by the Koivisto group in Jyväskylä [7] the beam emittance is also influenced by the plasma electrode position. In order to mach the plasma meniscus with the extraction electrode system at a fixed extraction voltage, it is important to have an adjustable puller electrode. It is very difficult to design a plasma electrode system remotely movable inside the plasma chamber. In the SuSI design we adopted a different approach. Because the plasma and the resonant zone location inside the plasma chamber is determined by the magnetic field structure, the other way to change the relative position of the plasma electrode from the plasma is to keep the plasma electrode fixed and move the axial magnetic field. This can be accomplished with two solenoids at each end of the ion source, INJ_1 and INJ_2 at the injection end, EXT_1 and EXT_2 at the extraction end. In order to have the magnetic field minimum easily adjustable, there is a third pair of solenoids between the injection and extraction ends, running electric currents in the opposite direction: MID_1 and MID_2. Each combination of INJ_i, MID_j and EXT_k (i, j, k =1, 2) is capable of producing the required magnetic field profile for optimum operation at 18 GHz.

Figure 1 represents the intermediate case when all six solenoids are powered and the axial magnetic field maximum at the extraction side is located at the position of the plasma electrode. In this case the distance between the two magnetic maxima is 408 mm. The length of the resonant zone for 18 GHz is 126 mm, for 14.5 GHz is 98 mm.

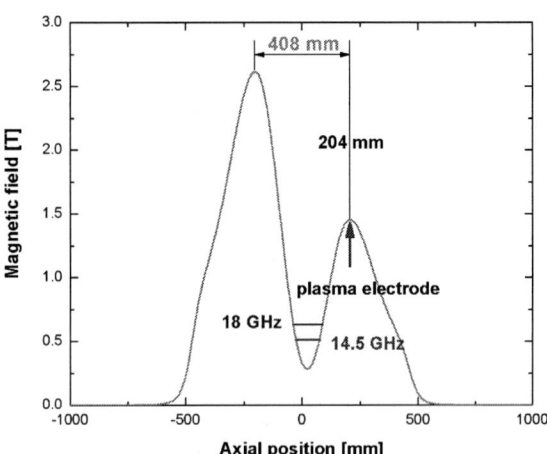

FIGURE 1. Axial magnetic field profile for intermediate distance between the two magnetic maxima.

Figure 2 represents the maximum case when only INJ_1, MID_1, MID_2 and EXT_2 solenoids are powered. In this case the plasma electrode is on the increasing side of the magnetic maximum at the extraction side. The distance between the two magnetic maxima is maximum, 460 mm. The length of the resonant zone for 18 GHz is 154 mm, for 14.5 GHz is 120 mm.

Figure 3 represents the minimum case when only INJ_2, MID_1, MID_2 and EXT_1 solenoids are powered. In this case the plasma electrode is on the decreasing side of the magnetic maximum at the extraction side.

The distance between the two magnetic maxima is minimum, 340 mm. The length of the resonant zone for 18 GHz is 102 mm, for 14.5 GHz is 78 mm.

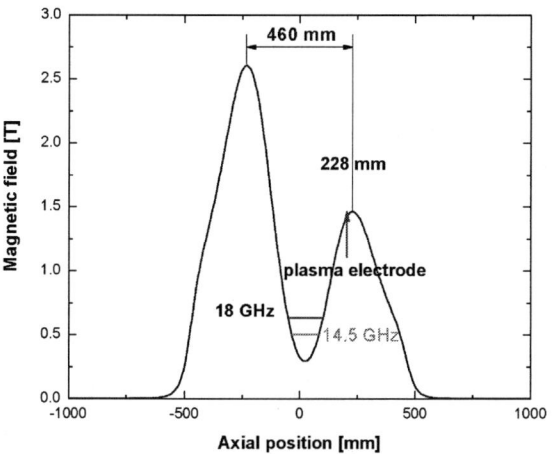

FIGURE 2. Axial magnetic field profile for maximum distance between the two magnetic maxima.

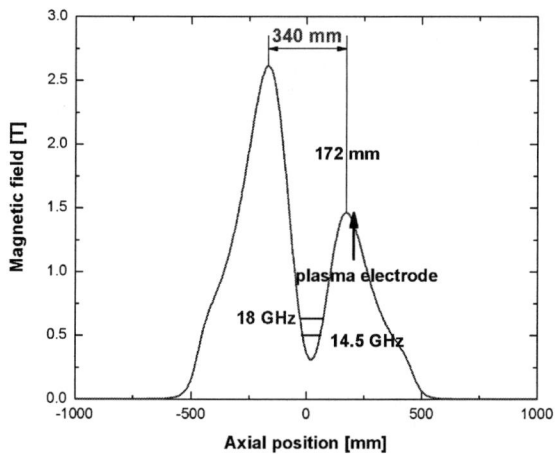

FIGURE 3. Axial magnetic field profile for minimum distance between the two magnetic maxima.

The currents in solenoids for the above mentioned three cases are tabulated in Table 1.

TABLE 1. Current densities in the SuSI solenoids

Coil	Intermediate case [A/mm^2]	Maximum case [A/mm^2]	Minimum case [A/mm^2]
INJ$_1$	74	120	0
INJ$_2$	74	0	170
MID$_1$	-60	-27	-100
MID$_2$	-60	-27	-100
EXT$_1$	61	0	150
EXT$_2$	61	96	0

All the magnetic field calculations were performed with two different codes: the Finite Element Method based TOSCA [8] and the Boundary Element Method based AMPERE [9]. The two codes gave similar magnetic field values.

The advantage of the above solenoid configuration relies in a great flexibility of shaping the axial magnetic field profile. Besides the three cases, a multitude of other situations can be easily obtained by tuning the current values in the solenoids. It is possible, for example to produce a flat-field magnetic configuration necessary for the volume-ECR type ion source [10]. The distance between the magnetic field maxima is variable in the range of 340 to 460 mm; the whole axial magnetic field profile can be shifted with fixed distance between the two magnetic maxima, equivalent with a plasma electrode movement of about 50 mm.

The magnetic field values on the surface of the superconducting coil system are shown in Figure 4 for the case when the distance between the two axial maxima is the smallest. The inner diameter of all solenoids is 300 mm; their width is 80 mm. The injection side solenoids have an outer diameter of 460 mm and 3200 turns of 2x1 mm superconducting wire with a Cu/SC ratio of 3. The extraction side and mid solenoids have an outer diameter of 400 mm and 2000 turns of the same wire. The spacing between the solenoids is 10 mm.

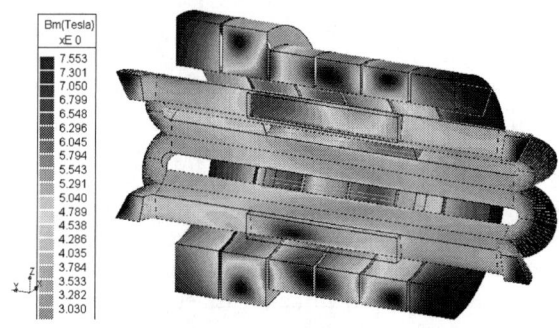

FIGURE 4. Magnetic field values on the surface of the superconducting coils.

The hexapole coils are wound from the same wire as the solenoids. Each hexapole coil is wound in layers around a 664 mm long metal pole which has the central 274 mm made from iron to increase the sextupole field in the center. The two end parts of each pole are made of aluminum. This solution was adopted in the VENUS hexapole construction in order to match the total axial thermal contraction of the pole with the contraction of coil at LHe temperatures [11]. The hexapole coils are 743 mm long; their turns are far away from the field of the solenoids, to minimize the forces acting on this part of the hexapole coils. One hexapole coil occupies 60° in the azimuthal direction; the cross-section of each side of the coil and that of the pole occupy 20° in azimuth. Figure 5 shows the radial magnetic field produced by the hexapole coils at the plasma chamber wall (R=50 mm), along the axis of the ion source. The central enhancement is due to the iron part of the metal pole. This iron helps also to reduce the necessary current in the middle two solenoids.

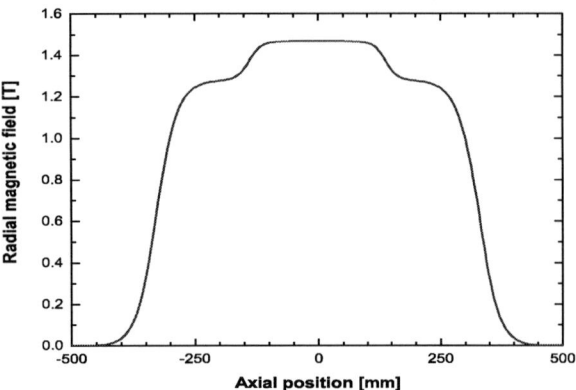

FIGURE 5. Radial magnetic field at R=50 mm along the axis of the plasma chamber. The current density in the hexapole coils is 175 A/mm^2.

In order to clamp down the hexapole coils, we will use the technology applied during the VENUS magnet construction [11]. The hexapole coils will be placed around a 160 mm OD stainless steel bore tube; the hexapole assembly will be inserted in the 244 mm ID inner bore of the solenoid bobbin. Thin stainless steel bladders will be inserted between each coil and they will be inflated with a low melting point indium alloy in order to provide azimuthal compression.

The total stored energy was calculated using the inductance matrix and it is the largest (350 kJ) in the case shown in Figure 4. Detailed thermal, stress and quench calculations will be presented elsewhere. The estimated heat load mainly due to the conventional current leads is 5 W, which will result in a 7-liter/h LHe consumption. For quench protection several cases where studied: all coils quench in the same time, or only one coil quenches then the others at various later times. Under any conditions the coils were found safe and self protected.

Present Status and Timeline

The first small solenoid winding was completed at NSCL on September 17, 2004. We also did a test winding of a hexapole coil to experiment the technique to be used to wind the real hexapole coils. The coil winding is expected to be finished by the end of March 2005. The design of the cryostat will be completed by the end of 2004. Successful tests were already made with the indium alloy inflatable bladders, this technology being new at NSCL. The design of the plasma chamber, injection box with hardware and the extraction box with a three-element accel-decel extraction electrode system will be presented in [12]. The first plasma is expected in early 2006. After extensive tests and optimization studies, SuSI will replace the SC-ECR.

ACKNOWLEDGMENTS

This work was supported by the National Science Foundation under grant PHY-0110253.

REFERENCES

1. F. Marti, P. Miller, D. Poe, M. Steiner, J. Stetson, and X.Y. Wu, in *Cyclotrons and their Applications 2001*, edited by F. Marti, AIP Conf. Proc. **600** (2001) p.64
2. T.A Antaya and S. Gammino, Rev. Sci. Instr. **65** 1723 (1994).
3. H. Koivisto, D. Cole, A. Fredell, C. Lyneis, P. Miller, J. Moskalik, B. Nurnberger, J. Ottarson, A. Zeller, J.DeKamp, R. Vondrasek, P.A. Zavodszky and Z.Q. Xie, in *Proceedings of the Workshop on the Production of Intense Beams of Highly Charged Ions*, Catania, Italy, 24-27 Sept. 2000, eds. S. Gammino and G. Ciavola, Italian Physical Society Conference Proceedings **72**, 135 (2001).
4. Z.Q Xie, and C. M. Lyneis, *Proceedings of the 13th International Workshop on ECR Ion Sources*, Texas A&M Univ., College Station, USA, 16 (1997).
5. D. Hitz, A. Girard, G. Melin, S. Gammino, G. Ciavola and L. Celona, Rev. Sci. Instr. **73**, 509 (2002).
6. Y. Higurashi, T. Nakagawa, M. Kidera, T. Aihara, M. Kase and Y. Yano, Rev. Sci. Instr. **73**, 598 (2002).
7. P. Suominen, O. Tarvainen and H. Koivisto, Rev. Sci. Instr. **75**, 1517 (2004).
8. www.vectorfields.com.
9. www.integratedsoft.com
10. G. D. Alton and D. N. Smithe, Rev. Sci. Instrum. **65** 775 (1994).
11. C. Taylor, S. Caspi, M. Leitner, S. Lundgren, C. Lyneis, D. Wutte, S.T. Wang, J.Y. Chen, IEEE Trans. Appl. Supercond. **10**, 224 (2000).
12. P.A. Zavodszky B. Arend, D. Cole, J. DeKamp, G. Machicoane, F. Marti, P. Miller, J. Moskalik, J. Ottarson, J. Vincent and A. Zeller, to be published in the *Proceedings of International Conference on Applications of Accelerators in Research and Industry*, Oct. 10-14 2004, Fort Worth, TX.

VII RADIOACTIVE ION BEAMS

Radioactive Ion beam production at GANIL : status and prospectives.

R. LEROY, C. BARUE, C. CANET, M. DUBOIS, M. DUPUIS, F. DURANTEL, W. FARABOLINI, J-L. FLAMBARD, G. GAUBERT, S. GIBOUIN, C. HUET-EQUILBEC, Y.HUGUET, P. JARDIN, N. LECESNE, P. LEHERISSIER, F. LEMAGNEN, J.Y. PACQUET, F. PELLEMOINE, M.G. SAINT LAURENT, O. TUSKE A.C.C. VILLARI

GANIL, BdH. Becquerel 14076 caen cedex 5, FRANCE

O. BAJEAT, S. ESSABAA, C. LAU, M. DUCOURTIEUX

IPN Orsay BP 1 F – 91406 Orsay, France

F.G. NIZERY

CEA/DAPNIA, Saclay, France

Abstract. Production of radioactive ions has started at GANIL on the SPIRAL facility since 2001 and numerous multicharged radioactive ion beams have been delivered for high energy nuclear experiments. This article makes an overview of the different beams that have been produced. In the mean time, an important R and D research program is continued in oder to produce new species of radioactive elements. A new concept of multicharged radioactive production that couples a monocharged ion source, based on the monolithe concept, to an ecr ion source like nanogan3 is under developments and is described The development of monocharged ion sources with high efficiencies is also motivated by a new big project that is under studies at GANIL : the SPIRAL 2 Project. The goal of this project consists in extending the disponible radioactive ion beams to very heavy elements created with a new method of production : while the spiral 1 facility uses the projectile fragmentation for radioactive nuclei production, the spiral 2 project is based on the fission of a Uranium carbide target induced by a neutron flow created by a high intensity deuton beam. The principle and an overview of the project is presented.

INTRODUCTION

The research and development strategy at GANIL is based on the short and long term needs of physicists. The need in producing enegetic radioactive ions has appeared in the years 95's and a big project consisting in producing and accelerating noble radioactive ions with short life times has been engaged and built. This facility called SPIRAL has been started in 2001 and delivers daily beams for nuclear experiments. The available beams are presently limited to noble gases or to gaseous elements that can be formed inside a target due to the technology that has been applied. Some new methods are under sudies that couple the production target to the ionisation system. The principle of production applied on SPIRAL gives acces to high intensities of neutron defficient light ions but is less efficient for neutron rich heavy ion production. It is the reason why a new project giving acces to heavy elements is under studies: the SPIRAL2 project.

THE RADIOACTIVE ION BEAM PRODUCTION ON THE SPIRAL FACILITY

The production process of the SPIRAL facility uses a high energy ion beam that hits a target and induces fragmentation reactions. The radioactive nuclei are stopped inside the target which is heated up to 2200°C to facilitate the migration of radioactive atoms to the surface. Usually the target is located at a short distance from an ion source and the radioactive atoms effuse via a transfer tube to the plasma region where they are ionised and then accelerated in a manner identical to that for stable ions. The number of created radioactive atoms depends on the primary beam intensity and on the integrated fragmentation cross section. However the creation rate of nuclei of interest is always low (from some particule per second to 10^{12} pps) and the major problem of the method is to be as efficient as possible in order to maintain a suitable radioactive ion beam intensity. This means that the system of production of the radioactive ion beam has to take into account all the losses process that can occur like sticking on the walls, leaks, chemical reactions, etc... The production time including diffusion out of the target, effusion, ionization, confinement, etc... has to be lower than the life-time of the nuclei of interest. This particularity implies that the production system has to be treated as an ensemble without disconnecting the production target from the ion source.

After production, the radioactive ion beam of interest is selected through a mass separator and accelerated by a cyclotron.

THE SPIRAL FACILITY

The Nanogan 3 configuration

In order to allow the futher acceleration by the cyclotron, a compact multicharged ion source has been developped and coupled it to a carbon target (see figure 1) The target is heated by the primary beam up to 2200°C. Such a temperature is a challenge for the target life duration and reliability and numerical calculations show that convenient temperatures can be achieved if the target presents a conical shape. In the case of a low beam power, an extra ohmic heating can be added through the axis of the target to maintain the diffusion of radioactive atoms.

After diffusion, the radioactive atoms effuse to the ion source through a cold transfer tube that makes a chemical selection as the main part of the non-gaseous elements sticks on the walls of the tube. The atoms then enter into the plasma of the ion source.

This source, called NANOGAN3, is a 10 GHz electron cyclotron resonance ion source with a magnetic field totally induced by permanent magnets (Figure 1). The source has been described in details in reference[i]. The choice of permanent magnets has been driven by the cost and the compacity of the source that limits the volume of radioactive waste after irradiation.

Total ionisation efficiencies –ie sum of the efficiency of each charge state- greater than 95% have been measured for different gases (Ar, Ne, Kr) with a calibrated leak during the presence and the absence of the primary beam. A comparison of the charge state distributions of stable argon during different moments of the production shows that after a short delay of outgassing the behaviour of the source is no longer affected by the presence of the hot target in its neighbourhood.

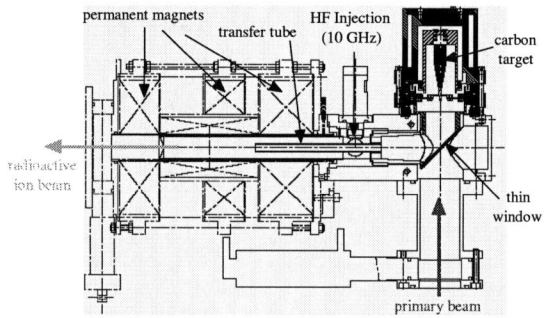

Figure 1:The Nanogan 3 configuration

A fifteenth of target ion sources have now been irradiated on the SPIRAL facility for the production of radioactive argon, neons and kryptons during 15 days with till 1.4 kW of primary beam power. A new target has been developed for ^{6}He and ^{8}He production which is divided into two parts because of the long range of He in carbon (see reference [ii]). In all cases, it has not been observed any decrease of the multicharged radioactive ion beam intensity during the long term irradiation that proves that the permanent magnets have not been damaged. The two first irradiated ion soures have been dismounted in order to change the target. The magnetic fields of these sources have been measured before and after irradiation and no significant change has been observed.

Table 1 gives the intensities of the beams before or after acceleration on the SPIRAL facility.

Radioactive beam (life time)	Intensity (pps)	Primary beam	Primary beam power (kW)
^{6}He$^+$ (0,8 s)	3.2 10^{7**}	^{13}C	1.2
^{8}He^{1+} (0,12s)	3 10^6	^{13}C	2,6

$^8He^{2+}$ (0,12s)	$1,5\ 10^5$	^{13}C	1,4
$^{18}F^{4+}$ (109mn)	$2\ 10^{5**}$	^{20}Ne	0,3
$^{20}O^{4+}$ (13,5s)	10^4	^{36}S	1,4
$^{21}O^{4+}$ (3,4s)	$1,3\ 10^3$	^{36}S	1,4
$^{18}Ne^{4+}$ (1,7s)	$3\ 10^6$	^{20}Ne	0,3
$^{23}Ne^{5+}$ (37s)	$5\ 10^6$	^{36}S	1,4
$^{24}Ne^{5+}$ (3,4mn)	10^6	^{36}S	1,4
$^{25}Ne^{4+}$ (0,6s)	$1,7\ 10^5$	^{36}S	0,8
$^{26}Ne^{5+}$ (0,23s)	$1,1\ 10^4$	^{36}S	0,8
$^{27}Ne^{5+}$ (0,032s)	$5,6\ 10^2$	^{36}S	0,8
$^{32}Ar^{9+}$ (0,098s)	10^3	^{36}Ar	1,1
$^{33}Ar^{8+}$ (0,173s)	$9,3\ 10^4$	^{36}Ar	1,1
$^{35}Ar^{8+}$ (1,78s)	$1,5\ 10^8$	^{36}Ar	1,1
$^{44}Ar^{9+}$ (11,9mn)	$1,4\ 10^6$	^{48}Ca	0,4
$^{46}Ar^{9+}$ (7,8s)	$7,7\ 10^4$	^{48}Ca	0,4
$^{72}Kr^{11+}$ (17s)	$2\ 10^2$	^{78}Kr	0,4
$^{74}Kr^{11+}$ (11,5mn)	$6\ 10^4$	^{78}Kr	0,4
$^{75}Kr^{11+}$ (4,3mn)	$5\ 10^5$	^{78}Kr	0,4
$^{76}Kr^{11+}$ (14,8h)	$3\ 10^6$	^{78}Kr	0,4
$^{77}Kr^{11+}$ (74,4mn)	$3\ 10^6$	^{78}Kr	0,4

Table 1: Radioactive ion beam intensities produced on the SPIRAL facility. The intensities marked with two stars are measured after acceleration while the other ones are measured before acceleration.

Further developments for the SPIRAL facility

The MONONAKE configuration for radioactive alkalii ion production

The present limitation of the NANOGAN 3 configuration is due to the cold transfer tube between the target and the plasma of the ion source. This configuration presents the advantage to make a chemical selection and to eliminate the isobaric atoms that can condense on the the tube before ionisation. However this advantage becomes a disadvantage for condensable radioactive ion beam production like for example for alcali elements. For multicharged alcali ion production, a solution consists in coupling a monocharged ion source to the NANOGAN 3 ion source. Figure 2 shows a design of the NANONAKE configuration. It consists in two sources. The first one is an upgrade of the MONOLITHE ion source that has been described in reference [iii]. The target is placed in a heater and the atoms can only escape through a hot tungsten tube where the alcali alements are ionised in the single charge state by surface ionisation process. The ions are then accelerated and injected into the NANOGAN 3 ion source with an energy just sufficient to drop the plasma potential barrier and to stop inside the plasma of the ECR ion source. The multi-ionisation process occurs and a multicharged ion beam can be extracted from the NANOGAN 3 source. In order to allow the injection of the 1+ ion beam in the ecr ion source, the HF injection has been changed and replaced by a circular waveguide that can be biased and is surrounded by a grounded electrode in order to decelerate the beam in the ecr source and to tune precisely the energy of the monocharged ion beam see Figure 2.

This configuration is constructed and the first off-line tests have began. The first test has consisted in checking that the ionisation of the multicharged ion source is not perturbated by the neighborhood of the 1+ ion source. In comparison with the nanogan3 configuration, the same charge state distribution with the same efficiency has been obtained for Ar. The second part of the off line test will consist in testing the charge breeding process by injecting a Lu powder in the production target, heating and ionizing the Lu vapor and injecting it in the ecr ion source. At least a third test is envisaged by injecting a radioactive ion beam inside the target and by measuring the different charge states of the radioactive element after the SIRa separator.

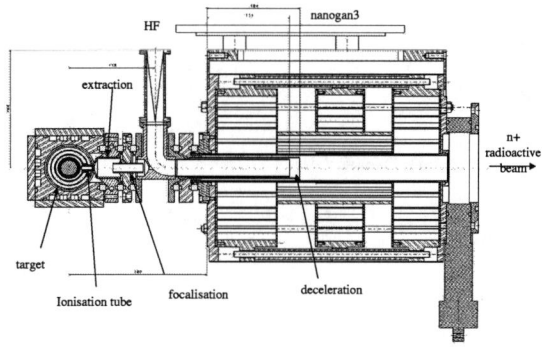

Figure 2: The nanonake configuration for alcali elements.

This method allows to obtain the overall efficiency of the system and has soon been used for measuring the diffusion properties of the SPIRAL targets. These tests will occur during the next year before implementation on the SPIRAL facility. It has to be noticed that the present design permits to be installed on SPIRAL without any modification of the production cave and can be removed with the existing remote controlled system.

THE SPIRAL2 PROJECT

To explore the production of heavy neutron-rich nuclei with suitable intensities, the method of production has been reviewed and it appears that the fission of uranium induced by neutrons with an energy of around 40 MeV can lead to interesting production rates for heavy neutron rich elements like Xe or Sn (see reference [iv]). In the same time, the production of light elements can be increased by increasing the intensity of the primary beam.

The SPIRAL 2 project deals with these two facts. It consists in building a new accelerator for the primary beam coupled to a new radioactive ion production area. The rare isotope beams are produced via the fission process, with the aim of 10^{13} fissions/s at least, induced either by fast neutrons from a C converter in a UCx target, or by direct bombardment of fissile material. The driver, with an acceleration potential of 40 MV, has to be upgradeable and versatile: it will accelerate deuterons (intensity up to 5 mA) and ions of mass-to-charge ratio A/q=3 (intensity up to 1 mA) and even higher A/q ions in a later stage (up to 6). The energy of the beam is fixed to 40 MeV for D^+ or to 14 MeV for elements with mass to charge ratio A/q=3 but could be upgraded to higher values (for example 95 MeV/A) in the future. After production, the radioactive ion beam is injected in a charge booster to increase the charge state and after is injected in the present CIME cyclotron for acceleration. A detailed description of the project including the accelerator is given in reference [v] and is available on the GANIL web site www.ganil.fr.

In terms of ion sources, the SPIRAL2 project presents ambitious performances.

For the stable primary ion beam production, different sources are under studies that can be divided into two main items: the deuteron ion sources and the A/q=3 ion sources. These sources are developed by CEA/DAPNIA in Saclay (France) and IN2P3/LPSC in Grenoble (France).

For the radioactive ion beam, the production system needs some sources that have to be efficient in term of ionisation and radiation resistivity. The sources are coupled to the uranium target placed in front of a rotating wheel that permits the creation of the fast neutrons needed for fission of the target.

At least an efficient charge booster has to be installed. This charge booster is based on the 1+/n+ technique. A PHOENIX ECR ion source developed at LPSC Grenoble is envisaged to be used.

The production system of radioactive ions

It has been seen previously that the rare isotopes of heavy elements are produced by the fission process of a uranium carbide target induced by fast neutrons. These neutrons are created by the interaction of an intense deuteron beam (5 mA) with an energy of 40 MeV in a carbon target. This represents a beam power of 200 kW deposited in a length of 6 mm. In order to sustain such high power deposition, thermal calculations have been made on the carbon target and shows that a rotating wheel presents a good compromize.

The effects of the diameter of the deuteron beam, of the rotating speed and of the radius of the wheel have been studied and show that the temperature of a 350 mm in radius wheel hitted by a beam of 40mm in diameter and rotating at a speed of 800 turn/mn does not excess 2100 K. The reader is encouraged to see reference [vi] for the details on the calculations.

The distance between the wheel and the uranium target has also been studied (see reference vi) and shows that the front of the target must be placed at a distance lower than 40mm from the entrance of the converter.

The uranium carbide targets

The target is designed to produce at least 10^{13} fissions/s with 5 mA of deuterons on the converter. Different geometries have been studied taking into account the number of fissions induced by the neutrons, the effusion time from the target and the technical feasibility. This target will be placed in an oven allowing a temperature of around 2000°C to maintain a suitable diffusion of radioactive elements.

In the same time a Research and Development program has been launched based on the use of high densities UCx target developed at PNPI in Saint Petersburg, Russia. A detailed description of the target-ion source system is given in see reference [vii].

The ion sources for the radioactive ion beam formation

As previously mentionned, the key word for an ion source in case of radioactive ion beam production is the efficiency. This efficiency is variable depending on the chemical element that has to be ionised and on the type of ion source that is used.

In case of SPIRAL 2, it looks clear that to fully profit from this facility, the largest choice of radioactive beams is wished. That is the reason why different ionization ways are considered depending on the chemical properties of the radioactive elements.

The ECR ion sources

This kind of sources is paticularly efficient for the production of gaseous elemens such a noble gases of gaseous compounds. At that time, no ECRIS tested on-line respects the features requested for SPIRAL 2, that means mainly a good ionization efficiency (more than 50%) for noble gases and some molecular compounds, and a life-time of 3 months under irradiation. A large development program has been launched in order to adapt a monocharged ion source to the SPIRAL2 conditions. A 2.45 GHz ion source called MONOBOB has been designed and constructed (Figure 3)

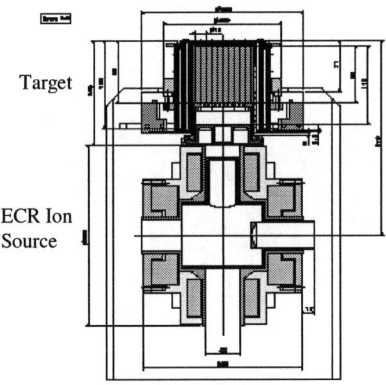

Figure 3: The monobob Ion source

This source presents the advantage to have large apertures that permits to place a target very close to the plasma suppressing by this way any transfer tube between the target and the ion source. The magnetic field is induced with four coils insulated with epoxy resin which supports a radiation dose of the order of 10^7 Sv. The result of the magnetic calculations is shown in Figure 4 The latests isomodule magnetic field surface is closed at 2500 Gauss. The RF power injection is made through a coaxial cable and an antenna.

The ionistaion efficiency has been measured and can attain 72% for Ar and 35% for Ne.

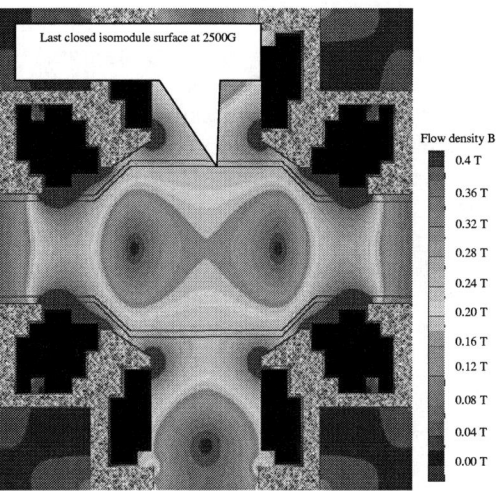

Figure 4: Magnetic field distribution in the monobob ion source

The FEBIAD ion source

The advantage of this kind of source is its small size that minimizes the delay time in the ionization chamber that alllows an access to short life time isotopes and that can be easily heated up to 1800°C. For instance, at ISOLDE- CERN, Switzerland or at IPN Orsay, France, FEBIAD ion sources at high temperature are connected to standard UCx targets and beams of isotopes with half life time lower than 300ms have been measured (for example ^{124}Ag or ^{94}Kr) (see reference [viii]). These prototypes work with an anode grid close to a hot cathode in order to extract the ionizing electrons. The ohmic heating of the cathode is about 1.5 kW and the extraction voltage about 150 V. Outside the vacuum chamber, around the ionization chamber, a magnetic coil is placed in order to optimize the ionization process. The magnetic field inside the chamber is about 300 Gauss. Furthermore a FEBIAD-kind prototype designed to work efficiently without magnet has been developed [ix] and ionization efficiency of about 35% for noble gases.

The SPIRAL 2 project raises two issues concerning such a type of ion source. The first one is the life of the source, which should be close to 3 months. The second one deals with the insulator required to apply the discharge high voltage. So far, the life of the MK5 units at ISOLDE-CERN is limited to one month. Under typical on-line conditions, the lifetime is rather twenty days. The main cause of the ionization efficiency collapse is not yet clearly identified.

Thermo-ionic ion source

The surface ion source consists simply of a refractory metal tube at a high temperature. For positive ion productions, the heated tube captures an electron from atoms with low ionization potential impinging on the surface. This ion source is the most efficient for the production of alkali ions. For the other elements, in particular alkali-earth and rare earth, it has been experimentally demonstrated that the ionization efficiency of the surface ion source is enhanced by orders of magnitudes when replacing the simple ionization tube by a hot cavity. The cavity has an extraction hole which area is small compared to the cavity inner surface At that time, only the surface ion source of TRIUMF demonstrated experimentally the possibility of working in conditions close to those foreseen for SPIRAL 2, in term of irradiation time and dose rate.

LASER ion source (Resonance Ionization by Laser Ion Source)

Different LASER IS have been developed for the production of radioactive ions of non volatile elements. This type of source is particularly efficient for the production of metallic ions. Furthermore, the ionization process is particularly selective.

The LASER IS will not be studied now but the constraints relative to its implantation and use will be taken into account, so as to allow the possibility to inject the laser beam through the low energy beam line and the implementation of a "LASER room" close to the production area.

The "PLUG" technology

The production of radioactive elements with the SPIRAL2 concept induces very high fields of beta, gamma and neutron radiation. This means that the equipment have to be protected against these radiation. The plug technology developed at TRIUMF, Vancouver, BC, Canada, will be adapted to the SPIRAL 2 production system. The principle of the plug consists in shielding all the element that are radio-sensitive and to permit a manual disconnection of the apparatus before handling for storing or dismantling. The size of the biololgical shield is under calculations. However the present size is fixed to 2 m as in Triumf and will be adjusted when the results of the dose rate calculations will be complete.The converter, target and ion source are placed in a box on the bottom on the plug and all the equipments (pumps...) are located on the top (see Figure 5). After irradiation the whole plug is removed and stored before its dismantling in a shielded cell. This type of functionning implies a high reliability on all the elements that equip the production system

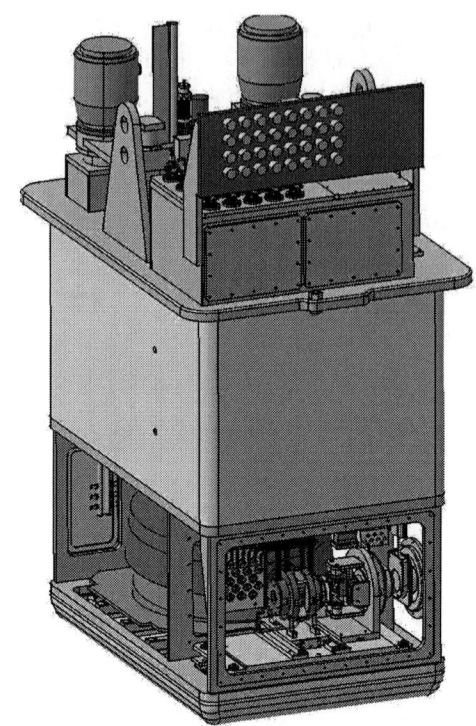

Figure 5: The PLUG for SPIRAL2

Acknowledgements

The authors are gratefull to all people from the SPIRAL 2 collaboration for their help and their fruitfull discussions.

References

[i] L. Maunoury et al, Proceedings of the 18th Int. Workshop on ECR Ion Sources, February 26-28, College Station, Texas USA (1997)
[ii] A.C.C. Villari et al Proceedings of. CAARI 2000, Denton, TX, USA (2000)
[iii] R. Leroy et al, proc. of the 9th ICIS, Rev. of Sci. Instr, **73**, p711 (Feb. 2002)
[iv] M.G. Saint Laurent et al, SPIRAL phase 2, European RTT, final report, www.ganil.fr/ research/developments/spiral2/index.htm
[v] LINAG phase 1 report, www.ganil.fr/ research/developments/spiral2/index.htm
[vi] F. Pellemoine, APD Spiral II, Thermal studies for the carbon converter, SST 220, GANIL report (June 2003).
[vii] V. Panteleev et al, these proceedings
[viii] S. Sundell et al, Nul. Instr. and Meth. **B 70** (1992) 160-164

[ix] C. Lau *et al.*, Nucl. Instr. and Meth. **B 204** (2003) 246-250

RECENT RESULTS WITH THE 2.45 GHZ ECRIS AT TRIUMF-ISAC[*]

Pierre Bricault, Keerthi Jayamanna, Dick He Ling Yuan, Miguel Olivo and Paul Schmor

TRIUMF, Vancouver, CANADA

Abstract. A 2.45 GHz ECRIS was built at TRIUMF for on-line applications at the ISAC facility. The ISAC facility utilizes a 500 MeV proton beam driver with intensity that can reach 100 µA on the isotope production target. In such an environment special care during the design phase had to be taken, all the ECRIS components especialy the two coils had to be radiation hard. Following a disappointing run with a Ta foil target intensive studies were carried out to determine the relationship between the ionization efficiency and the gas pressure in the ion source. The Neon ionization efficiency decreased from 2% to 0.01% when the tank pressure surrounding the ECR increased from 1.5×10^{-6} to 3×10^{-6} Torr. This pressure increase corresponds to approximately 10^{14} Kr/s and 5×10^{13} Xe/s. In May 2004 additional on-line tests with a SiC target were undertaken. The goal was to produce ^{18}Ne for a high precision half-live determination. Again when operating the ECRIS with the same operating parameters as the ones obtained off-line the ionization efficiency decreased when the target was bombarded with more that 5µA. After modifying the ECRIS operating conditions by injecting only Ar as a support gas and the proton beam could be increased up to 30 µA, which is the nominal for a SiC target.

INTRODUCTION

The ISAC radioactive ion beam facility began operation in November 1998. UntilMay 2004 it was operating using only a surface ion source. The ISAC facility utilizes a 500 MeV proton beam driver with intensity that can reach 100 µA on the isotope production target [i, ii]. In such an environment special care during the design phase had to be taken, all the ECRIS components especially the two coils had to be radiation hard [iii].

A 2.45 GHz ECRIS was built at TRIUMF for on-line applications at the ISAC facility. The source cavity is a single mode TE111 that operates at 2.45 GHz. In order to avoid large containment time the plasma volume is limited using a 12 mm diameter quartz tube located at the centre of the cavity [iv]. Figure 1 shows a drawing of the ECR ion source and target assembly. The quartz tube lies in the centre of the plasma chamber. The Electron Cyclotron Resonance (ECR) ion source has the following characteristics:

- 2.45 GHz operating frequency,
- Single mode cavity,
- Radial injection of the Radio Frequency (RF) power.
- No radial magnetic field confinement,
- The axial magnetic field confinement is very shallow. (The magnetic field longitudinal distribution is superimposed onto the mechanical drawing in fig. 1.)

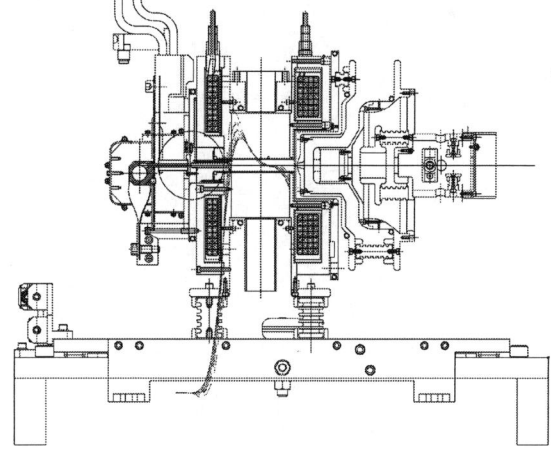

Figure 1-- Drawing of the ECR and target assembly as installed on the tray at the bottom of the target module.

---

[*] TRIUMF receives federal funding visa a contribution agreement through the National Research Council of Canada

Off-line optimization found that the 2.45 GHz ECRIS gives the best ionization efficiency and source lifetime when using He as support gas.

The ECR was installed on-line during the Fall 2002 and stable beam commissioning commenced. At the time the ECR ion source had a 5 mm diameter exit hole. The extracted beam intensity was such that we experienced large copper sputtering on all insulators in the exit module optics. After rebuilding the optics system with new shields to prevent deposition on the insulators the ECRIS was reinstalled in April 2003 with a smaller extraction hole of 3 mm. Two collimators were also added to limit the beam losses in order to reduce the sputtering caused by the extracted ion beam. On-line tests started in May 2003. A tantalum target 22 g/cm^2 was installed in order to measure the production yield of the various noble gasses elements from Helium to Xenon. During the run we observed a lower yield of noble gasses by a factor from 10 to 100 with respect to the observed yield at ISOLDE/CERN using a FEBIAD ion source. ISLOLDE operated with 1 µA of 600 MeV protons onto a 122 g/cm^2 Ta target. We operated with 5 µA of 500 MeV protons onto a 22 g/cm^2 Ta target and expected a similar yield.

Following the disappointing run with the Ta foil target we decide to conduct intensive studies of the relation between the ionization efficiency and the pressure. Tests were conducted with a well-known quantity of gases of Kr and Xe that were injected into the ECR while monitoring the ionization efficiency.

NEON IONIZATION EFFICIENCY VERSUS PRESSURE AND GAS FLOW

The optics of the exit modules was tuned to obtain the best transmission through the pre-separator slit (IMS:SLIT0). The ISAC target station and mass separator is described in the reference [v]. The ECR was tuned for the best Neon ionization efficiency. We performed all the runs at an extraction voltage of 30 kV. With this ECRIS we found that the tuning range for the coils and the gas is quite limited. Reference [vi] gives a more detailed description of the Neon ionization efficiency as a function of the coils, input RF power and support gas flow. To study the effect of the pressure increase inside the ECR on the ionization efficiency Kr and Xe were injected into the ECR through the diaphragm valve while Neon was coming from a calibrated leak. We recorded the ^{22}Ne and ^{20}Ne ion current, the pressure, the Kr^{1+} and Xe^{1+} beam current and the total extracted beam current. The pressure was then gas-type corrected. The beam current was measured at the focal plane of the first stage of the mass separator [5] IMS:FC0. The total extracted current beam was deduced from the high voltage power supply minus the leakage current when the ECRIS was off. We also measured the emittance as a function of the Xe flux injected into the ECR.

Figure 2 and 3 show the Neon ionization efficiency as a function of the pressure in the tank and injected gas flow. The efficiency from ^{22}Ne and ^{20}Ne differs because we had to use Argon to ignite the ECR ion

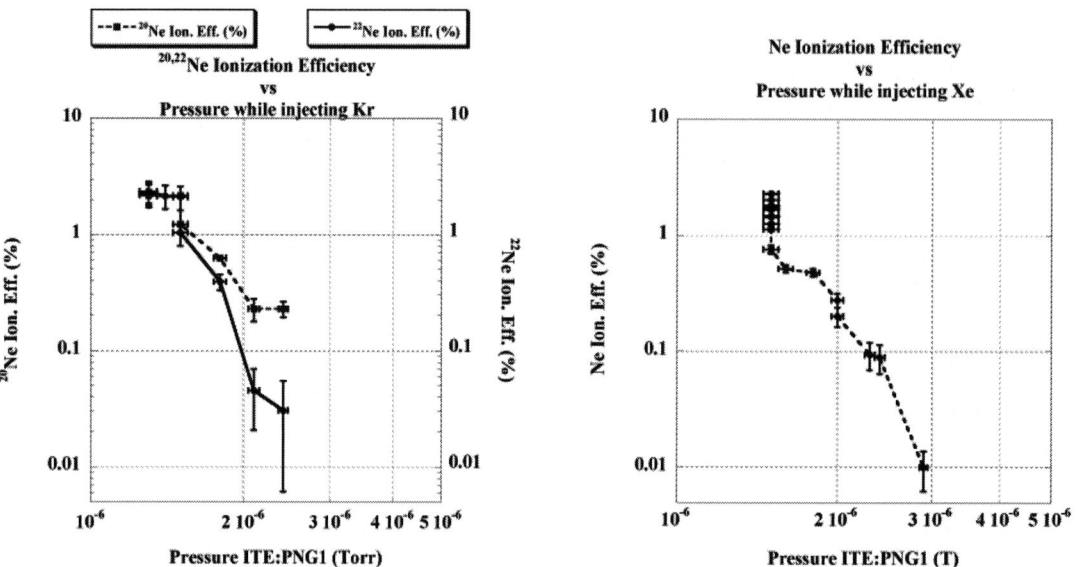

Figure 2--The left plot shows the ^{20,22}Ne ionization efficiency with respect to the pressure into the containment box surrounding the ECR while injecting Krypton. The plot on the right shows the ^{20}Ne ionization efficiency with respect to the pressure into the containment box while injecting Xenon.

source in that case. The ^{20}Ne values were contaminated by ^{40}Ar^{2+} beam. We can see from those two plots that the Ne ionization efficiency shows a decrease with the pressure increase. The ionization efficiency drops by more than two orders of magnitude with the pressure going from 1.5×10^{-6} to 3.0×10^{-6} Torr. Figure 3 shows the Neon ionization efficiency versus the Kr and Xe flux. We can see that the efficiency starts decreasing at about 5×10^{13} Xe atoms/s and 10^{14} Kr atoms/s.

BEAM EMITTANCE AS A FUNCTION OF THE XENON FLUX

One of the biggest unknowns for efficient beam transport was the emittance of the beam from the ECRIS under on-line conditions. The beam was tuned to the mass separator and the beam emittance was measured at the focal magnet plane as a function of the Xe flux. Figure 4 shows the emittance as a function of Xe flux injected into the ECR. Looking in detail at the emittance plot we can see clearly that the emittance is increasing with the flux, which may be the result of collision of ions on the residual gas atoms at the extraction.

^{18}NE ON-LINE RESULTS

In May 2004 on-line tests with a SiC target were undertaken. The goal was to produce ^{18}Ne for a high precision half-live determination. The ECRIS was operated with the same tune developed during the off-line tests, where a 2% Neon ionization efficiency was obtained. A flow of 1 SCCM Helium was used as the support gas. The ^{18}Ne yield was measured with 1 μA to 5 μA and followed the proton beam. Above

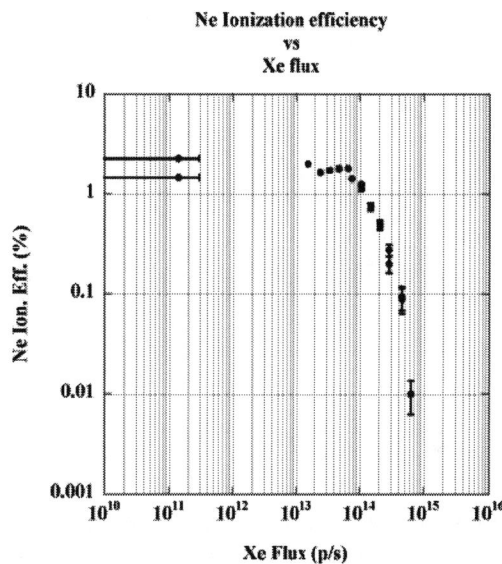

Figure 3 -- The left plot shows the ^{20,22}Ne ionization efficiency with respect to Krypton flux. The plot on the right shows the ^{20}Ne ionization efficiency with respect to Xenon flux.

7 μA the yield did not increase and at 10 μA it was lower than at 5 μA. The on-line Neon ionization efficiency was measured and it decreased from 1% at 5 μA to 0.05% at 10 μA.

Furthermore, due to carbon coating of the quartz liner in the plasma chamber, the ECRIS had to run with an Ar-He mixture. The Ar provides a way to reduce the carbon coating on the quartz tube. Unfortunately, the addition of Ar also reduces the ionization efficiency by a factor 3. Figure 5 shows the trend of the ^{18}Ne yield as a function of the proton beam current on target. The dots and the squares represent the measured ^{18}Ne yield when the ECRIS was operating with He. The sudden drop at 6 μA corresponds to the addition of the Ar with He as support gas. Then we try to reduce the amount of injected gas into the ECR. We then change the support gas from He to Ar. In those conditions we were able to operate the ECRIS with a proton beam up to 30 μA on the SiC target and the ^{18}Ne yield just scale with the proton beam. The open dots and squares in figure 5 show the ^{18}Ne yield when we were operating the ECRIS with only 0.003 SCCM Ar.

CONCLUSION

The goal of this study was to determine the operating range in pressure of the ECR-1 installed at

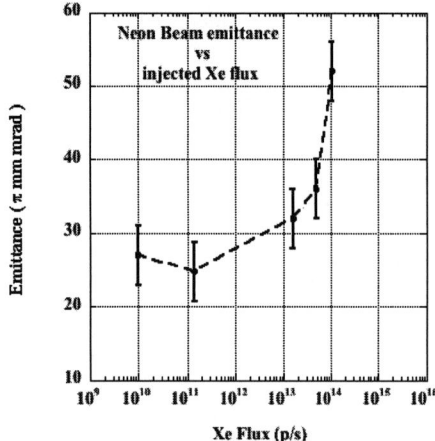

Figure 4-- 4RMS plot of the beam emittance with respect to the Xe flux.

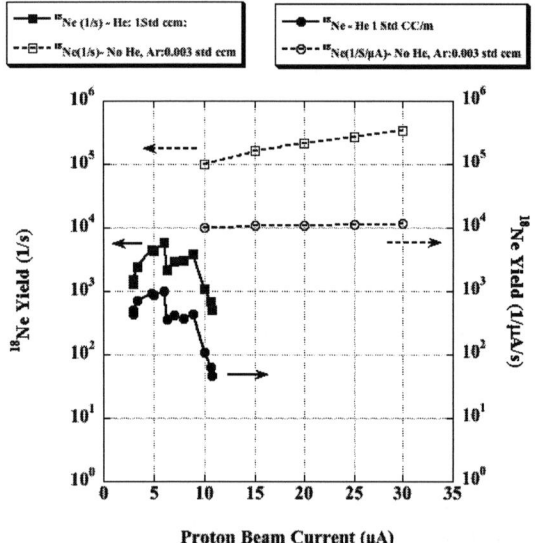

Figure 5-- ^{18}Ne yield as a function of the proton beam current. Solid lines show the ^{18}Ne yield operating the ECRIS with the nominal parameters. The dashed lines show the yield operating the ECRIS without He and only Ar as support gas.

the ISAC Target Station. We were able to reproduce the Neon, Kr and Xe, ionization efficiency of 2.6 %, 25.9% and 62.1%, respectively, at the ISAC target station, that was observed on the off-line test stand. Kr and Xe gases were injected into the ECR and we have measured the ionization efficiency of Neon, Kr and Xe as a function of the flux of gas. The measurements show also that the emittance increases with the gas flux injected into the ECRIS and we observed a low energy tail formation, which is a possible sign of collision on gas during the acceleration.

We can conclude that the source is stable over an external pressure range between 1×10^{-6} to 1.5×10^{-6} Torr. Above that pressure the ionization efficiency starts dropping. That corresponds to an estimate flux of 10^{14} atoms/sec. If we want to run the ECR on-line we have to make sure that we stay within those limits if we want to maintain the 2% ionization efficiency.

During the on-line run we observed that the ionization efficiency was dropping significantly when the proton beam was increase to 10 μA. We change the support gas from 1 SCCM of He to 0.003 SCCM of Ar only and we were able to maintain the ionization efficiency relatively constant up to 30 μA, which is the nominal beam current we can run on SiC target.

ACKNOWLEDGMENTS

We would like to acknowledge the assistance of several people who made that work possible, Clint Laforge, Mike McDonald and Raymond Dubé for the installation of the ECRIS, John McKinnon for the remote handling, Guy Stanford for the engineering, Marik Dombsky for radioactive ion beam yield measurement, Andy Hurst and ISAC operation crew for beam delivery. We also would like to thank the physicist crew of the GPS-1 and 8π experiment who assist in the yield measurement.

REFERENCES

[i] P. Bricault, m. Dombsky, P. W. Schmor, G. Stanford, Nucl. Instr. And Meth. B 126 (1997) 213.

[ii] Pierre Bricault, Marik Dombsky, Paul Schmor, Guy Stanford, Ian Thorson and Jaroslav Welz, *Proceeding of the 1999 Particle Accelerator Conference, New York, USA,* 1999.

[iii] D. Yuan, K. Jayamanna, M. Dombsky, D. Louie, S. Kadantsev, R. Keitel, T. Kuo, M. McDonald, M. Olivo, and P. Schmor, *Rev. Sci. Instrum.* **71**(2) 643 (01 Feb 2000)

[iv] K. Jayamanna, D. Yuan, M. Dombsky, P. Bricault, M. McDonald, M. Olivo, P. Schmor, G. Stanford, J. Vincent, and A. Zyuzin, *Rev. Sci. Instrum.* **73**(2) 792 (01 Feb 2002)

[v] Pierre Bricault, Paul Schmor, Guy Stanford, Clive Mark, Marik Dombsky, Mike Gallop, Lutz Moritz and Lorne Udy, "Target and mass separator concept for the RIB facility", Proceeding of the XV international conference on Cyclotrons and their Applications 1998, Caen, France. p. 347.

[vi] "Study of the Ionization Efficiency of ECR-1 at ISAC", Pierre Bricault, (2004), TRIUMF Design Note, TRI-DN-04-17.

Charge State Breeding with an ECRIS for ISAC at TRIUMF

F. Ames, K. Jayamanna, D.H.L. Yuan, M. Olivo, R. Baartman, P. Bricault, M. McDonald, P. Schmor

TRIUMF, 4004 Wesbrook Mall, Vancouver BC V6T 2A3, Canada

T. Lamy

LPSC UJF-IN2P3-CNRS, 53 Av. Des Martyrs, 38026 Grenoble, France

Abstract. For the acceleration of radioactive ions the usable mass range is limited by the A/q acceptance of the first accelerator stage. Since an efficient primary ion source normally produces singly charged ions, charge state breeding is necessary if higher masses are to be accelerated. At TRIUMF an ECR source has been chosen as a breeder due to its potential high efficiency in producing intermediate A/q values. To minimize the necessity for further stripping an A/q around 6 is desirable. A 14 GHz "PHOENIX" booster from Pantechnik has been set up on a test bench. The singly charged ions are produced from different ion sources, which can be mounted in a standard ISAC target-ion-source set-up. For the first tests an ECR source to produce noble gas beams has been chosen. The aim of the measurements at the test bench is to find the optimum operation conditions of the charge state booster and the injection and extraction ion optics. Working with radioactive ions always means that the system should aim for high efficiency, as the production of such species is limited. Therefore, special emphasis has to be put on the highest yield for the production of the desired charge state. A second point is the analysis of the extracted beam quality in order to optimize mass separation and transport efficiency. The paper shows the status of the set-up and reports on first results of the charge breeding of Ar, Ne and Xe. With Xe a total efficiency of 22.5 % has been achieved.

INTRODUCTION

With the development of radioactive beam facilities at different laboratories charge state breading has become an important issue especially in connection with post acceleration. At an ISOL facility the ions are produced typically in a 1+ state at an energy of several 10 keV. But in order to allow the acceleration to higher energies an increase in charge state is essential. There are several methods being used so far. The first one involves the stripping of the singly charged ions after an initial acceleration to some 100 keV/u. This method is limited to low masses, where the first acceleration is possible. At higher masses the only way is to increase the charge state before acceleration. In order to do so they are injected into an ion source for highly charged ions. There are two types of ion sources being used so far. An electron beam ion source EBIS has been used at the post accelerator REX-ISOLDE at CERN [1]. It can deliver beams of highly charged ions up to an efficiency of about 10% for a single charge state. The disadvantage of such a system is that it requires a pulsed beam at a small emittance for the incoming ions. That means a cooling and prebunching of the beam is necessary, which limits the total efficiency and the intensity to be used. If the source should run in a continuous mode the only solution so far is the use of an ECR [2], [3], [4]. The efficiency for the production of intermediate charge states has been reported as high as about 10 % especially if noble gas ions are used and the system can be run also at high intensity. A disadvantage is the sometimes high background due to the ionization of the support gas or residual gas. Therefore a small emittance of the extracted beam is essential. This will reduce the background due to neighboring high intensity A/q values or scattered ions in the beam lines.

THE CHARGE STATE BOOSTER AT THE TRIUMF ION SOURCE TEST STAND

At TRIUMF radioactive ions are produced at the ISAC facility by means of a high intensity 500 MeV proton beam hitting a solid target. To produce singly charged ions a surface ion source and an ECR have been used so far. The maximum A/q ratio of 30 for the post acceleration is given by the acceptance of the first linear RFQ accelerator. To increase this number it is planned to introduce charge state breeding (CSB) with an ECR source. Therefore a 14 GHz "PHOENIX" booster from Pantechnik has been set up first on a test bench. It can be equipped with several ISAC type ion sources. The 1+ section has already been described in detail in [5]. Some space in front of the CSB has been reserved to allow for the insertion of a beam cooling device. Following the CSB a double focusing mass spectrometer consisting of a 90° magnetic and two 45° electrostatic benders has been installed. Two electrostatic quadrupole doublets in front of the magnet ensure the focusing of the beam from the CSB to the object point of the spectrometer. The emittance of the incoming and outgoing beam can be measured at several points by means of inserting emittance meters [6]. The entire set-up is shown in figure 1.

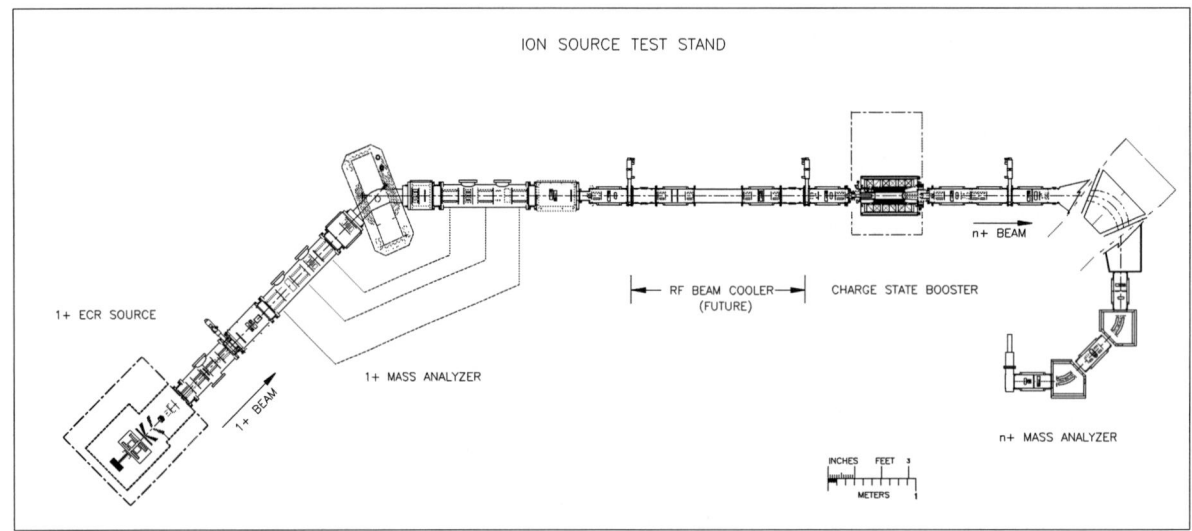

FIGURE 1. Set–up of the TRIUMF ion source test bench.

Set-up of the n+ ECR source

To meet the velocity acceptance of the ISAC RFQ accelerator with a mass to charge ratio around 7.5 the source has to be run at a potential of about 15 kV. The set-up of the PHOENIX ion source has been slightly changed for the requirements at TRIUMF. In order to limit the total emitted current the extraction aperture in the plasma electrode has been reduced from originally

8 mm diameter to 3.5 mm. The extraction is done with one electrode on ground potential 12 mm in front of the extraction aperture. In order to protect the window, which insulates the source vacuum from the waveguide, from being damaged by the plasma, it has been placed outside the stray magnetic field of the source. Additionally, water cooling of the waveguide has been installed and the portion of the waveguide at high vacuum has been equipped with small holes and connected through a high voltage insulator to the vacuum chamber in front of the source for additional pumping. This guarantees a high vacuum inside the waveguide and thus minimizes the risk of a plasma discharge in it, which could destroy the window. The support gas is fed in directly into the plasma chamber at the injection side. On each side of the source pumping with both a 500 l/s turbo pump and a 2000 l/s cryo pump ensures a pressure below some 10^{-7} Torr.

MEASUREMENTS AND RESULTS

In first commissioning measurements the CSB has been run without injecting ions in order to optimize the extraction and analyzing ion optics. Ar or He has been used as a support gas. The efficiency and charge state distribution for the ionization of Ar for gaseous injection has been determined. The charge state distribution when using He as support gas and injecting Ar via a calibrated leak is shown in figure 2. The maximum in the distribution is at 7+ with an efficiency of 2.7 %. Summing up the efficiency for all charge states gives a total efficiency of 12%. For this measurement a microwave power of 100 W and a He flow of about 10^{-3} sccm have been used. This efficiency value includes the transmission of the extraction and analyzing system. Especially in the line in front of the magnet, the high beam current causes losses due to space charge effects. As the vacuum in the analyzer for these first measurements is only about $1 \cdot 10^{-6}$ Torr, charge exchange for the higher charge states will occur, resulting in an estimated loss of up to about 20%.

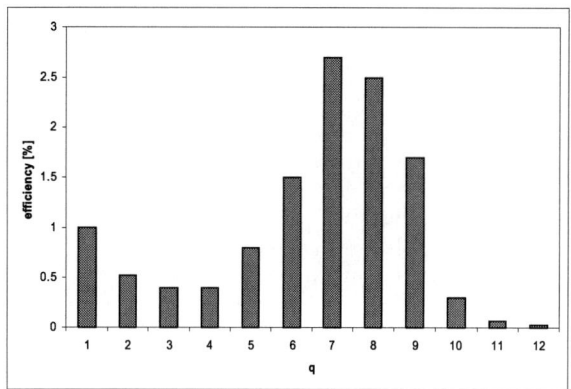

FIGURE 2. Charge state distribution of Ar when injected as a gas via a calibrated leak.

The emittances for the different charge states of Ar have been determined at the end of the analyzing system. Values up to 20 π mm mrad at 15 kV source potential have been found for 86% of the beam enclosed. Within the measurement accuracy of ±10% no significant dependence on the charge state could be found.

For the charge breeding measurements an ECR [6],[7] source has been used to produce beams of singly charged Ne, Ar and Xe ions. The CSB has been run with a low flow of He gas or with no additional support gas. First the potential of the CSB relative to the 1+ source has been changed to find the optimum deceleration parameter. An example of such a measurement is shown in figure 3. As has been reported from other groups, a broad structure can be found for noble gases with some saturation behavior at low potentials. This indicates the possibility for noble gases of being reemitted from the plasma chamber walls if their initial energy is too high for being captured by the plasma.

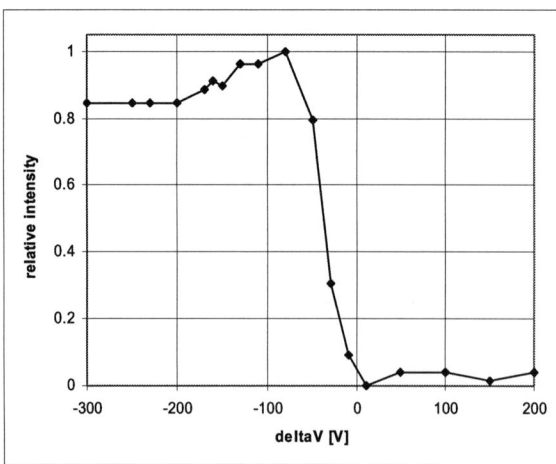

FIGURE 3. Relative intensity of the ^{22}Ne^{4+} signal as function of the CSB potential with respect to the 1+ ion source.

After this the efficiencies for the different charge states have been determined. An example is shown in figure 4 for Xe. In this case the maximum was at 13+ with 3.7 %. Summing up all the charge states a total efficiency of 22.5 % can be found. This measurement has been done with a microwave power of 250 W and no support gas. In the case of Ar and Ne the maximum in the distribution was at 8+ and 4+ respectively, with a smaller total efficiency, but at these measurements a microwave power of only 100 W could be used.

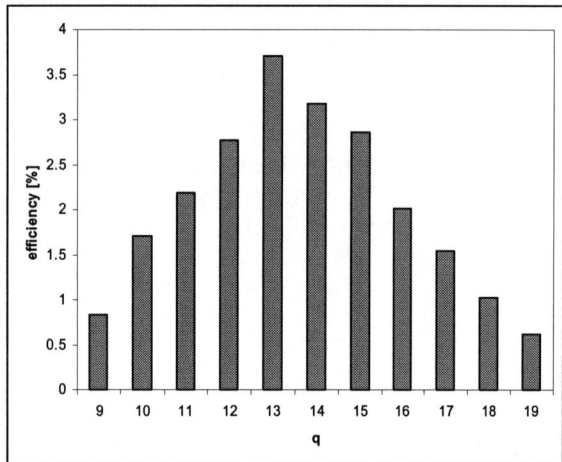

FIGURE 4. Efficiency of the charge state breeding of Xe for different charge states.

CONCLUSION AND OUTLOOK

The ECR charge state booster has been put successfully into operation at a test beam facility at TRIUMF. The first measurements could already demonstrate its capability of producing ions of intermediate charge states from singly ionized injected beams. The total efficiency is 22.5 % for Xe and we expect this to improve. More studies on its dependence on the support gas or the composition on the residual gas in the plasma chamber will be done. Additionally the influence of the reduced extraction hole in the plasma electrode compared to the original design has to be studied in more detail with simulations and experimentally. Here a compromise between high extraction efficiency and a small emittance, which guarantees high transmission through the rest of the set-up, has to be found. In order to extend the measurements to non noble gas ions a surface ion source will be installed in the 1+ line. In principle the efficiency can be increased by reducing the emittance of the incoming beam. This will be studied with the addition of a gas filled RFQ cooler [9] in front of the CSB. Such a device is being presently constructed at TRIUMF. Especially the overall efficiency with this system and the reliability for an on-line operation together with the CSB has to be studied.

ACKNOWLEDGMENTS

TRIUMF receives federal funding via a contribution agreement through the national research Council of Canada.

REFERENCES

1. Ames F. et al., Rev. Sci. Instr. 75 1607-1609 (2004)
2. Sortais P. et al., Nucl. Phys. A 701 537-549 (2002)
3. Lamy T. et al., Rev. Sci. Instr. 75 1624-1626 (2004)
4. Jeong S.C. et al., Rev. Sci. Instr. 75 1631-1633 (2004)
5. Dombsky M. et al., Nucl. Instr. Meth. B 126 50-54 (1997)
6. Yuan D. et al. Rev. Sci. Instr. 67 1275-1276 (1996)Bricault P., these proceedings
7. Jayamanna K. et al, Rev. Sci. Instr. 75 1621-1623 (2004)
8. Bricault P., these proceedings
9. Herfurth F., Nucl. Instr. Meth. B 204 587-591 (2003)

Radioactive Beams Using the AECR-U and the 88-Inch Cyclotron

M.A. McMahan, D. Leitner, J. Powell and C. Silver

Lawrence Berkeley National Laboratory
Berkeley, CA

Abstract. The high ionization efficiency of an Electron Cyclotron Resonance (ECR) ion source combined with the mass resolution of a cyclotron is ideal for the generation of some ISOL-type radioactive ion beams (RIBs). In two separate projects at the 88-Inch Cyclotron at LBNL – BEARS and the Recyclotron – we have developed techniques to efficiently ionize and accelerate beams of gaseous species of ^{11}C ($t_{1/2}$ = 20 min), ^{14,15}O ($t_{1/2}$ = 70 sec, 2 min) and ^{76,79}Kr ($t_{1/2}$ = 14,35 hours). Measurements of the ionization efficiency and hold-up times are discussed, along with issues of source contamination and poisoning encountered in running both RIBs and high-intensity stable beam experiments using the same ion source, the LBNL AECR-U. Methods used to tune clean RIBs through the Cyclotron with high efficiency are also discussed, including the use and limitations of analog beams.

INTRODUCTION

In two separate collaborations between 88-Inch Cyclotron Operations staff and researchers at LBNL and other institutions, a few ISOL-type radioactive ion beams have been developed in batch mode using the 88-Inch Cyclotron as a post-accelerator. The high charge states and efficiency of the LBNL AECR-U makes it ideal for this application, particularly when the radioactive species can be produced in gaseous form. In this paper we describe the techniques developed to optimize production, ionization and acceleration of light carbon and oxygen beams in the BEARS project [1] and krypton isotopes in the Recyclotron project [2]. The beams developed thus far are given in Table 1.

Each of the five steps - production, transport, trapping, ionization and acceleration – must be optimized in order to obtain a flux of ions on target which makes nuclear structure and astrophysics measurements feasible. The optimization and the 'leaks in the pipeline' are different for the two projects, mainly because of the half-life differences. Radiation safety issues also differ between the two projects and are an integral part of the development process.

THE 88-INCH CYCLOTRON FACILITY

The 88-Inch Cyclotron – shown schematically in Figure 1 - is a sector focused K=140 cyclotron which accelerates both light and heavy ions. Of the three ECR sources, two of them – the LBNL-ECR and the AECR-U – are presently coupled to the Cyclotron through the axial injection line. The third source, VENUS, is being commissioned and will be coupled to the Cyclotron in the future. For RIBs, the AECR-U, which presently generates the highest intensity and efficiency for high charge states heavy ions, is used. The parameters of this ion source are given in Table 2. [3] The AECR-U is the workhorse of the facility, extensively used for the two largest Cyclotron programs: 1) heavy element studies, which needs high intensity heavy ions, and 2) radiation effects testing of microelectronics, which uses cocktails of beams as heavy as bismuth. Because of this, it is essential that the ion source be kept uncontaminated so it can be maintained for day-to-day operation. This requirement has driven the development of RIBs at the Cyclotron to restrict beams to those whose half-lives and those of the daughter decays are on the order of a few days or less.

TABLE 1. Radioactive ion beams developed at the 88-Inch Cyclotron

BEARS Beams			
Species	Half-life (sec)	Reaction	Yield (atoms/μA-sec)
^{11}C	1200	$^{14}N(p,\alpha)$	3×10^9
^{14}O	70	$^{14}N(p,n)$	1.5×10^8
^{15}O	120	$^{15}N(p,n)$	3×10^9
Recyclotron Beams			
	Half-life (hrs)		Yield (atoms/μA-hr)
^{76}Kr	14.8	$^{74}Se(\alpha,2n)$	4×10^{12}
^{79}Kr	35	$^{76}Se(\alpha,n)$	5×10^{12}

The beam is extracted from the ion source at 15,000 Volts, bunched, and injected axially into the Cyclotron. The emittance and transmission through the various sections of the injection line are very well understood. The overall transmission is 33%. [4]

TABLE 2. Parameters of the LBNL AECR-U

Magnetic Field: Ampere-Turns	317,000
Magnetic Field: Peak Field	1.7 T
Microwave Frequency	10 GHz + 14 GHz
Microwave Total Power	2600 W
Extraction High Voltage	15 kV

DEVELOPMENT OF THE RIBS

I. Production and Transport

The radioactive carbon and oxygen are produced at the LBNL Biomedical Isotope Facility (BIF), a commercial 11 MeV, 40 μA proton cyclotron made be CTI. An internal nitrogen gas target is purged periodically and the radioactive species sent 300 m down the hill in a gas transport line to the 88-Inch Cyclotron. Transfer takes about 15 sec. The radioactive carbon naturally reacts with oxygen in the target to make CO_2, the chemical form needed for trapping with LN_2. Unfortunately for radioactive oxygen, this turns out not to be the case; the oxygen preferentially sticks to the walls unless there is enough hydrogen in the target for it to form water. Therefore, an automated chemistry step takes place between the target and the transfer line, converting the water to CO_2. This adds some delay to the process, which affects the final amount of oxygen transferred.

The krypton isotopes are produced using (α,xn) reactions. For ^{76}Kr, an enriched ^{74}Se target is used. The target thickness was chosen carefully in order to maximize the energy range of the $^{74}Se(\alpha,2n)$ reaction while minimizing the production of ^{75}Br from the $(\alpha,2np)$ reaction and ^{75}Se from the $(\alpha,n2p)$ reaction, both of which lead to long-lived ^{75}Se, which is to be avoided for radiological health and safety concerns. The production period was approximately one half-life.

II. Trapping and Transfer to Ion Source

The LN_2 trap is warmed up and unloaded into a reservoir next to the AECR-U. For BEARS, the target purge, transfer, trapping and transfer is automated on a 1-5 minute cycle. For the Recyclotron, the transfer takes place over a couple of hours.

III. Ionization

The efficiency of the AECR-U for carbon and oxygen has been optimized and measured with a stable CO_2 leak. The optimized charge state distributions with absolute efficiencies are given in Figure 1a for carbon and 1b for oxygen. For ^{11}C and ^{14}O, the efficiency is substantially lower, as can be seen in the figures. This is partially due to the difficulty in tuning a beam which is fluctuating in intensity on a 1-5 minute time scale as decay occurs. The primary loss in efficiency, however, can be attributed to a hold-up time in the ion source which is 4-5 times greater for the RIB than for stable CO_2. This is most likely due to the long transfer line between the reservoir and the source. The hold-up time for ^{11}C is given in Figure 2. The fast component is 24 seconds, followed by a long tail; for stable CO_2, we have measured the hold-up time to be 5.6 sec without a slow component.

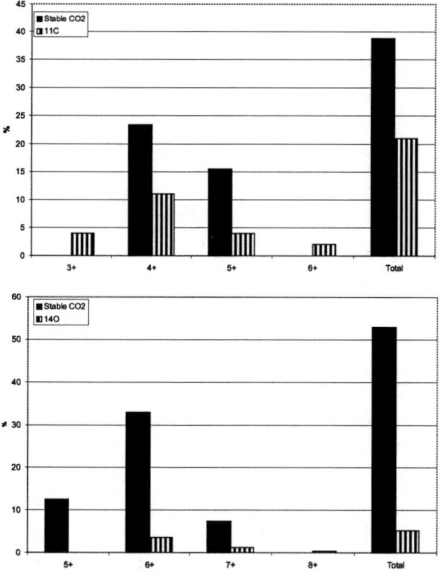

FIGURE 1. Absolute ionization efficiency for a) ^{11}C and b) ^{14}O compared to carbon and oxygen from stable CO_2 leak.

For the radioactive Kr beams, we also used a stable Kr leak to optimize the ionization efficiency, and at the target position to measure the time and efficiency of the transfer and trapping steps. Figure 3 gives the optimized Kr efficiencies using the standard leak. It is possible to get more than 10% of the krypton into one charge state around q= 13-20. Figure 4a and 4b show the effect of the mixing gas on the efficiency and the hold-up time of the krypton. As expected, high pressures in the ion source are to be avoided.

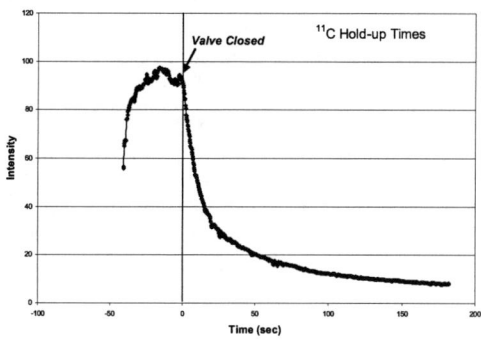

FIGURE 2. Hold-up time for ^{11}C. The fast component is 24 sec.

IV. Acceleration and Beam Transport

The carbon and oxygen beams are tuned using analog beams: ^{22}Ne and ^{11}B for ^{11}C and ^{14,15}N for ^{14,15}O. The frequency resolution of the 88-Inch is 2 kHz, which corresponds to a mass resolution of 1/3000. If the mass of the analog beam and RIB are too close they cannot be separated. This is the case for ^{14}O/^{14}N, for example. Also, when one beam is much higher intensity than the other, then the tail of the large beam overlaps, and the frequency resolution is worse. In order to guarantee a clean RIB, we use a thin nickel foil in the beamline to fully-strip the carbon and oxygen. Then the analyzing magnets in the beamline will allow delivery of an uncontaminated beam.

The choice of charge state tuned from the ion source depends on the desired energy for the experiment. The AECR-U is most efficient for the C^{+4} and O^{+6} charge states. Either increasing or decreasing these will lower the overall efficiency. In addition, the percentage of beam which is fully stripped in the foil decreases with energy. For ^{11}C, we have run experiments at several energies requiring charge states of +2, +3 and +4. At the lowest energy measured, 2 AMeV, the beam intensity was four orders of magnitude less than the typical intensity obtained around 5-10 AMeV. An additional problem is encountered if attempting to run RIBs at the highest energies of the 88-Inch Cyclotron. These require using fully-stripped carbon and oxygen from the ion source, for which no analog beams are available. Various options are being considered to tune these beams, including putting a silicon diode detector inside the cyclotron to pick out the beam during acceleration.

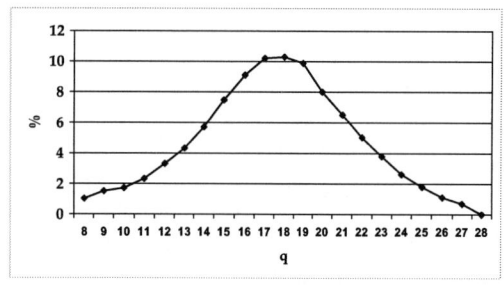

FIGURE 3. Optimized krypton charge state distribution

TABLE 3. Intensities and Reliability of RIBs at the 88-Inch Cyclotron

BEARS Beams			
Species	Half-life (sec)	Flux (ions/sec)	Reliability
^{11}C	1200	5-10 x 10^7	Very good
^{14}O	70	1-3 x 10^4	marginal
^{15}O	120	5-10 x 10^5 (expected)	Under development
Recyclotron Beams			
	Half-life (hrs)	Total Ions /batch	Efficiencies(prod/ion +accel (%)
^{76}Kr	14.8	6-7 x 10^{11}	12-25/1-3
^{79}Kr	35	7-9 x 10^{11} (expected)	12-25/1-3 (expected)

There are many possible analog beams to the radioactive krypton isotopes, depending on energy and charge state. If the frequencies are too close, e.g. ^{76}Kr/^{76}Se, the analog beam is tuned from the ECR. Even charge states are avoided for ^{76}Kr because of ^{38}Ar contamination, which is always present in the ion source due to small air leaks.

CONCLUSION

The radioactive beams developed in batch mode are summarized in Table 3, with final efficiencies and fluxes under the best operating conditions for the Cyclotron and ion source.

For the three BEARS beams, the reliability and flux on target is directly related to the half-life, with 11C being the most reliable and intense, and 14O the least. 15O is still under development and hasn't yet been accelerated.

For the Recyclotron beams of krypton, in particular ^{76}Kr, we have measured efficiencies at several stages in the process. The combined ionization, acceleration and beam transport efficiency was 2.2% in the best case, giving a total of 7 x 10^{11} atoms on target for a 20 hour production/acceleration cycle. The ^{79}Kr beam has been produced and trapped but not yet accelerated through the Cyclotron.

ACKNOWLEDGMENTS

The authors wish to acknowledge the other members of the BEARS collaboration, led by Prof J. Cerny III, of University of California at Berkeley and LBNL, and the Recyclotron collaboration, especially Jeff Cooper of LLNL. None of this work would have been possible without the dedication and expertise of the 88-Inch Cyclotron Operators: Michael Beaudreau, Tom Gimpel, Aran Guy, Mike Johnson, Jim Morel, Reba Siero, and Ray Thatcher.

This work was performed under the auspices of the US Department of Energy Office of Science by the University of California Lawrence Berkeley National Lab under Contract No. DE-AC03-76SF0098.

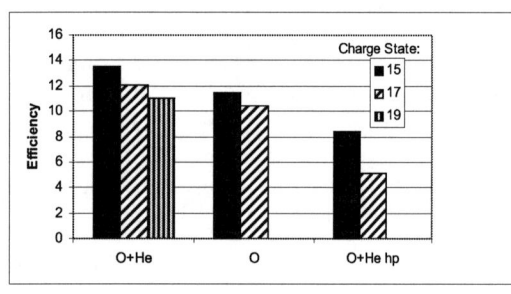

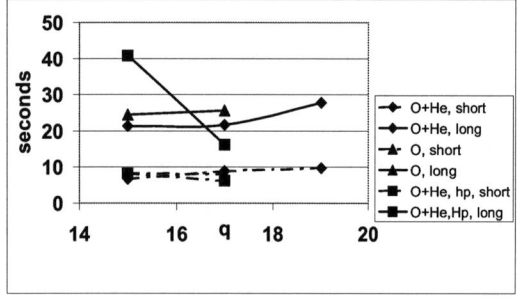

FIGURE 4. a) Effect of gas mixture and pressure on absolute efficiency (a) and hold-up time (b) for q = 15,17 and 19 for stable krypton leak.

REFERENCES

1. J. Powell et al, *Nucl Instr Meth* **A455**, 452 (2000).
2. J. Cooper et al., *Nucl. Instr Meth*, in press (2004)
3. Xie, Z.Q., *Rev. Sci. Instrm.* **69**. 625 (1998).
4. D. Leitner, et al., LBID-2349, LBNL Internal Report, (2000).

VIII POSTERS

Design Study of a Hybrid ECRIS

D. Hitz[1], A. Girard[1], D. Guillaume[1], L. Guillemet[1], P. Seyfert[1], J.M. Poncet[1]
L.T. Sun[2]

[1]*CEA-Grenoble, Département de Recherche Fondamentale sur la Matière Condensée, Service des Basses Températures, 17 Rue des Martyrs, 38054 Grenoble Cedex 9, France*
[2]*Institute of Modern Physics (IMP), Chinese Academy of Sciences, Lanzhou, 730000 China*

Abstract. A new ion source to be installed on a compact high voltage platform is under study. The purpose of this machine is to deliver very high charge states, which requires long ion lifetimes and then a large plasma volume. On the other hand, to deliver intense beams, the highest frequency compatible with the magnetic confinement has been chosen. Due to the room available for this source, the magnetic system is chosen hybrid: radial field is created by permanent magnets, while axial field is produced by 4K superconducting coils placed in a cryostat filled with LHe. This article gives the major parameters of this ion source, with a special emphasis on the magnetic system and its cryogenic features.

INTRODUCTION

It is now well recognized that ECRIS with a very good confinement and long ion lifetimes have the capability to deliver very high charge states [1-2]. Depending on its future situation, a new ECRIS will have its magnetic confinement made of room temperature or superconducting coils and/or permanent magnet. In order to be installed on a small high voltage platform at Hahn Meitner Institut Berlin, on has designed a hybrid ECRIS whose goal is to deliver at least 100 pnA of elements with M/q =4.7. After a short review of the main trends of the project, this new ECRIS will be described in details.

SHORT REVIEW OF EXISTING SCALING LAWS

To reach very high charges, it is absolutely necessary to have a good plasma confinement. A high frequency is of course of interest, in order to maximize the electron density, but it is not always possible to have both a high confinement and a high frequency, since it was proved [3] that there is relationship between the frequency of the wave and the quality of the magnetic confinement. It is now worldwide admitted that a good confinement of the plasma is achieved when the value of the hexapolar field at the wall B_{hex} over the resonant field B_{res} is of the order of 2. The axial field at injection should be in "high-B" mode and minimum-B value, which defines the last closed magnetic surface, is related to the hexapolar field. On the other hand, gradient at resonance must be steep to control the electron energy. Other parameters are also to be taken into account:
- plasma chamber size: a large diameter leads to a better the confinement
- plasma length (defined by the length of the hexapole)
- resonance zone length
- minimum-B field value in the mirror

These experimental scaling laws had been progressively derived first in the MSU source at 6.4 GHz, and then with the SERSE source at frequencies of 14, 18 and 28 GHz. They have been adopted in the design of the Grenoble Test Source [4] and led to optimum performances of this ECRIS for the production of high charge states: the GTS source has produced the largest currents of Ar^{17+}, firstly at 14 GHz, then at 18 GHz [2]. On the other hand, thanks to its optimum confinement, GTS was able to work without support gas even for Ar^{17+}..

Moreover, the production of heavy metallic ions has been demonstrated with the GTS source, via a sputtering technique. A biased probe has been used and its material sputtered in the GTS source: with this technique still never reached charges of tantalum ions have been produced.

As a conclusion of this chapter, it is clearly demonstrated that the application of the design rules to the GTS source lead to a very successful source, which delivers the highest charges ever produced in an ECRIS. It is therefore of primary importance to fit with these design rules for the hybrid ECRIS

DESIGN OF THE MAGNETIC STRUCTURE

Due to the constraints on available space on the HV platform it is not possible to adopt the technology of a fully superconducting source at 28 GHz (GYROSERSE type). Only this technology allows working at 28 GHz for the production of high charges. Therefore, it is necessary to reduce the value of the working frequency: lower frequency transmitters exist at 18 GHz (CPI, Thalès) and 24 GHz (Gycom, CPI). Therefore we designed a hybrid source (ie: superconducting coils for the mirror field and permanent magnets for the hexapolar field) able to work in the range 18-24 GHz. Several hybrid ion sources are in operation at lower frequency, they however are set at ground potential [5]. Even if they are attractive regarding the cryogenic aspect, the use of HTS coils was not chosen in our design: their short range capability would lead to set injection and extraction coils at hexapole ends, with coils inner diameter equivalent to hexapole inner diameter, like in a Caprice type source. On the other hand, one would be forced to suppress middle coils which are absolutely necessary to get the best performances of an ECRIS.

This hybrid ECRIS is a compromise between SERSE and GTS whose parameters are indicated in ultra-high vacuum technology to minimize charge exchange processes.

In the following the description of the magnetic design is given; all computations were performed firstly with POISSON, and then confirmed with TOSCA.

Mirror Field

Due to the long resonance length and in order to have a better tuning of the field gradient at resonance, and of the minimum-B field at the midplane, a set of four coils is proposed, both central coils being supplied with a counter current as compared to the other two coils located at the injection and extraction sides. Coils are made of NbTi, using the same conductor as for SERSE, which proved very reliable. Soft iron pieces are also added close to the source axis to obtain a larger magnetic field. Finally, because of the large inner radius of the warm bore, the axial magnetic field is enhanced with two thick iron yokes set at both the injection and the extraction sides of the source. Figure 1 shows the magnetic field line contours of the system.

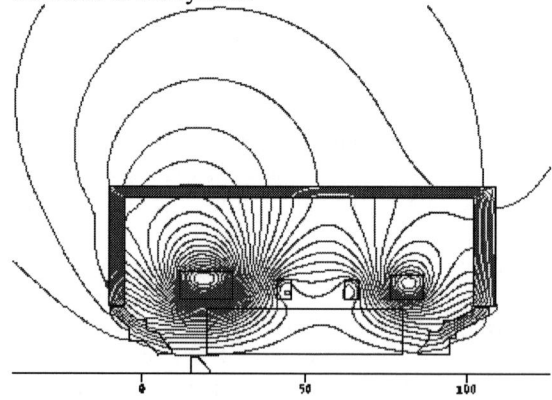

Figure 1: magnetic field flux line contours obtained from Poisson code

TABLE 1. Main parameters of the hybrid ECRIS in comparison with one fully superconducting and one room temperature source

	SERSE	**GTS**	**Hybrid ECRIS**
Frequency	14 – 18 GHz	14 – 18 GHz	18 – 24 GHz
Resonance length	14 GHz : 50 mm 18 GHz : 65 mm	14 GHz : 95 mm 18 GHz : 145 mm	18 GHz : 70 – 160 mm 24 GHz : 90 – 233 mm
Plasma chamber	L : 500 mm Ø : 130 mm	L : 300 mm Ø : 80 mm	L : 600 mm Ø : 100 mm

Table 1. A large plasma chamber diameter allows long ion lifetimes which are necessary to get high charge states. A long resonance zone would also enhance the extracted current. Finally the whole source is based on

Each coil has its own power supply, and then several functioning points are possible. Most of them are presented in table 1. In Figure 2 are presented different magnetic field profiles corresponding to different modes of table 2 while figure 3 presents the whole axial profile of the magnetic field

TABLE 2. possible coil currents giving magnetic profiles presented in Figure 2

Mode	Injection coil (A/mm^2)	Central coils (A/mm^2)		Extraction coil (A/mm^2)
		Coil 1	Coil 2	
A	96	-86	-86	80
B	96	-78	-78	80
C	96	-84	-84	98
D	96	-63	-63	98
E	76	-70	-70	75
F	76	-60	-60	75

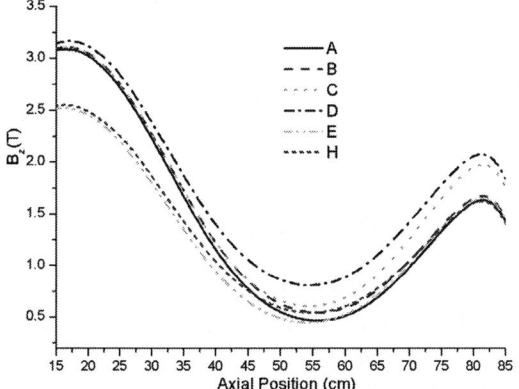

Figure 2 : Possible magnetic field distributions

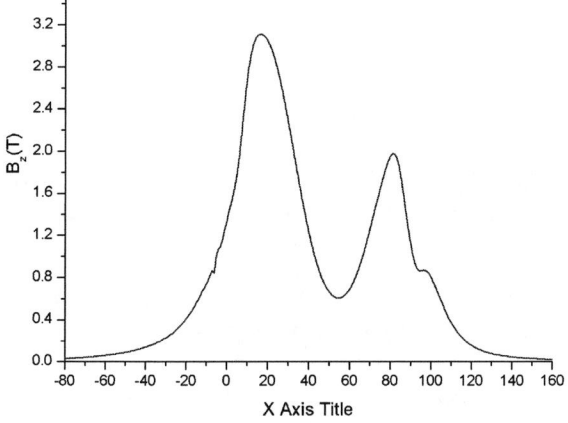

Figure 3 : axial magnetic field of the hybrid ECRIS

Furthermore, with an adjustment of the position of both iron yokes at injection and extraction side, the unbalanced electromagnetic forces between these iron yokes (the outer magnetic field flux circuit yoke and both additional yokes) and the solenoidal coils can be minimized. These yokes also work as the link parts between the hexapole and the other components. It must be noticed that coils are calculated to be operated within a 20 % safety margin.

For all possible operating modes listed in Table 1, unbalanced forces are 6.5 tons (A), 6.3 tons (B), 1.4 ton (C), 0.8 ton (D), 1.9 ton (E), 2.9 tons (F). Then C, D, E and F are the safest modes which ensure a source operation without quench.

Hexapolar Field

A superconducting hexapolar structure would require a too large cryostat, which could not fit with the constraints on the high voltage platform. Therefore a hybrid system with a permanent magnet hexapole has been chosen. Dimensions of this magnetic system are: ID= 114 mm, OD = 390 mm, length = 650 mm. This system made of 36 segments is shown in Figure 4. Because of the small size of the inner parts of the magnet and because of the strong opposite radial field created by coils and of the large axial field, some parts of the hexapole need to be built with magnets having a very strong coercive force. And then 3 different magnet qualities are used. At plasma chamber wall, calculated radial field is 1.55 T which is in agreement with our magnetic scaling laws at 18 GHz.

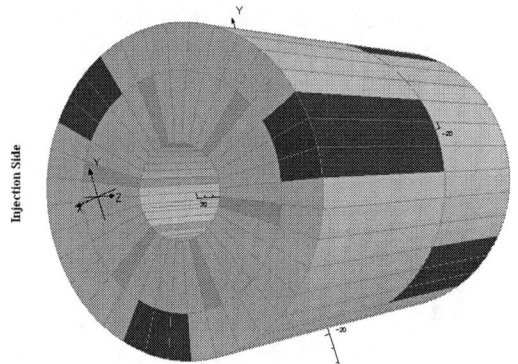

Figure 4 : 36 poles hexapolar system. light grey magnets, B_r =1.44 T, H_{cJ} = 1195 kA/m. black magnets, B_r = 1.14 T, H_{cJ} = 2865 kA/m. For dark grey magnets, B_r = 1.28 T, H_{cJ} = 1830 kA/m

At 18 GHz, the last closed surface is more than twice the resonant field, which gives an excellent confinement. At 24 GHz the last closed surface is approximately 1.7 T, which still offers a confinement suitable for the production of high charges. It is worth noting that operation of this source at 28 GHz, although possible (since the resonance at 28 GHz is closed), is not optimized for the objective of the ion source. The whole magnetic design is presented in Figure 5.

Figure 5 : whole magnetic assembly showing 4 coils (dark), part of the hexapole, iron plug and iron yoke.

CRYOGENIC STUDIES

As the ion source has to be installed on a HV platform, we cannot use a cryostat cooled by cryogenic fluids from main storage through metallic insulated pipes. The cryostat must therefore be autonomous and two techniques are available for that purpose:
- With liquid circulation: LN_2 and LHe Dewars have to be placed near the cryostat and have to be changed of course before being empty.
- With cryocoolers: the cryostat will have an autonomous operation until a failure of the cryocooler occurs. For high reliability operation, however, it could be necessary to have a liquid circulation available.

The advantages and drawbacks of these two solutions are summarized in Table 3.

TABLE 3. cryo-cooling techniques		
	Advantages	drawbacks
Liquid circulation	- Simple - Low cost manufacturing	- High cost running - Ion source stopped during Dewar exchange
Cryocooler	- Autonomous running - Low cost running	- High cost manufacturing

Cryostat Schematic Design

The cryostat, like the SERSE one, uses liquid helium for the cooling down of the coils, and liquid nitrogen for the thermal shields. This cryostat can be operated either with tanks for the supply of the cryogenic liquids or with cryocoolers. This choice has some consequences on the reliability and availability of the ion source. Regarding the 4K cryocoolers: 2 different technologies are possible:
- Gifford Mac Mahon (GM): it is a simple and reliable technology and the most powerful available cooler of this type is a 1.5W@4.2K
- Gifford Mac Mahon+Joule-Thomson technology where reliability has to be proved: a 3.5W@4.2K JT type device is available

In a first step we have designed a cryostat with only one GM. It works with a natural gravity cooling loop. An open circuit LHe working facility is added (for cooling down and cryocooler redundancy).

For the 77 K cryocoolers, several GM-type devices are possible. The cryostat works with a natural gravity cooling loop. As for LHe we add an open circuit LN_2 working facility (cooling down and cryocooler redundancy). Another cooling down loop is necessary for to cool down the coils 300 K to around 100 K.

CONCLUSION

A new ECRIS whose goal is to deliver high charge states is under study. This machine takes benefits of all recent progresses done in ion source understanding and technology as well. Its design will also allow some flexibility in axial magnetic configuration. Finally to enhance its capability to deliver high charge states, a double frequency (18 – 24 GHz) will be used.

ACKNOWLEDGMENTS

The design of this hybrid ECRIS has been performed under contract CEA-HMI #GR-768152.

REFERENCES

1. S. Gammino et al, Rev. Sci. Inst. 72 (2001) 4090..
2. D. Hitz et al., "Production of Highly Charged Ions with the Grenoble Test ECT Ion Source", Proc. 10th Int., Conf. Ion Sources, Dubna, 2003, Rev. Sci. Inst., 75 5 (2004) 1403.
3. D. Hitz et al. Rev, Sci. Inst. 73 (2002) 509
4. D. Hitz, D. Cormier, J.M. Mathonnet, Proceedings of 8th European Particle Conference, Paris 3-7 June 2002.
5. T. Nakagawa et al, Electron Cyclotron Resonance Ion Sources development in RIKEN, Proc. 10th Int., Conf. Ion Sources, Dubna, 2003, Rev. Sci. Inst., 75 5 (2004) 1394.

Characterisation And Performance Of The CERN ECR4 Ion Source

C. Andresen, J. Chamings, V. Coco, C.E. Hill, D. Küchler, A. Lombardi, E. Sargsyan and R. Scrivens

AB Department, CERN, 1211 Geneva 23, Switzerland

Abstract. To optimise the heavy ion injector for the LHC, a good knowledge of the parameters of the ECR4 ion source and the beam transport in the Low Energy Beam Transport (LEBT) for a lead ion beam is necessary. Results of the emittance measurements of the full beam (O + Pb) leaving the source using a scanning slit and profile monitor will be presented. Furthermore, the emittance of a single charge state after the source has been measured using a solenoid scan coupled to a guillotine and spectrometer. The results for last year's operation for the SPS fixed target physics with indium will also be presented.

INTRODUCTION

Following the success of the light ion fixed target programme at the SPS, an international collaboration was formed to build a heavy ion injector [1] for the CERN complex and most especially for the SPS. This installation became operational in 1994 [2] providing Lead ions to physics. The programme came to an end in 2003 with a final operational period using Indium ions instead of the more usually requested Lead ions.

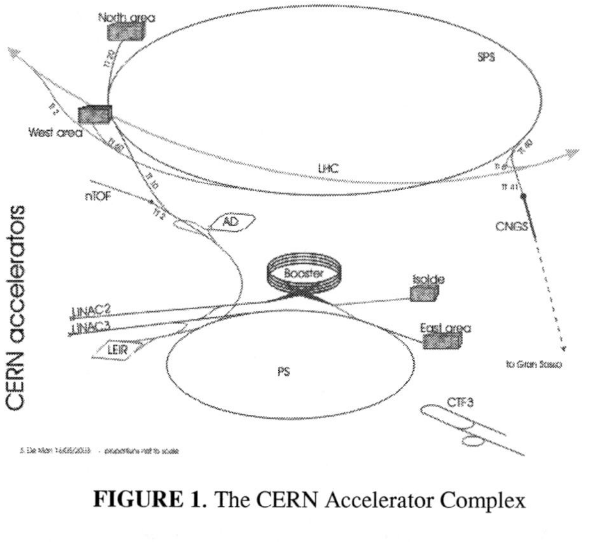

FIGURE 1. The CERN Accelerator Complex

Within the CERN accelerator complex (Fig. 1) the current major accelerator project is the Large Hadron Collider (LHC) which is being constructed in the border zones between France and Switzerland to collide 2 * 7 TeV proton beams deep underground [4]. Although the emphasis for the project has been on protons, Lead ion beams at 2.76 TeV/u have been requested by the physics community. To prepare for this new operating regime a programme of measurements on the ion beam were made to see if there were any fundamental limitations in the linac.

IONS FOR LHC

Using the present system of multiturn injection of the ions into the PS Booster (PSB), there would be a factor of 30 missing in beam brightness. This could be overcome by increasing the source current, but this would result in an unacceptable emittance growth in the beam which would be incompatible with the very tight emittance budget of the LHC injection chain. In view of CERN's considerable experience in phase space cooling, experiments demonstrated that it would be possible to stack and phase space cool heavy ions in a storage ring. By this means it proved possible to accumulate at least 25% of the required number of ions in the ring [4] with the linac running at 2.5 Hz instead of 0.8 Hz. The stacking was limited by the balance between recombination losses due to the vacuum perturbations induced by beam losses and the injected beam. Thus by doubling the linac ion current

CP749, *Electron Cyclotron Resonance Ion Sources–16th International Workshop on ECR Ion Sources* edited by M. Leitner and doubling the repetition rate, the goal could be achieved.

The majority of the components of the linac had been designed for 10 Hz operation so the repetition rate increase would only require the replacement of a few power supplies. The current increase was more problematic within the funding made available. It was felt that an increase in microwave frequency at the source from 14 to 18 GHz, the acceleration of Pb^{25+}, relaxing the beam stability as the source could be tuned at each programmed LHC filling together with finding and compensating aperture restrictions, the equivalent of doubling the current could be achieved. Thus a campaign of beam measurements was started.

SOURCE EMITTANCE

During the design phase of Linac3 the emittance of the beam from the source in the pulsed afterglow mode was assumed to be the same as that as in d.c. operation. Inadequate time resolution of the measuring equipment and a lack of time during source commissioning meant that this was never remeasured. Measurements on the GANIL test bench which included a solenoid after the source indicated 100% emittances of <120 mm.mrad at a waist [1].

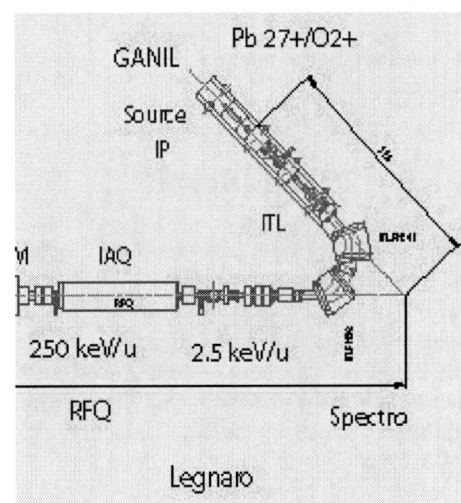

FIGURE 2. Low energy end of the ion linac, Linac3.

Following the end of physics operation, opportunity was taken to install a slit/collector emittance meter 600 mm downstream of the source extraction to investigate the source emittance. Fig. 2 shows the low energy end of the linac and the meter was installed in place of the solenoid after the source. This device had a small integration time (400 μs) with acquisition at a user defined instant in time. The source was felt to be sufficiently stable that emittances did not change during the measurement or during the day. These measurements of the total (O + Pb) [5] beam extracted from the source on a typical operational afterglow beam indicated a 90% geometrical (1 rms) during the afterglow of :- H 142 mm.mrad (22.0), V 192 mm.mrad (29.8). Fig. 3 is a overlay of the fitted 4 rms ellipses for 8 measurements in both planes which demonstrates the stability of the source and its reproducibility.

Of major interest was the evolution of the emittance during the afterglow Fig. 4 shows the evolution with time in the horizontal plane over the typical beam length used by the accelerator. The emittance can be seen to reduce by about 25% during the useful pulse. If a similar phenomenon were to be exhibited in the vertical plane, then there would be a substantial increase in beam brightness. However there was no evidence of an increase in beam intensity out of the linac during the beam pulse which would seem to indicate that there are no major acceptance limitations along the machine.

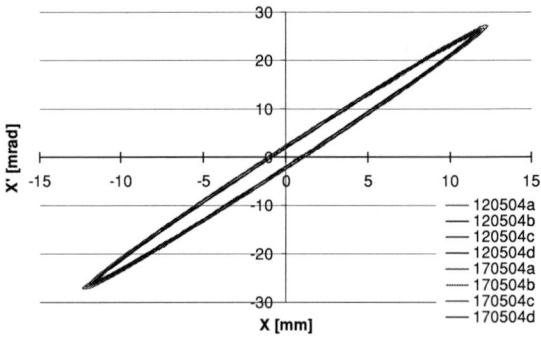

FIGURE 3. 4 rms emittances of 8 measurement (12xx H plane, 17xx V plane)

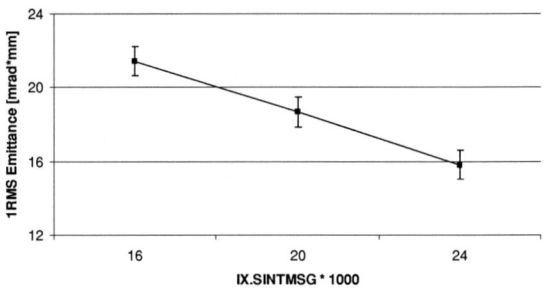

FIGURE 4. Evolution of horizontal emittance during the afterglow. The initial point is at the start of the afterglow. The last 800 μs later. 20 corresponds to PSB beam injection.

Normal operation of the source used an afterglow that maximised both stability and intensity over the injectable part of the beam. Alternate settings could give a very smooth afterglow but at reduced intensity. The rms H,V emittances measured for an operational beam were typically 20.8 mm.mrad, 35.6 mm.mrad whereas for a smooth beam these became 16.9 mm.mrad, 22.3 mm.mrad. The operational beam was adjusted to maximise beam out of the stripper.

The density distribution of the particles in the beam has long been a subject of conjecture and screens in the PSB injection line gave some evidence that the beam, at the end of the linac, was of a strange form. A further observation with a slightly more sensitive screen attached to a parasitic experiment seemed to confirm this view. The images suggested that the beam had a triangular form with three hot spots in the corners of the triangle. The most probable source of this form would be the plasma density distribution due to the hexapolar field in the plasma in the extraction area. Figure 5 is an emittance plot of a typical beam showing the ion density distribution. The beam was observed to consist, typically, of ten separated islands of varying density whose position remained constant over a series of measurements but changed position as source parameters changed. This might suggest that the non uniform distribution already exists at the source exit. Staining and surface erosion on the puller electrode supports this conjecture. However, there is no scintillator screen at this point to verify this possibility

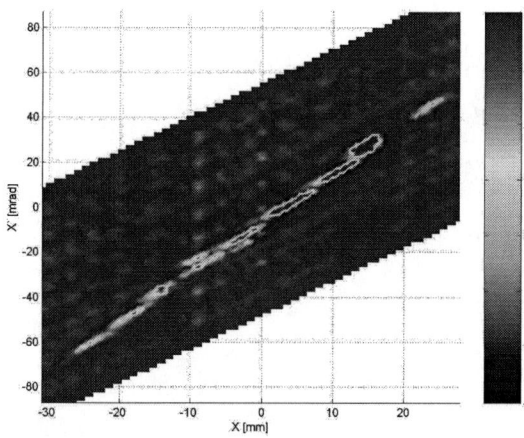

FIGURE 5. Emittance plot showing density islands.

ION BEAM EMITTANCE

A set of slits exists in a diagnostic box just upstream of the first spectrometer magnet and a Faraday cup downstream of the second magnet. Opportunity was taken to use this slit collector system in conjunction with a Charge State Distribution program and solenoid current scanning to measure the width of the beam [6]. Using Trace validated Mathcad routines, the emittance of a single charge state could be computed from this information. These measurements gave a geometrical emittance of 100 mm.mrad for Pb^{27+} and 120 mm.mrad for In^{21+} at the source extraction in the horizontal plane This technique which does not require any additional installation in the existing beam line should prove invaluable later when the linac is upgraded for LHC operation.

INDIUM BEAMS

Physics requested an intermediate mass ion beam for their last period of ion physics. The element selected would need to be mono-isotopic as the mass differences between isotopes would give capture difficulties in the following accelerator, thus wasting ions, and should have melting and vapour pressure qualities similar to Lead. For these reasons Indium was chosen.

Initial tests showed that the optimum charge state from the ECRIS was In^{21+} which resulted in an extraction voltage of approximately 15 kV to meet the injection energy requirement for the RFQ of 2.5 keV/u. This ion was stripped to In^{37+} after the linac. Typically 70-90 eµA of In^{21+} was injected, after mass analysis and matching, into the RFQ

PROBLEMS

To obtain a vapour pressure of around 10^{-5} mbar in the plasma chamber, the indium sample must be heated to around 650 °C. Unfortunately, molten indium is capable of wetting many glasses and ceramics, a phenomenon accentuated by the vacuum environment inside the source. This wetting action caused the formation of short circuits inside the micro-oven used to evaporate the sample which resulted in considerable instabilities in the beam. Indium oxide charges were tried but the sample lifetime proved too short to be useful (<24 hours).

A solution to the problem was found in regulating the power delivered to the oven. For a given operating temperature the power needed to maintain this temperature is essentially constant. Variations may occur as emissivity changes with time but this is a slow process. Thus to a first order, the oven temperature should be constant for a fixed heating power. For lead operation, constant voltage control of the oven heating was sufficient to maintain source stability. For Indium this was not sufficient due to the varying impedance of the oven and thus a power controller was developed [7]. Once this had been installed the beam the beam intensity was generally reproducible to ± 10% (as compared to ± 5% for lead).

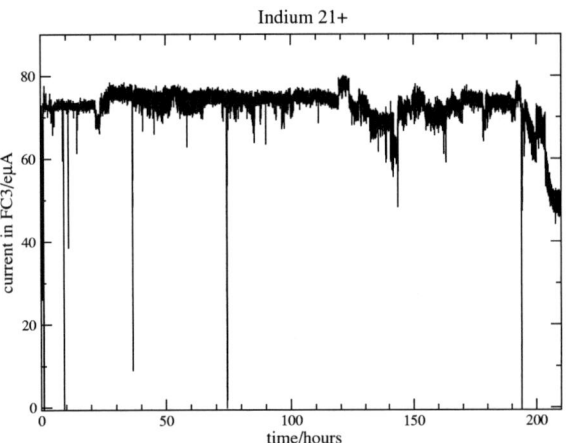

FIGURE 6. Typical long term stability of the Indium beam after the RFQ

Running the oven at a higher temperature resulted in a much reduced lifetime of the oven filament. Power control helped in this respect but tests showed that an operational life of 8 ± 4 days could be expected.

PERFORMANCE

The Indium beam was scheduled for a total run time of 1141 hours continuous. Of this time 28 hours were lost due to machine faults and another 28 for oven replenishments. Opportunity was taken from the presence of weekly LHC proton beam study sessions on the SPS to change the oven and sample at these times. This also ensured the presence of qualified personnel for the operation. Over the period in question the ion beam was maintained around 80 eµA which was more than sufficient to satisfy the requirements of physics [8].

REFERENCES

1. D. Warner, Editor. "CERN Heavy-Ion Facility Design Report", CERN 93-01, 1993.
2. H. Charmot et al. "Operational Experience with the CERN Hadron Linacs", Proc. 18th Int. Linac Conf., Geneva, 1996, CERN 96-07, 360, 1996.
3. O. Brüning et al .(Eds) "LHC Design Report, Vol 1, The Main Ring", CERN 2004-003, 2004.
4. J. Bosser et al. "Experimental Investigations of Electron Cooling and Stacking of Lead Ions in a Low Energy Accumulation Ring" Particle Accelerators, 63, 171, 1999.
5. C. Andresen et al., "Measurement of the ECR4 Ions Source Emittance", CERN AB-ABP, to be published.
6. J. Chamings, "Measurements on Pb^{27+} Sources for the CERN Heavy Ion Injection Chain", M. Phys. Thesis, University of Surrey, GB (AB-Note-2004-44 (ABP), 2004).
7. C.E. Hill, D. Küchler, M. O'Neil, "New Beam for the CERN Fixed Target Heavy Ion Programme", Proc. 15th Int. Workshop on ECR Sources, Jyväskylä, 2002, JYFL Research Report 4/2002, 2002.
8. J. Chamings et al., "New Operational Beam for the CERN Heavy Ion Programme", Proc. 10th Int., Conf. Ion Sources, Dubna, 2003, Rev. Sci. Inst., 75,. 2004.

A 2.45 GHz Singly-charged ECR Ion Source for RIB Production

H. Y. Zhao [1,2*], X. Z. Zhang [1], H. W. Zhao [1], Z. M. Zhang [1],
L. T. Sun [1], H. Wang [1], B. H. Ma [1]

[1] *Institute of Modern Physics (IMP), Chinese Academy of Science, Lanzhou, 730000, P. R. China*
[2] *Graduate School of the Chinese Academy of Science, Beijing, 100039, P. R. China*

Abstract. The stripping mode for production of multi-charged radioactive ion beams is discussed, and its advantages are also discussed. A 2.45 GHz ECRIS serving as the primary ion source was designed for production of singly charged radioactive ion beams. This source is compact, reliable, and economical. The magnetic field of the source provided by two permanent magnet rings has been calculated with POISSON code. Additionally, the structure of the source is presented.

INTRODUCTION

In recent years there has been an increasing interest in investigating the properties of atomic nuclei and nuclear reaction. It is well known that the use of secondary beams of radioactive nuclei enables these researches. For the purpose of development of RIBLL (Radioactive Ion Beam Line of Lanzhou) in IMP, the production of secondary beams of radioactive nuclei is one of our necessary and important goals. In order to produce multi-charged radioactive ion beams, two ionization steps, which have been proposed by some laboratories [1, 2, 3], are recommended: a compact 2.45 GHz ECR ion source is used to produce singly charged radioactive ions, the ion beams of the wanted species are selected by an analyzing magnet, then the analyzed ion beams are decelerated and injected into a high charge state ion source (ECRIS or EBIS) for further ionization.

The first step of the project is to build a 2.45 GHz ECR ion source, which should be reliable, inexpensive, and flexible. The magnetic field of the ion source provided by two permanent magnet rings has been calculated. Since its radial structure is considerably simple, the ion source can be located as close as possible to the radioactive atom production target to save transport time.

THE PRODUCTION OF MULTI-CHARGED RADIOACTIVE IONS

To obtain radioactive ion beams, there are varieties of approaches, including external target mode, internal target mode, stripping mode, etc [4]. Comparing these methods, we decided to use the third one. That is to say, a primary ion source (here, a compact all-permanent magnet ECR ion source), which is adjacent to the radioactive atoms production target, is dedicated to the production of singly charged radioactive ions, then after being selected and decelerated, the ion beam is re-injected into another ion source (ECRIS or EBIS) for multiple charge state ionization (as shown in **Figure 1**).

Considering that the yield of radioactive atoms is very low, the ion source for the production of singly charged radioactive ion beams is expected to have high ionization efficiency. And it is well known that the ionization efficiency of electron cyclotron resonance ion source (ECRIS) is relatively high, which is one of

---

* Corresponding author.
E-mail address: zhaohuanyu@tsinghua.org.cn

the main reasons why we chose an ECR ion source for the production of RIB.

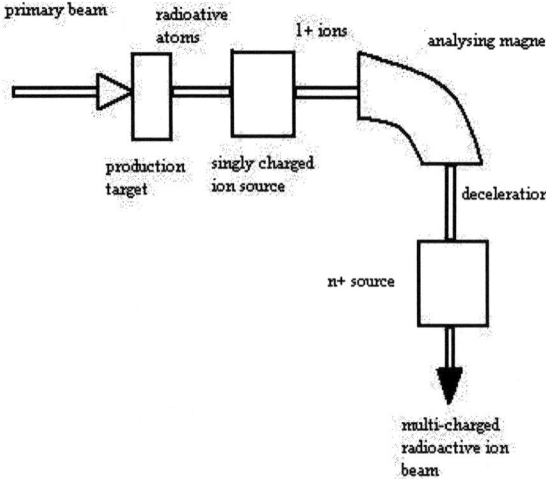

FIGURE 1. Schematic plot of the stripping mode.

As for the production of radioactive ions, there is another problem to be considered. Since the life time of radioactive elements is limited, the ion source should be placed as close as possible to the target in order to save diffusion time. But on the other hand, it seems to be better to have the ion source far from the target if one takes the high radiation environment into account. It is obvious that the stripping mode has advantage to solve this dilemma: it combines the safety of the ion source producing multi-charged ions (now placed outside of the production cave) with the fast transfer of the radioactive atoms, which can be transformed into singly charged ions very close to the target by a low-cost ECR ion source.

DESIGN OF THE 2.45 GHZ ECRIS

It has been mentioned that the first step of the stripping mode is to produce singly charged ions. For this purpose, a 2.45 GHz ECR source is being developed in IMP. Compared with an ECR source running at higher frequency, a 2.45 GHz ECR source would create a lower ionization factor and at the same time lower electron temperature, thus suppressing the yield of the high charge state ions, and enhancing the yield of singly charged ions [2]. In addition, as the emittance and energy spread increase proportionally to the axial magnetic field at the extraction region and in the plasma where ions are created [5], the emittance and energy spread growth are smaller than those in ECR sources working at higher frequency.

The magnetic configuration of the ECRIS, which is provided by two permanent magnet rings, has been calculated through POISSON program. The magnetization direction of the magnet ring which is close to extraction region is parallel to the main axis of the source, and that of the other ring is slanted. The calculated magnetic field distribution is shown in **Figure 2**. One of the features of this structure is that there is enough room between the two magnet rings, which makes the possibility that the source is close to the production target. Additionally, the simple structure of the source can also considerably decrease the cost. In order to protect the permanent magnet rings from demagnetization due to high temperature, a water-cooled plasma chamber has been adopted.

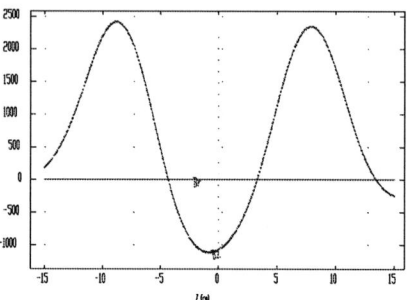

FIGURE 2. The magnetic field provided by two permanent magnet rings.

The schematic structure of the ion source is shown in **Figure 3**. Microwave power is fed into the plasma chamber through a ridged waveguide with a three-stub tuner, which can concentrate microwave power at the center of the microwave window [6]. The matching of microwave can be adjusted through the tuner.

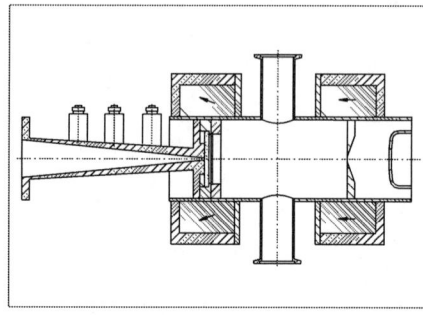

FIGURE 3. Mechanic sketch the 2.45 GHz ECR ion source.

The extraction system consists of two electrodes. The diameter of the plasma electrode hole, which is located at the point of 1700 G magnetic field, is 6 mm.

The plasma chamber has a diameter of 90mm and length of 148mm. It is obvious that the source is very compact and flexible. These characteristics make it suitable to produce singly charged radioactive ions. At the same time, because of its simple structure its cost is low. Considering the high temperature and irradiation environment, the life time of the source of any type is limited, thus it is reasonable to use one more inexpensive.

EXPERIMENTS IN FUTURE

It has been pointed out that this 2.45 GHz ECRIS is designed for the production of singly charged radioactive ion beams, so the ionization efficiency of singly charge ions is one of the issues we concern about. Therefore the ionization – especially singly ionization efficiency is planed to be measured, and the measured results will be compared with those of a 2.45 GHz intense proton source, which is also under construction.

Considering that the ion beams provided by this source need to re-injected into another source for charge breeding, the emittance is expected to be as small as possible. Thus the measurement of emittance is also one of our future plans.

ACKNOWLEDGMENTS

The support of the key projects of Pilot Project of the Knowledge Innovation Program launched by the Chinese Academy of Sciences through the contract KJCX1-09 and the National Scientific Fund for Outstanding Youth through the contract 10225523 is gratefully acknowledged.

REFERENCES

1. P. Jardin, W. Farabolini, G. Gaubert, J. Y. Pacquet et al, Nucl. Instr. and Meth. B 204 (2003) 377.
2. D. Yuan, D. Jayamanna, M. Dombsky et al, Rev. of Sci. Instr. 71 (2000) 643.
3. A. Efremov, V. Bekhterev, S. Bogomolov, G. Gulbekyan et al, Nucl. Instr. and Meth. B 204 (2003) 368.
4. A. C. C. Villari, Nucl Instr. And Meth. B126 (1997) 35.
5. W. Krauss-Vogt, H. Beuscher, H. L. Hagedoorn, J. Reich, and P. Wucherer, Nucl. Instr. and Meth. A 268, (1988) 5.
6. J. Sherman, A. Arvin, L. Hansborough, D. Hodgkins, et al, Rew. Sci. Instr. 69 (1998) 1017.

The Latest Results of the All Permanent ECR Ion Source LAPECR2

L. T. Sun, H. W. Zhao, Z. M. Zhang, W. He, H. Wang, B. W. Ma, X. W. Ma

Institute of Modern Physics (IMP), Chinese Academy of Sciences, Lanzhou, China

Abstract. A high performance all permanent Electron Cyclotron Resonance Ion Source (ECRIS) LAPECR2 (Lanzhou All Permanent ECR ion source No. 2) has been under construction in IMP for one year. This ECRIS is running at 14.5GHz. The magnetic field configuration is designed according to the famous magnetic scaling laws: B_{inj} = 1.4T (2.2T with an iron plug at the injection side), B_{ext} = 1.1T, B_{min} = 0.43T and B_{rad} = 1.2T. According to the final design, the source body is φ650mm×560mm in dimension and about 900kg in weight. The source turns out to be the largest all permanent ECRIS in the world. To fabricate such an ion source is really very difficult, however we have succeeded in overcoming the many difficulties and the source has been set up recently. In this article, the typical parameters of the constructed ion source are presented.

INTRODUCTION

A 300kV HV (High Voltage) platform is now under construction in IMP. This HV platform is supposed to deliver intense high charge state ion beams and medium charge state ion beams for atomic physics, material surface research and biology research. A high performance ECRIS is meant to put on this platform. Considering the virtues like simplicity, easy handling, electricity-free (except for the power supply for microwave generator), no need for high pressure cooling water, low running expense and long time running stability, an all permanent ECRIS LAPECR2 was then determined as the choice. The source LAPECR2 should be able to deliver tens of to hundreds of eμA medium charge state ion beams such as C^{4+}, O^{6+}, Ar^{8+}, Xe^{20+} and at least several eμA high charge state ion beams such as Ar^{14+}, Xe^{27+} and etc.

The IMP ECR group has been devoted to ECRIS research for about 17 years [1]. In recent years, with the development and necessity of atomic physics research and material research, compact, simple and easy-handling ECRIS is in great demand. In 2003, we successfully developed an all permanent ECRIS-LAPECR1 for atomic physics research [2]. This source has been put on a 50kV HV platform to deliver intense low charge state ion beams (such as He^+, He^{2+}, B^+, C^+, C^{2+}, O^+, O^{2+}, N^+, N^{2+}, Ar^+...) and medium charge state ion beams like C^{4+}, O^{6+}, Ar^{8+}. The outer dimension of this source is Ø204mm×294mm. **Fig. 1** gives the photo of this source. With the experience obtained from the fabrication of LAPECR1, we can do the job to build a huge all permanent ECRIS. Meanwhile, permanent material in China is comparatively very cheap. Then, we decided to build a huge all permanent ECRIS which will be discussed in this paper.

FIGURE 1. The LAPECR1 ion source in IMP.

THE DESIGN OF THIS SOURCE

This source is supposed to be able to produce intense medium charge state ion beams and high charge state ion beams. Empirically, LAPECR2 should be working at high-B mode. Because of the inflexibility of the magnetic field of permanent magnets, the magnetic field configuration of the source should be carefully designed. The ECRIS Scaling Laws [3, 4] are the good criterions to follow. For optimal operation at 14.5GHz, the axial magnetic field peaks should be no less than 2.1T and 1.05T respectively, and the maximum radial magnetic field at the inner wall of the plasma chamber should be $B_{rad} \geq 1.15T$. Thanks to the development of NdFeB permanent material, N type NdFeB materials with high remanence and good working characteristics are now available. High magnetic field provided by permanent magnetic rings is possible.

Large volume plasma chamber is always favored for high performance ECRIS. The plasma chamber of LAPECR2 is intentionally designed to be as large as possible. After some compromise of the magnetic field and the plasma chamber volume, Ø67mm is determined to be the inner diameter of the plasma chamber, which is a double-wall (3.5mm thickness) structure one for good cooling of the hexapole. With such a large diameter plasma chamber, the magnetic field is hard to get high enough. In our design, N45M type NdFeB material (B_r=1.35T, H_{cb}=-12.7kOe) is adopted for the axial magnet rings, and N45M and N42SH (B_r=1.28T, H_{cb}=-12.0kOe) type NdFeB materials are adopted for hexapole magnet. After some careful calculations, the axial magnetic field is designed to be B_{inj}=1.42T, B_{ext}=1.12T and B_{rad}= 1.22T [5].

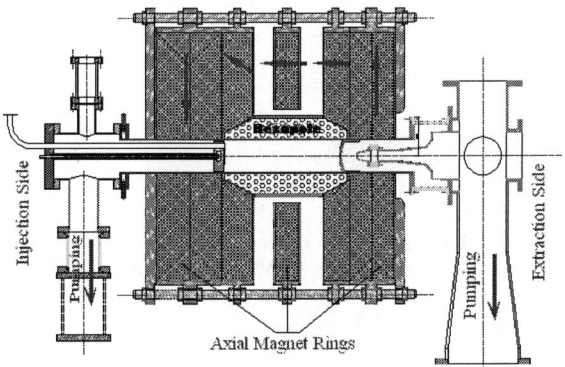

FIGURE 2. Schematic plot of LAPECR2.

As is shown in **Fig. 2**, the pumping system of the source is designed to be simple and effective. The required base vacuum condition of this source should be better than 1.0×10^{-7}mbar. Especially, for the extraction side, higher vacuum condition is necessary for the effective extraction of high charge state ion beam, and thus a 1100L/m pump is used at the extraction side. A rectangular microwave guide is directly inserted to the working region to let effective and stale RF power feeding [6]. The stainless steel (SS) plasma chamber is covered with an aluminum liner at the inner wall to enhance the secondary electron emission efficiency. Other special techniques like biased-disc, triode extraction system, gas mixing and micro-oven are adopted to enhance the performance of the source.

THE PRESENT STATE OF LAPECR2

We have successfully set up the magnetic body of LAPECR2, despite of the strong magnetic forces between the magnetic rings (maximum 100kN). The outer dimension of the source is Ø650mm×560mm, and the weight is high up to 950kg. This source is probably the largest all permanent one in the world. The axial magnetic body is mainly composed of three magnetic rings: the injection magnetic ring, the middle magnetic ring and the extraction magnetic ring. The injection and extraction magnetic rings both consist of 3 rings: two main magnetic rings and a supplementary ring to help increase the magnetic field and enhance the magnetic gradient. We have measured the axial magnetic field, and the distribution is shown in **Fig. 3** The injection magnetic field is 1.28T which is about 0.14T lower than expected, but thanks to an iron plug at the injection side, it will enhance B_{inj} to 2.2T. The extraction magnetic field is 1.07T that is 0.05T lower than expected. The minimum magnetic field is 0.43T which is exactly what we want.

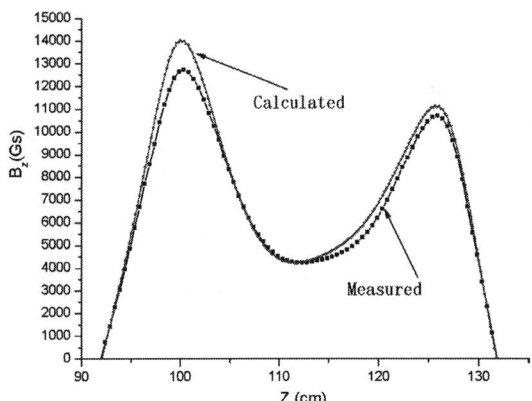

FIGURE 3. Axial magnetic field distribution.

The hexapole is a Halbach structure 36-segmented magnet. To ensure high enough the axial magnetic field and also a good radial magnetic confinement, the hexapole is shaped to a special configuration as shown in **Fig. 2**. The radial magnetic field distribution at the pole is given in **Fig. 4**. Easy to see B_{rad}=1.21T. The distribution of radial magnetic field at the plasma chamber inner wall along the symmetric axis is also measured and the result is shown in **Fig. 5**. According to **Fig. 5**, a good radial magnetic field confinement is available at the ECR region. The B_{rad} at different poles are different according to our measurement, but the difference is less than 5% which is good enough.

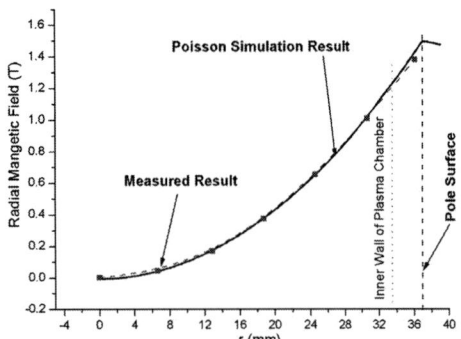

FIGURE 4. Radial magnetic field distribution.

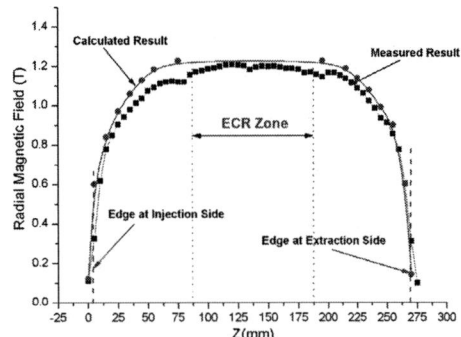

FIGURE 5. Radial magnetic field distribution along the symmetric axis at r=33.5mm.

In **Table 1**, the typical parameters of LAPECR2 are given. Comparing with the parameters of LECR2 [7], we can find that the typical parameters are approximately the same as or better than the ones of LECR2. One can expect that LAPECR2 is able to produce good results for the 300kV HV platform.

TABLE 1. The Comparison of the Typical Parameters of LAPECR2 and LECR2.

Parameters	LAPECR2	LECR2
Magnetic mirror (T)	1.28 (2.2*), 1.1	1.5, 1.1
B-minimum (T)	0.43	0.39
B_{rad} (T)	1.2	1.0
RF	14.5	14.5
RF Feeding mode	Rectangular wave guide	Coaxial
Mirror length	255mm	300mm
Resonance length	100mm	86mm
Plasma chamber ID	67mm	70mm
Chamber material	316L SS	316 SS
HV	25~30kV	25kV

* with an iron plug at the injection side

CONCLUSION

We are constructing a 14.5GHz all permanent ECRIS in IMP. The magnetic body is ready now, which is a Ø650mm×560mm dimension and 950kg weight cylindrical magnet. This source is supposed to deliver intense medium charge state ion beams as well as high charge state ion beams for IMP 300kV HV platform. Many latest techniques of ECRIS are adopted on this source to enhance the performance. The other components of LAPEC2 source like the vacuum tubes, plasma chamber and etc are under processing, and they will be ready at the end of November this year. Some preliminary results are expected to be given at the end of this year.

ACKNOWLEDGMENTS

The support of the key projects of Pilot Project of the Knowledge Innovation Program launched by the Chinese Academy of Sciences through the contract KJCX1-09 and the National Scientific Fund for Outstanding Youth through the contract 10225523 is gratefully acknowledged.

REFERENCES

1. H. W. Zhao, B. W. Wei, Z. M. Liu et al, R. S. I. Vol. 71, No. 2, Feb. 2000, pp. 646-650.
2. L. T. Sun, H. W. Zhao, Z. M. Zhang et al, *Brief Review of Multiple Charge State ECR Ion Sources in Lanzhou*, Proceedings of 12th Int. Conf. on the Physics of Highly Charged Ions, 6-11 September, 2004, Vilnius Lithuania.
3. D. Hitz, A. Girard, G. Melin et al, Proceedings of 15th Int. Workshop on ECR Ion Sources, Jyäskylä, Finland, pp. 100-103.
4. S. Gammino, G. Ciavola, L. Celona et al, R. S. I., Vol. 72, No. 11, Nov. 2001, pp. 4090-4097.
5. L. T. Sun, H. W. Zhao, Z. M. Zhang and D. Hitz, The Design of A High Charge State All Permanent ECR Ion Source, R. S. I., Vol. 75, No.5, May 2004.
6. Z. M. Zhang, H. W. Zhao, X. Z. Zhang et al, Proceedings of 15th Int. Workshop on ECR Ion Sources, Jyäskylä, Finland, pp. 126-127.
7. H. W. Zhao, X. Z. Zhang, Z. W. Liu et al, Proceedings of 14th Int. Workshop on ECR Ion Sources, Geneva, Switzerland, pp. 216-219.

New Design of the ECRIS Plasma Chamber Using a Modified Multipole Structure

P. Suominen, O. Tarvainen, P. Frondelius, V. Nieminen and H. Koivisto

Department of Physics, P.O. Box 35, FIN-40014 University of Jyväskylä (JYFL), Finland

Abstract. It has been found that the radial magnetic field strength plays an important role for the production of highly charged ions with an Electron Cyclotron Resonance Ion Source. Remarkable improvement in the ion source performance has been achieved due to the upgrade of the permanent magnet hexapoles. Unfortunately, methods to increase the radial magnetic field strength are often limited. We have studied a new method, known as the Modified Multipole Structure (MMPS) in which the magnetic field is increased only at the locations where the plasma flux hits the plasma chamber wall. A MMPS plasma chamber for the JYFL 6.4 GHz ECRIS has been designed in order to experimentally study the new idea. This article presents the results of simulations, along with the design and features of MMPS plasma chamber.

INTRODUCTION

The need of highly charged and intense ion beams requires the maximum performance of Electron Cyclotron Resonance Ion Sources (ECRIS) [1]. It has been found that the most straightforward way to improve ECRIS performance is to use a higher microwave frequency. However, this requires a stronger magnetic field in order to satisfy the scaling rules [2, 3] for the optimum magnetic field configuration. In most cases the radial magnetic field strength of a conventional ECRIS is too weak for the production of highly charged ion beams. For example, a remarkable improvement in the performance of the ANL 14 GHz ECRIS2 (Argonne National Laboratory) was obtained after upgrade of the hexapole [4].

The magnetic field is usually designed so that the multipole field strength at the plasma chamber wall is quite homogenous. This work presents simulations and a plasma chamber design in which the radial component of the multipole field is much stronger than the azimuthal component, i.e. the multipole field is not homogenous. According to the simulations described in references [5] and [6], a small-cross section high-permeability material block can significantly increase the radial magnetic field at the magnetic pole, i.e. the location where plasma flux intersects the chamber wall. With a special design (described in reference [6]) a small iron block can boost the radial magnetic field inside the plasma chamber to a value of over 2 T while the magnetic field elsewhere stays relatively homogenous, at around 1.3 T. The effect of iron is very local, only a few millimeters wide, and strongly depends on the shape and the size of the iron block. The aim of this method is to improve the magnetic confinement of the plasma by increasing the radial mirror ratio (B_{rad}/B_{ecr}). Experiments with a fully superconducting ECRIS [2, 3] have shown that the strength of the radial magnetic field should be about 2.0 – 2.2 times the resonance field B_{ecr}. As all these experiments are made using a homogenous multipole field the effects of local magnetic peak at the magnetic poles is not yet known. To study these effects a new MMPS plasma chamber has been designed for the JYFL 6.4 GHz ECRIS.

1. MODIFIED MULTIPOLE STRUCTURE – JYFL-MMPS

The idea of the JYFL-MMPS [5, 6] is to guide the magnetic field lines into the plasma chamber using a material of high permeability, usually iron. The most special feature in our plasma chamber design is the inclusion of 4 mm × 20 mm × 491 mm iron bars at the magnetic poles. In the ECRIS, the optimization of the design has to be made using a 3D-simulation as a relatively high solenoid field is also present, which causes saturation of the iron. Simulations were performed using the Radia-code [7]. Some of the simulations can be compared with 2D-simulations performed with the FEMM-code [8]. According to FEMM and Radia simulations the iron poles can boost the radial magnetic field from 0.5 T up to 0.9 T for the design shown in figure 1. One special feature is also that the position of the iron poles can be changed online allowing the strength of the radial magnetic field to be adjusted without breaking the vacuum.

2. PLASMA CHAMBER DESIGN

The new design of the plasma chamber consists of two parts. The outer part holds the magnet array and the inner part is a separate vacuum vessel. The inner chamber is made of a 140 mm aluminum tube (see fig. 1), which can be easily detached (for example to clean the plasma chamber mechanically). The most critical part of the inner chamber is at the iron poles where the aluminum wall thickness is only 1 mm. Almost the whole heat load is concentrated at these locations due to the plasma flux. In order to keep magnets as cool as possible twelve cooling pipes are embedded into the inner part of the chamber. According to the 2D heat transfer simulations performed with FEMLAB [9], the cooling pipe close to the iron pole is so efficient that the temperature of the outer surface of the inner part rises only a few degrees. Consequently, there is no need for separate cooling of the magnet array. However, if the iron pole is in contact with inner chamber, heat conduction through the iron pole could heat up the magnet holders and damage the magnets.

Sensors will be attached to the magnet holders to avoid damage and to monitor the magnet temperature.

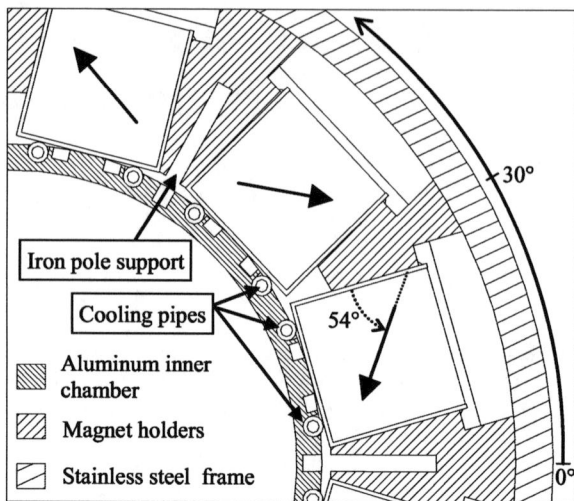

FIGURE 1. Plasma chamber design.

The hexapolar magnetic field is formed by arranging the magnet bars as shown in figure 1. The permanent magnet material is NdFeB of grade N48. The magnets are glued into stainless steel boxes, which are supported by holders. The magnet holders are mounted to a stainless steel frame (cylinder), which is the main supporting structure of the outer part. The magnet holders enable the radial position of each magnet to be adjusted within a range of 8 mm, which allows the strength of the multipolar magnetic field inside the plasma chamber to be changed between 0.4 T and 0.55 T. The direction of magnetization is 54° (see fig. 1), which gives maximum homogeneity of the magnetic field at the plasma chamber wall without the iron poles.

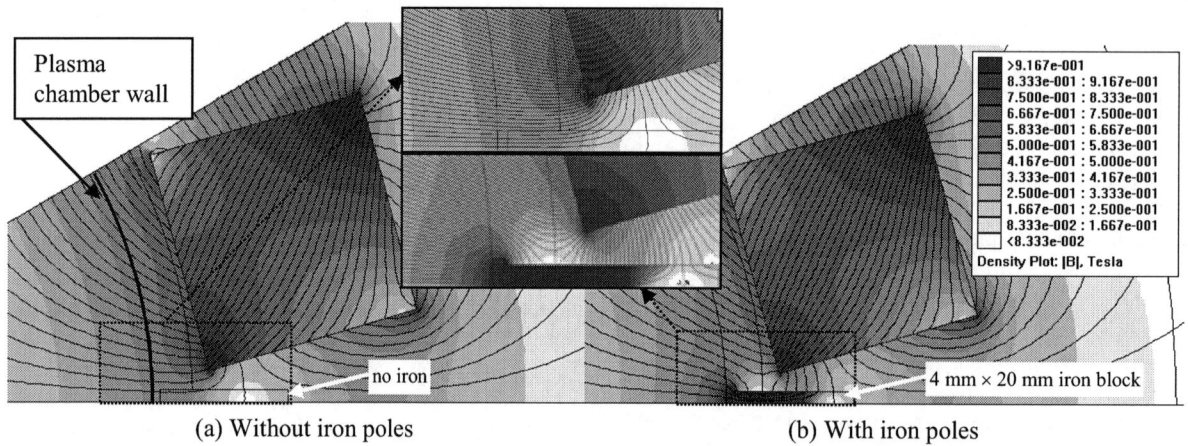

FIGURE 2. The magnetic field simulated with FEMM (2D, no solenoid field). Part (a) shows the typical structure, part (b) shows the MMPS with the additional iron block.

All dimensions in this article are given in cylindrical coordinates as well as the magnetic field components, i.e. $B_{tot} = \sqrt{B_r^2 + B_t^2 + B_z^2}$. In this notation r, t and z correspond to the radial, azimuthal and axial components of the magnetic field. Due to the simulation input geometry, a value of z = 25 mm corresponds to the height where the solenoid field has only a B_z-component. Consequently, the injection and extraction z-coordinates for the JYFL 6.4 GHz ECRIS are z_{inj} = 280 mm and z_{ext} = -190 mm, respectively. The plasma chamber inner radius is R_{pc} = 70 mm. The notation shown in figure 1 is used for the azimutal coordinates ($\theta = 0°$ at the magnetic pole).

Figures 2, 3 and 4 show that the effect of the iron pole is very local, only a few millimeters wide. The work described in reference [10] shows that the ion flux is concentrated into the center of the plasma flux at the plasma chamber wall. Because ions tend to follow electrons in the quasi-neutral plasma flux it can be assumed that the electron flux is also concentrated into a small area and even a narrow magnetic field peak at the magnetic pole can reflect a remarkable proportion of the electrons back to the plasma.

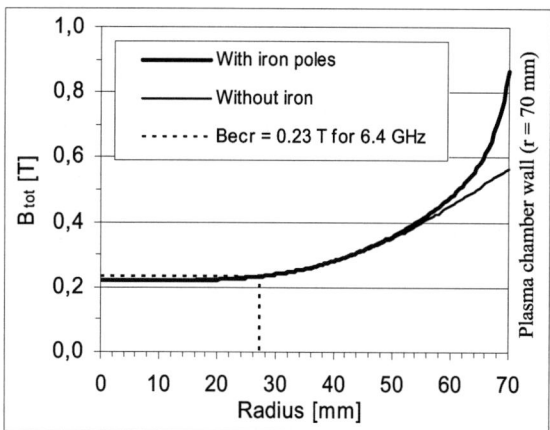

FIGURE 4. Effect of the iron pole on the magnetic field profile (z = 0 mm, θ = 0°) simulated with Radia (3D, with solenoid field).

In the new plasma chamber design the radial position of the iron poles can be adjusted so that the distance between the iron and the plasma chamber inner wall is between 1 and 12 mm. Figure 5 shows the influence of the position of the iron block on the magnetic field at the magnetic pole (at the plasma chamber wall). When the distance is greater than 5 mm the iron pole does not affect the magnetic field inside the plasma chamber.

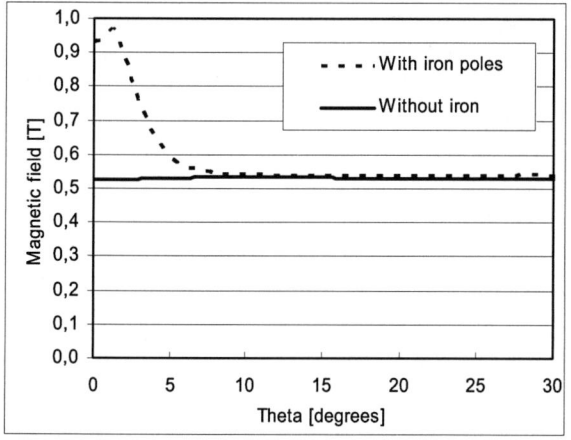

FIGURE 3. The influence of iron poles on the magnetic field at the plasma chamber wall (r = 70 mm, z = 0 mm) simulated with Radia (3D, with solenoid field).

Figure 4 shows that the electron cyclotron resonance with a 6.4 GHz microwave frequency is at a radius of about 30 mm, where the iron pole does not affect the magnetic field. Therefore the size and the shape of the plasma should not be changed. Rather, the electron density in the plasma should increase due to the higher mirror ratio and lead to an enhancement in the production of highly charged ions.

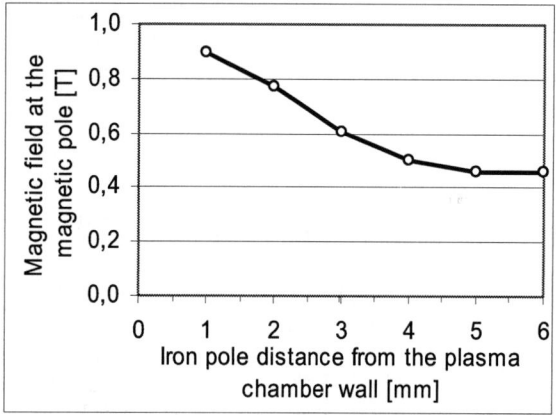

FIGURE 5. The influence of iron pole position on the magnetic field at the magnetic pole simulated with FEMM (2D, no solenoid field).

The general effect of the iron poles can be seen from figure 6 where the total magnetic field B_{tot} and the radial magnetic field B_r are plotted with and without the iron poles at the plasma chamber wall (at the magnetic pole) as a function of the z-coordinate. The axial component B_z is also plotted and is the same in both cases.

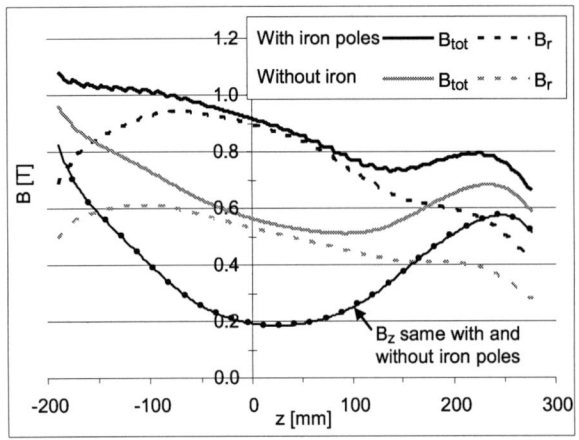

FIGURE 6. Effect of the iron pole on the different magnetic field components at the plasma chamber wall (at the front of the magnetic pole, r = 70 mm, θ = 0°) simulated with Radia.

The simulation shows that the shape of the magnetic field is maintained and the positive effect of the iron poles is very clear along the plasma chamber wall. The average increase of the radial magnetic field B_r is 56 %. Results presented in reference [6] show that the impact spots of the plasma flux are clearest at the height where the value of B_{tot} is at a minimum (at z = 100 mm in fig. 6). This indicates that the strength of the total magnetic field minimum at the plasma chamber wall may also be an important parameter in ECRIS design. By using the JYFL-MMPS B_{tot} increases 50 % and the radial mirror ratio (B_{rad} / B_{ecr}) increases 68 % (from 1.9 to 3.2) at z = 100 mm.

3. SUMMARY AND FUTURE PLANS

Construction of the new plasma chamber has already begun. The chamber should be ready by the end of 2004. Before the new chamber is assembled in the JYFL 6.4 GHz ECRIS the magnetic field and the effect of the iron poles will be measured. After installing the chamber the total magnetic field when the axial field-producing solenoids are energized will be measured. The next step will be to run the ECRIS and study how the new aluminum plasma chamber behaves without iron poles (the original plasma chamber is made of copper). The final step is to study the effect of the iron poles, especially on the production of highly charged ion beams. The first results of the effects of the iron poles are expected in spring 2005. The new MMPS plasma chamber makes it possible to perform new types of measurements and hopefully will reveal new information how the magnetic multipole in an ECRIS should be designed.

ACKNOWLEDGMENTS

This work has been supported by the Academy of Finland under the Finnish Centre of Excellence Programme 2000–2005 (Project No. 44875, Nuclear and Condensed Matter Programme at JYFL).

REFERENCES

1. R. Geller, Electron Cyclotron Resonance Ion Sources and ECR Plasmas, Institute of Physics Publishing, London, 1996, ISBN 0 7403 0107 4.
2. S. Gammino, G. Ciavola, R. Harkewicz, K. Harrison, A. Srivastava, P. Briand, Rev. Sci. Instr. 67 (1996) 4109.
3. D. Hitz, A. Girard, G. Melin, S. Gammino, G. Giavola, L. Celona, Rev. Sci. Instr. 73 (2002) 509.
4. R. C. Vondrasek, R. Scott, R. C. Pardo, Rev. Sci. Instr. 75 (2004) 1532.
5. H. Koivisto, P. Suominen, O. Tarvainen, D. Hitz, Rev. Sci. Instr. 75 (2004) 1479.
6. P. Suominen, O. Tarvainen, H. Koivisto, Nucl. Instr. and Meth. B 225/4 (2004) 572.
7. Radia magnetic field simulation code, version 4.098, copyright of ESRF, Grenoble France also portions copyright Wolfram Research Inc. <http://www.esrf.fr/machine/groups/insertion_devices/Codes/Radia/Radia.html>.
8. D. Meeker, Finite Element Method Magnetics, Version 3.2, Build 3Dec02. Available from <http://femm.foster-miller.com>.
9. FEMLAB, The COMSOL Group, Version 2.3 <http://www.comsol.com/products/femlab>
10. O. Tarvainen, P. Suominen, H. Koivisto, I. Pitkänen, Rev. Sci. Instr. 75 (2004) 1523.

Design and Calculations for the New ECRIS at KVI

H. R. Kremers, J.P.M Beijers, S. Brandenburg, I. Formanoy, J. Mulder, J.Sijbring, H. Koivisto[¶], K. Rantilla[¶]

Kernfysisch Versneller Instituut, Zernikelaan 25, 9747 AA Groningen, The Netherlands.
[¶]*Department of Physics, Accelerator Laboratory, University of Jyväskylä, P.O. Box 35, FIN-40351, Finland.*

Abstract. In this paper a brief description is given of the on-going upgrade of the CAPRICE-type ECRIS injector of the K=600 AGOR cyclotron at KVI. This upgrade is motivated by the new TRIμP program, which requires a significant increase of available beam intensity by up to two orders in magnitude. The upgrade follows the AECR design of the university of Jyväskylä, which was originally pioneered at LBNL (USA). We will discuss the mechanical design and magnetic field calculations of the solenoidal and the permanent magnetic hexapole fields.

INTRODUCTION

Since 2002 a new radioactive beam facility is under construction at KVI, called the TRIμP facility [1]. The aim of this new facility is to produce, separate, decelerate and trap short-lived radioactive ions or atoms for various fundamental physics studies. The radioactive nuclei of interest will be produced and separated by heavy-ion reactions on suitable targets in an ISOL device at collision energies between 6 and 50 MeV/amu. In order to produce the short-lived radioactive nuclei at sufficient rates a significant increase of projectile beam intensity by up to two orders of magnitude is necessary. This is beyond the reach of our present CAPRICE-type ECRIS injector. We have therefore chosen to upgrade the present ECRIS according to the well-known AECR design of LBNL [2], which is also in use at several other accelerator laboratories, e.g. Argonne National Laboratory, Michigan State University, Texas A&M and the university of Jyväskylä. Our upgrade is a close copy of the Jyväskylä ECRIS [3]. The new plasma chamber is actually built by the university of Jyväskylä.

In the following we will discuss the mechanical design in more detail including the modifications with respect to the Jyväskylä design. Also magnetic field calculations of the solenoids and permanent magnetic hexapole will be discussed. The various parts are now under construction and we hope to ignite the first plasma in the first part of the next year.

THE UPGRADE DESCRIPTION

Figure 1 shows a cross section of the new ECRIS. Several subsystems of the present source are being reused, including both injection and extraction solenoids with their 1000 A power supplies, the 2 kW 14 GHz microwave generator and the extraction chamber with electrostatic extraction lenses. All other parts are changed or modified, the most significant change being a new aluminum plasma chamber (the plasma chamber of the CAPRICE source is made of stainless steel) housing the open hexapole configuration characteristic of the AECR design. The new plasma chamber has a much larger volume (76.2 mm diameter and 320 mm length) compared to the CAPRICE source (60 mm diameter and 200 mm length). Furthermore, because of the six radial access ports between the hexapole bars the new plasma chamber can be pumped much better than the closed chamber of the CAPRICE source. Vacuum pumping is done by one 170 l/s turbomolecular pump and two 500 l/s turbomolecular pumps. The new source will be equipped with a moveable biased disk located on axis.

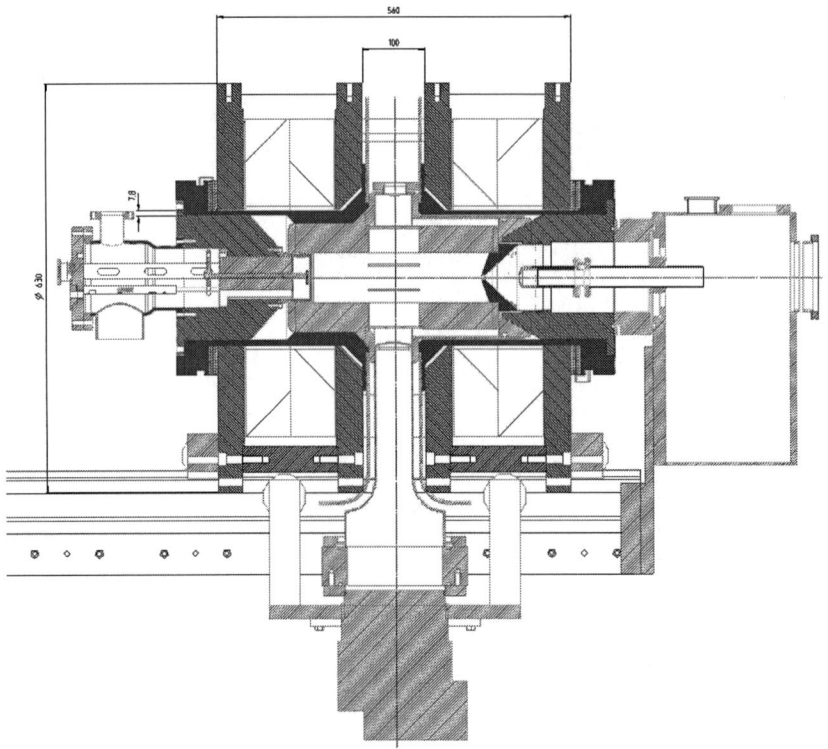

FIGURE 1. The KVI-AECR source, for details see text.

The microwave power is injected via a rectangular waveguide connected off-axis to the plasma chamber. We use the same cooling scheme of the hexapole bars as adopted by the Berkeley and Jyväskylä sources which requires encasing the bars in 0.8 mm stainless steel cans to protect them for corrosion [4,5].

We made a few modifications to the Jyväskylä design, in particular to improve the high-voltage insulation between the inner and outer parts of the source. A better insulation is needed because our source must be able to operate at a maximum voltage of 40 kV. Another modification is made to the extraction side where the hollow iron extraction plug is now on high voltage instead of on ground potential. The reason for this is again to have a longer insulating path length between high voltage and ground potential. Another advantage of this construction is that the plasma electrode is directly mounted on the iron plug with a M104 thread instead of to the aluminum plasma chamber. We can thus change its position without breaking the vacuum.

MAGNETIC CONFIGURATION

The existing injection and extraction solenoids of the CAPRICE ECRIS will be used for the new source. These solenoids have an inner bore diameter of 208 mm, large enough to fit the new plasma chamber. Each solenoid has a total number of 144 windings yielding a maximum of $1.44 \cdot 10^5$ ampere-turns. In order to increase the magnetic mirror ratio two iron plugs are inserted on the injection and extraction sides, respectively. These iron inserts are on high voltage and are separated from the solenoids by the 40 kV insulation. The injection plug is identical to the one used in the Jyväskylä ECRIS, but the extraction plug with its vacuum and high-voltage insulation was redesigned as explained above. The axial magnetic field distribution as calculated with FEMLAB is shown in Fig. 2.

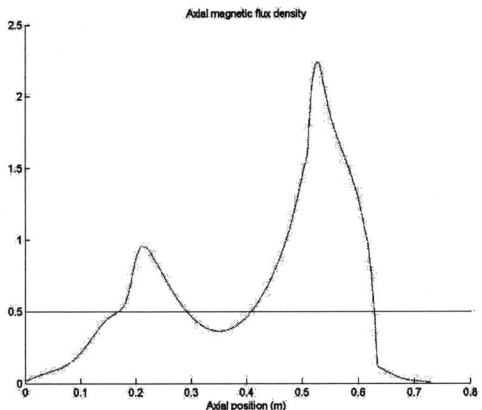

FIGURE 2. Axial magnetic field profile.

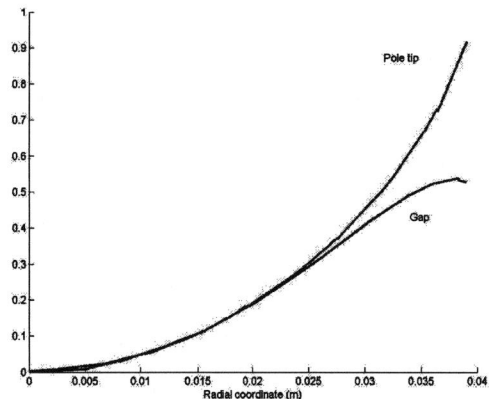

FIGURE 4. FEMLAB calculation of the magnetic field profile along the radius through the pole tips and through the gaps of the open hexapole structure.

The new permanent hexapole consists of 6 Nd-Fe-B magnet bars having a length of 295 mm. Each bar is build up of two segments glued together at the interface. One such segment is shown in Fig. 3. Following the ANL design, the easy axis is set at 40° [6]. When these bars are set in a hexapole configuration they will produce a maximal pole tip field of 0.93 T and a minimum of 0.53 T, see Fig. 4.

Figure 5 shows a three-dimensional picture of the magnetic hexapole. Full 3d calculations of the magnetic structure with the FEMLAB code are in progress. As an example the magnetic field of the hexapole along the plasma chamber wall at the position of the pole tip is shown in Fig. 6. As can be seen the hexapole field extends approximately 5 cm beyond its edges in the axial direction.

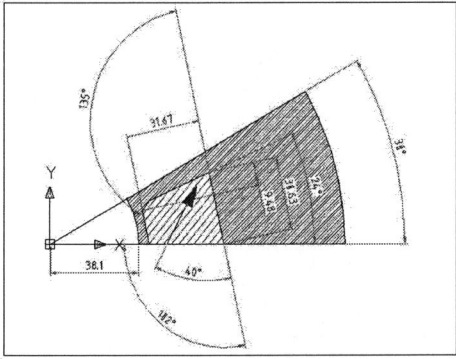

FIGURE 3. One segment of the magnetic hexapole with easy axis set at 40°.

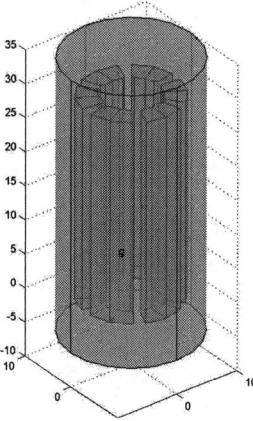

FIGURE 5. The magnetic hexapole structure.

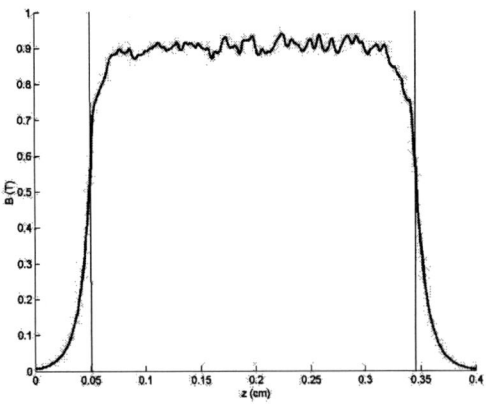

FIGURE 6. The magnetic field of the hexapole along the plasma chamber wall (at radius r=3.81 cm) at the position of the pole tip.

ACKNOWLEDGMENTS

This work is part of the research programme of the "Stichting voor Fundamenteel Onderzoek der Materie" (FOM) with financial support from the "Nederlandse Organisatie voor Wetenschappelijk Onderzoek" (ZWO). It is supported by the Rijksuniversiteit Groningen (RuG) and by the European Union through the Large-scale Facility program LIFE under contract number ERBFMGE-CT98-0125. The authors also gratefully acknowledge support and help from the H. Koivisto, K. Ranttila and J. Ärje from the university of Jyväskylä and R. Vondrasek from Argonne National Laboratory.

REFERENCES

1. Berg, G.P. et al., *Nucl. Instr. Meth. in Phys. Res. B* **204**, 532-535 (2003).
2. Xie, Z.Q. and Lyneis, C.M., *Rev. Sci. Instr.* **65**, 2947 (1994).
3. Koivisto, H. et al., *Nucl. Instr. Meth. in Phys. Res. B* **174**, 379-384 (2001).
4. Xie, Z.Q. and Lyneis, C.M., *Proc. of the 13th International Workshop on ECR sources,* College Station, 1997, p. 16.
5. Koivisto, H., DeKamp, J. and Zeller, A., *Nucl. Instr. Meth. in Phys. Res. B* **174**, 373-378 (2001).
6. Vondrasek, R., private communication (2003).

Permanent Magnet Microwave Source For Generation Of EUV Light

S. K. Hahto, K-N. Leung, J. Reijonen and Q. Ji
LBNL, Berkeley, USA

D. Schneider
LLNL, Livermore, USA

R. Bruch, S. Kondagari and H. Merabet
University of Nevada, Reno, USA

Abstract. A permanent magnet 6.4 GHz microwave plasma generator has been designed and constructed at Plasma and Ion Source Technology group at Lawrence Berkeley National Laboratory for applications in Extreme Ultraviolet Lithography (EUVL). In order to produce 13.5 nm EUV light, Xenon plasma was formed with the goal of producing Xe^{10+} ions, which are associated with the formation of 13.5 nm radiation. The goal was to diagnose the source plasma by extracting Xe- ions from the source plasma and by measuring the EUV light spectrum with a grazing incidence monochromator. 13.5 nm light was observed in the measurements indicating that Xe^{10+} existed in the plasma.

INTRODUCTION

The Extreme Ultraviolet Lithography (EUVL) is considered to be the most favored technique for manufacturing integrated circuits at 65 nm node and beyond[1]. Industry standard has been set on 13.5 nm EUV light. Xenon ions with charge state of ten are found to be responsible for the production of the 13.5 nm line in electromagnetic spectrum. To test if a reasonable concentration of Xe^{10+} ions could be obtained with a simple plasma source, a permanent magnet microwave plasma generator using 6.4 GHz microwave frequency has been developed by the Plasma and Ion Source Technology Group at Lawrence Berkeley National Laboratory (LBNL).

MAGNETIC FIELD SIMULATIONS

To produce a microwave discharge with 6.4 GHz frequency, a simple axial magnetic field structure consisting of two permanent magnet blocks was designed. The goal was to have a resonance volume with 2200 Gauss magnetic field in the middle of the source. Figure 1 shows OPERA-3D[2] simulation of the geometry.

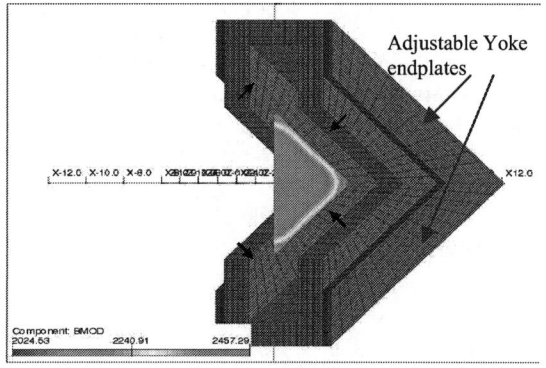

Figure 1. OPERA-3D simulation of the magnetic field created by the two $NdFeB_{38}$ magnet blocks. The magnetization direction of the magnets is indicated by arrows. Half geometry is shown

Each permanent magnet block consisted of four individual magnets glued together so that the magnetization direction of each magnet was radial. The magnetic field could be fine tuned with 8 adjustable yoke endplates so that the resonance magnetic field in the middle of the source could be optimized. Simulations showed that quite large volume of nearly uniform magnetic field strength could be achieved with this simple geometry.

THE CONSTRUCTED ION SOURCE

Figure 2 shows a schematic of the constructed ion source and the completed source at the test stand.

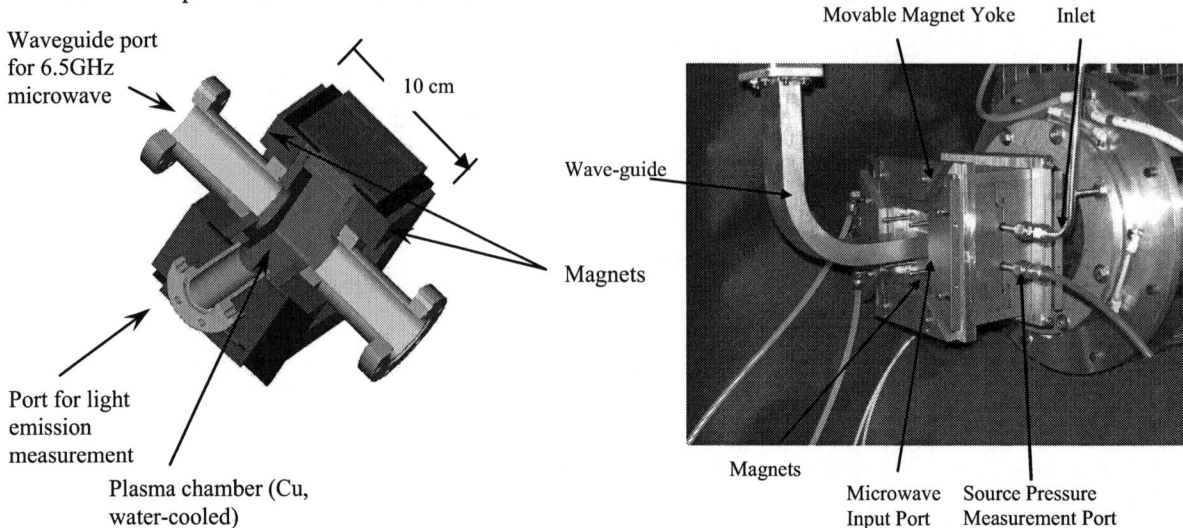

Figure 2. (left) a schematic of the permanent magnet microwave source, (right) the completed source installed at a test stand

The plasma chamber was a square copper chamber which has three vacuum ports. The two magnet blocks are placed on opposing sides of the chamber so that each magnet has a vacuum port going through the center of the block. The third vacuum port is placed transversely to the source axis. Microwaves are injected into the source parallel to the magnetic field lines. The magnetic field was measured in the middle of the source chamber at different movable yoke positions. Figure 3 shows the measured B-field at 0 mm, 10 mm and 15 mm yoke displacements. At 0 mm displacement the yokes are all the way in.

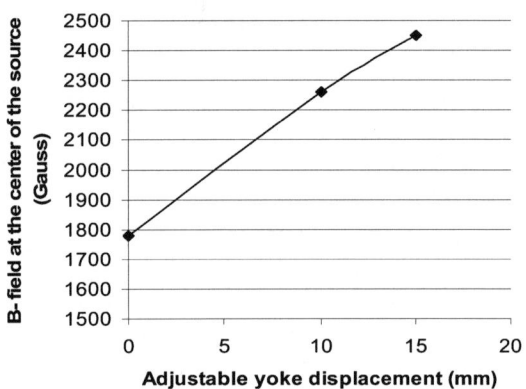

Figure 3. Effect of the moveable yoke displacement to the measured B-field at the center of the ion source.

MEASUREMENTS

The source was operated typically at 250 – 500 W microwave power. Stable plasma was achieved at source pressures of 0.2 mTorr and above. Xenon ions were extracted from the vacuum port opposite to the microwave feed port with a 2-electrode extraction system. The extracted ion beam was analyzed with 60° magnetic dipole mass separator. Figure 4 shows a typical measured Xenon charge state distribution.

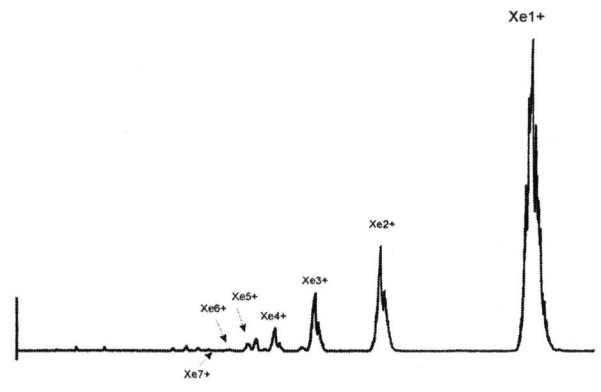

Figure 4. Xenon charge state distribution.

The fractions of xenon charge states Xe^{2+} and higher were not affected noticeably by the microwave power level. With decreasing source pressure the amount of higher charge states increased, as indicated by figure 5. The extracted total current density from the source decreased rapidly at pressures below 0.2 mTorr.

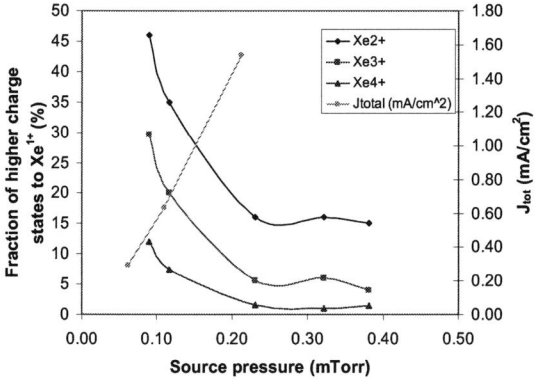

Figure 5. The fractions of some xenon charge states to Xe^{1+} and the total current density extracted from the source through a 2 mm diameter aperture at 250 W RF power.

Xe^{7+} was the highest charge state reliably detected in the extracted ion beam. Any potential Xe^{10+} was hard to detect due to overlapping of the peak with some minor impurity peaks and the low level of the current.

To diagnose the emitted EUV light from the permanent magnet microwave source, a measurement with Acton Research 1.5 grazing incidence monochromator[3] was conducted in collaboration with the University of Nevada, Reno. The monochromator was installed to the transverse vacuum port so that the light emitted from the center of the plasma chamber could be measured. Figure 6 shows the analyzed spectrum over 12.8 – 14 nm range[4].

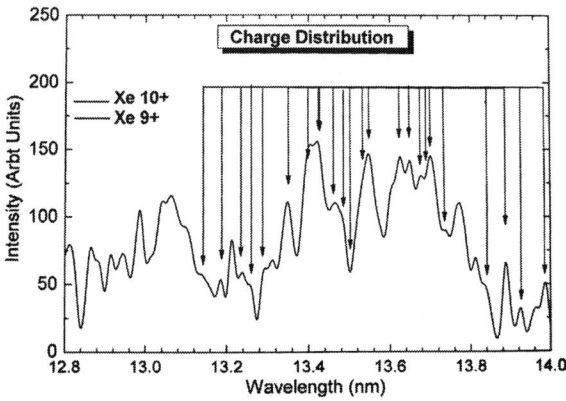

Figure 6. Measured EUV spectrum over 12.8 – 14 nm range. Several peaks were identified as originating from Xe^{9+} and Xe^{10+}.

The EUV measurements confirmed the existence of Xe^{10+} ions in the central region of the plasma. The fact that the Xe^{10+} could not be reliably identified in extracted ion beam is likely due to the fact that the source pressure was relatively high. Also the axial plasma confinement was not optimized for the source leading to loss of plasma particles, especially the fast electrons, to the ion source endplates.

FURTHER DEVELOPMENT OF THE SOURCE

In order to increase the Xe^{10+} fraction the plasma confinement along the magnetic field lines will have to be improved. This would also enable lower pressure operation which would be less detrimental to the high charge states. A simple way of creating a confining magnetic field is by placing straight permanent magnet bars with varying magnetization direction into the front and back plates. Figure 7 shows an OPERA-3D simulation of the added cusp magnets.

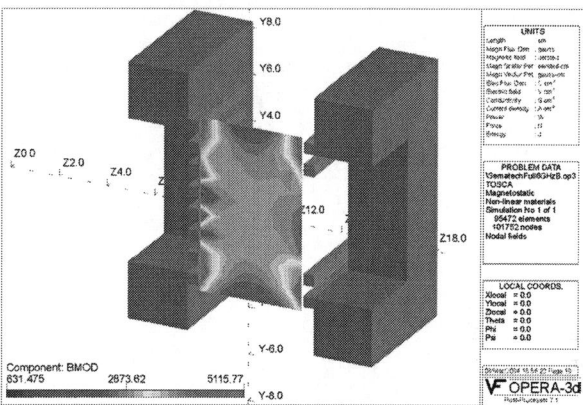

Figure 7. OPERA-3D simulation of the cusp magnets installed into the front and back plates of the source.

CONCLUSIONS

Compact permanent magnet microwave plasma generator was constructed to study the production of 13.5 nm EUV light in Xenon plasma. The existence of 13.5 nm light was verified with a grazing incidence monochromator. Xenon charge states up to Xe^{7+} could be measured in the extracted ion beam. Higher charge states could not be detected due to low peak intensity and overlapping with impurity peaks. In order to improve the high charge state production further optimization of the source, such as gas mixing and improved axial confinement of the plasma, should be implemented.

ACKNOWLEDGEMENTS

This work was supported by International Sematech under contract No. 81FK and Department of Energy under Contract No. DE-AC03-76SF00098. The authors thank Dr. Claude Lyneis of LBNL for the loan of the 6.4 GHz microwave generator and Prof. Reinhardt Bruch of Univ. Nevada, Reno, for providing the EUV measurement instrumentation and analyzing the EUV spectrums.

REFERENCES

1. http://www.intel.com/technology/itj/q31998/articles/art_4.htm

2. OPERA-3D v. 9.0 User Guide, Vector Fields Inc., Aurora IL, USA

3. Operation Instruction Manual, ARC Model GIMS-551.5, Acton Research Corporation, Acton MA, USA

4. S. Kondagari, MSc thesis, Univ. of Nevada, Reno (2004)

Tests of New NIRS Compact ECR Ion Source for Carbon Therapy

M. Muramatsu[1,2], A. Kitagawa[2], Y. Sakamoto[2], S. Sato[2], Y. Sato[2], Hirotsugu Ogawa[2], S. Yamada[2], Hiroyuki Ogawa[3], Y. Yoshida[4], A.G. Drentje[5]

[1] *Graduate school of Mechanical Engineering, Toyo University, Japan*
[2] *National Institute of Radiological Sciences, 4-9-1 Anagawa, Inage, Chiba 263-8555, Japan*
[3] *Accelerator Engineering Corporation, Ltd., 2-13-1 Konakadai, Inage, Chiba 263-0043, Japan*
[4] *Department of Mechanical Engineering, Toyo University, 2100 Kujirai, Kawagoe, Saitama 350-8585, Japan*
[5] *K.V.I, University of Groningen, 9747 AA Groningen, The Netherlands*

Abstract. Ion sources for medical facilities should have characteristics of easy maintenance, low electric power, good stability and long operation time without maintenance (one year or more). Based on the performance of the proto type compact source, a 10 GHz compact ECR ion source with all permanent magnets has been developed. Peak values of the mirror magnetic field along the beam axis are 0.59 T at the extraction side and 0.87 T at the gas injection side, respectively, while the minimum B strength is 0.25 T. The source has a diameter of 320 mm and a length of 295 mm. The result of beam tests showed that a C^{4+} intensity of 530 µA was obtained under an extraction voltage of 40 kV. This paper describes the experimental results for the new source.

INTRODUCTION

Heavy-ion cancer treatment is being carried out at Heavy Ion Medical Accelerator in Chiba (HIMAC) with 140-400 MeV/u carbon ions[1]. A more compact accelerator facility for cancer therapy is now being designed at the National Institute of Radiological Sciences (NIRS)[2]. In order to reduce the size of injector, it is effective to increase the injection energy to the post-accelerator Linac. Also, an ion source requires a higher charge-to-mass ratio and a lower electric power for easy installation of the source on the high voltage platform. A compact ECRIS with all permanent magnets is one of the best types for this purpose. The required beam intensity and emittance for the source are on the order of 500 µA and 0.6 π mm mrad (normalized) for C^{4+}.

At first, prototype compact ECRIS with all permanent magnet (Kei-source) was designed to study its performance in producing C^{4+} [3]. The fixed magnetic field profile of the Kei-source was copied from that of the 10 GHz NIRS-ECR source[4] in the HIMAC, which has already proven to be reliable and stable for the production of C^{4+}. This particular field profile seems to be suitable for the production of C^{4+}, but not for C^{5+} or C^{6+}. Kei-source has satisfied the basic requirements of the medical application. Our previous articles reported the detail of the Kei-source[5,6]. Based on the results of Kei-source, we have designed a new all permanent magnet ECRIS for practical use (Kei2-source), in which the magnetic field and the extraction region are modified for improving the performance.

SPECIFICATIONS OF KEI2-SOURCE

The transmission efficiency in the low energy transport depends strongly on the injection energy due to the space charge effect. It is effective to increase the injection energy to the Linac; an ion source requires a high extraction voltage. Based on the experience on 18 GHz NIRS-HEC[7] designed with the extraction voltage of 50 kV, the Kei2-source has basically a similar structure in this point of view, as shown in figure 1. In order to realize such high voltage, there are two improvements. First, the extraction electrode is cooled directly by water. This is very effective to reduce the outgas from the electrode and to keep good vacuum around the extraction region. Second, all components of the source, including plasma chamber, permanent magnets, all power supplies, and vacuum system, are put into the high voltage box at the extraction voltage. This is necessary to keep a good insulation. These two improvements allow up to significantly improve reliability of the source.

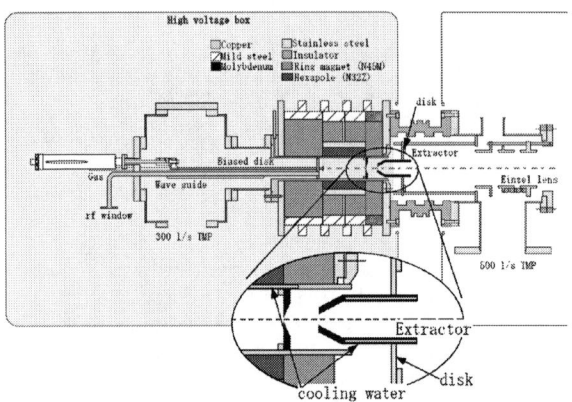

FIGURE 1. Schematic view of the compact ECR ion source (Kei2-source).

General structure of the Kei2-source (size of magnets, vacuum chamber, gas injection, waveguide, plasma chamber, and extraction system) is almost the same as the previous Kei-source. The source has an outer diameter of 320 mm and a length of 295 mm. The mirror magnetic field is set to 0.87 T, 0.59 T, and 0.25 T at the injection side, extraction side, and minimum B region, respectively. The size of the magnets and their arrangement were determined in a way that both peak and minimum values in the mirror field become close to those of the NIRS-ECR source. The magnetic field was calculated using POISSON/SUPERFISH code[8]. The sextupole magnetic field on the plasma chamber surface is 1.1 T. The commercial model N45M (made in Shin-Etsu Co.) is employed as a permanent magnet. A Traveling-Wave-Tube (TWT) amplifier is used in the new source. The TWT amplifier was employed in order to find the optimum frequency under the fixed and uncontrollable magnetic field. Microwave power is fed into the plasma chamber through a rectangular wave guide from axial direction. The TWT amplifier is frequency of 8-10 GHz and can be driven both in cw and pulse mode with the maximum output power of 300 W. A biased disk was also used for optimizing. Many groups reported that the biased disk method is useful for increasing the beam intensity of highly charged ion (HCI)[9-11], see also the review in ref. 12. The diameter of the disk was 8 mm, which was made of molybdenum. The disk position is movable between 7 mm to upstream and 25 mm to downstream the direction from the peak of mirror field (injection side). From our experience this method seems suitable for such compact ECRIS, since the disk does not need a large space for the installation. The plasma chamber is made of copper for a good cooling efficiency, in order to avoid a decrease in the magnetic field due to the high temperature. The plasma chamber is 50 mm inner diameter and 120 mm in length. The vacuum pressure of the gas injection side and beam extraction side are 7.0E-6 Pa and 8.0E-5 Pa, respectively. The estimated vacuum in the plasma chamber is around 1.0E-4 Pa.

BEAM TESTS OF KEI2-SOURCE

The ion source is being operated at a test stand, which consists of an analyzing magnet, four monitor boxes for Faraday cups, horizontal slits, an emittance monitor, and two vacuum pumps (both 500 /sec turbo molecular pump). The analyzing magnet has bending angle of 90 degree, edge angles of 28.8 degree for both ends, and radius of curvature of 0.5 m. The maximum magnetic rigidity is 0.13 Tm. A couple of horizontal slit and a Faraday cup are installed at the object point of the analyzing magnet (FC1) and the other at the focusing point (FC2). The emittance monitor is also installed downstream the FC2. In order to study the basis performance of the source, beam test is being done using CH_4 gas.

In order to obtain the best performance for medium charged ion (such as C^{4+}), it is necessary to know some dependences of the operation parameters. The tuning parameters of the source are the gas flow, microwave power, microwave frequency, biased disk voltage, and its position. Three operation parameters of microwave power, frequency and extraction voltage were examined for optimizing C^{4+} yield under the extraction voltage of 35kV. Figure 2 shows the variation of intensity of C^{4+} at the FC2 versus microwave power of the TWT amplifier. The C^{4+} intensity increased with increasing the microwave power. Second, microwave frequency dependence was investigated. The variable microwave frequency is useful to find a good condition for C^{4+} under the fixed magnetic field. Figure 3 shows the variation intensity of C^{4+} versus microwave frequency. We can find that the optimal frequency for C^{4+} is 9 or 10 GHz. Third, the extraction voltage dependence was studied under the fixed frequency of 10 GHz. The result is shown in figure 4.

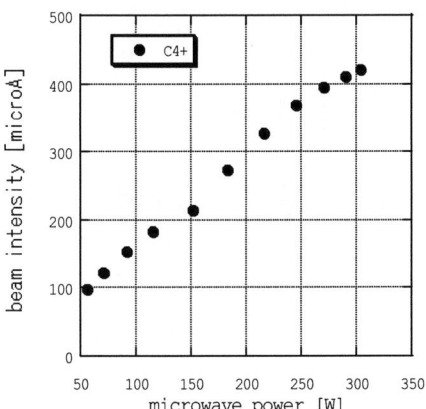

FIGURE 2. Microwave power dependence.

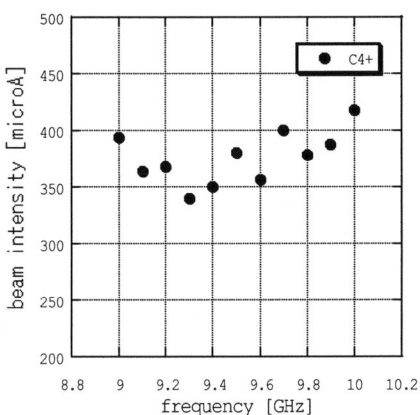

FIGURE 3. Microwave frequency dependence.

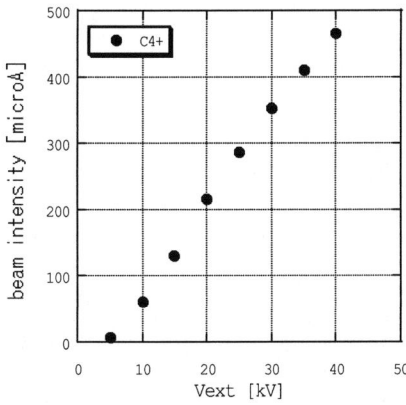

FIGURE 4. Extraction voltage dependence.

The effect of the biased disk was investigated. Figure 5 shows the comparison of the CSD of the carbon ion, comparison between with and without the disk. Operation parameters for with and without disk were optimized for production of C^{4+}, respectively. Operation parameters are as follows. The vacuum pressure at the extraction side of the cases with and without disk were 2.7E-4 and 4.7E-4 Pa, respectively. The microwave frequency of the cases with and without disk were 10.067 GHz and 10.000 GHz, respectively. The microwave power was 300 W. The extraction voltage was 40 kV. In the case with disk, the disk was connected electrically with the plasma chamber and its potential was equal to the chamber wall. Under these conditions, the beam intensity of C^{4+} with and without biased disk were 530 and 360 µA, respectively. This again shows that the biased disk is effective for the HCI production.

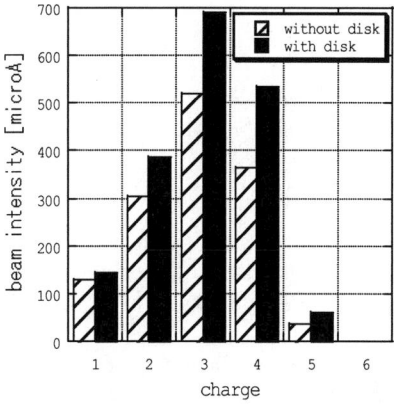

FIGURE 5. CSD of the carbon ion. Comparison between with and without disk.

In any case the beam stability is very important for therapy applications. It is desirable for the source to be operated for one year without serious troubles. A beam stability test has thus been made under the pulsed mode (2Hz, 4 msec) with optimized source parameters for C^{4+}. At the initial beam intensity of 280 µA, the amount of scatter in measured values, this is the difference between minimum and maximum values, is within 6% during 90 hours without adjustment of the operation parameters under the extraction voltage of 30 kV as shown in Figure 6. In this beam test, there was continuous unwanted discharge in the puller at high voltage extraction (above 40 kV). Start of the discharge, it causes of short distance between the end of the plasma chamber and the puller of 5 mm. In the case of the low extraction voltage (around 30 kV), there is no unwanted discharge. At the high voltage extraction, it causes of vapor from Cu puller because of unwanted discharge. Then, the vacuum pressure in the extraction region would be worse. To avoid this problem, beam stability test was being done under the extraction voltage of 30 kV.

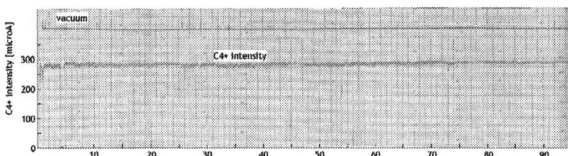

FIGURE 6. Beam stability of the C^{4+} under the extraction voltage of 30 kV.

In addition, He, CH4, O2, and Ar have been tested. Table 1 summarizes the beam performance of the Kei2-source. Beam intensity of 530 µA at the C^{4+} using CH_4 gas was obtained. This result is better than previous compact source (Kei-source). It is clear that the higher magnetic field and the higher extraction voltage were effective to increasing the beam intensity.

TABLE 1. Beam performance of the Kei2-source [µA]

	1+	2+	3+	4+	5+	6+	7+	8+	9+	10+	11+
He	>1000	**540**									
C		335	550	**530**	60						
O		800	815	580	385	210	**77.5**				
Ar			445	295	225	135	80	75	**88**	30	2.5

Bold: the ECRIS was tuned to these charges

In order to increase the beam intensity, a biased cylinder method was tested. This method is effective for increasing the beam intensity of HCIs[13]. Figure 7 shows the photo of a comb shape cylinder. The cylinder has an outer diameter of 48 mm, inner diameter of 47 mm and a length of 85 mm. The cylinder is made from aluminium. The cylinder was installed into the plasma chamber and electrically insulated by ceramic tubes in-between. The comb electrode covers the pole areas. Operation parameters were optimized for Ar^{7+} under the argon gas with oxygen as a mixing-gas. Result of the beam test is shown in figure 8. It is clear that the biased cylinder is effective for increasing the intensity of HCIs under the positive cylinder voltage of 70 V. However, beam intensity of HCIs has not yet reached the best record. It seems that the sufficient microwave power did not enter the ECR plasma, due to the small inner diameter of cylinder (i.d.=38 mm at the end of cylinder).

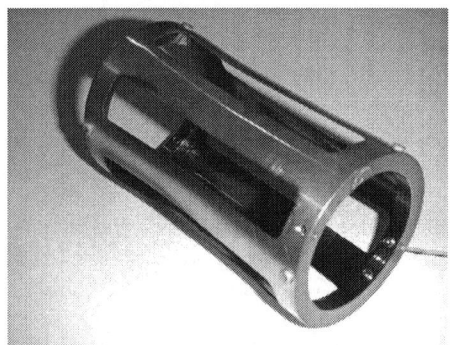

FIGURE 7. Photo of a biased cylinder.

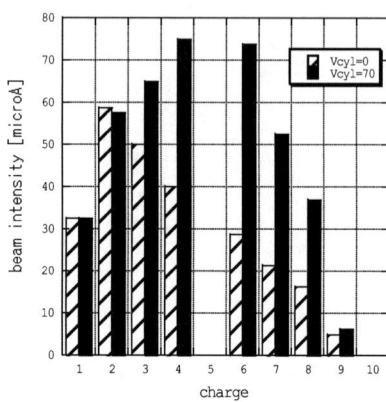

FIGURE 8. CSD of the argon ion. Comparison between the cylinder voltage 0 V and 70 V.

CONCLUSION

We developed a new compact ECR ion source with all permanent magnets for high-energy carbon ion facility. Source parameters were optimized for production of C^{4+} based on the experience of the 10 GHz NIRS-ECR ion source and previous compact source. Beam intensity of C^{4+} was obtained 530 µA at extraction voltage of 40 kV. Beam stability was better than 6% at C^{4+} of 280 µA during 90 hours with no adjustment of the operation parameters. However, extraction voltage has not yet reached the requirement values. It is necessary to modify the puller.

REFERENCES

1. Y. Hirao, Proc. of the Int. Conf. on Cyclo. and their Appl., East Lancing, May 2001
2. K. Noda et al., proceedings of EPAC2004, to be published.
3. M. Muramatsu et al., Rev. Sci. Instrum. 71, 984 (2000)
4. A. Kitagawa et al., Rev. Sci. Instrum. 65, 1087 (1994)
5. M. Muramatsu et al., Rev. Sci. Instrum. 73, 573 (2002)
6. M. Muramatsu et al., Proceedings of the 15th International Workshop on ECRIS, 2002, p, 59.
7. A. Kitagawa et al., Rev. Sci. Instrum. 69, 674 (1998)
8. Reference Manual for the POISSON/SUPERFISH Group of Codes, LA-UR-87
9. G. Melin et al., Proceedings of the 10th International Workshop on ECRIS, 1990, p, 1.
10. S. Gammino et al., Rev. Sci. Instrum. 63, 2872 (1992)
11. S. Biri et al., Proceedings of the 14th International Workshop on ECRIS, 1999, p, 81.
12. A. G. Drentje, Rev. Sci. Instrum. 74, 2631 (2003)
13. A. G. Drentje et al., Rev. Sci. Instrum. 75, 1399 (2004)

Traveling Wave Vs. Cavity RF Injection For The ORNL HRIBF Volume-Type ECR Ion Source

Y. Liu, F. M. Meyer, J. M. Cole

Oak Ridge National Laboratory, P.O. Box 2008, Oak Ridge, TN 37831-6368, USA

Abstract. The performance of the ORNL HRIBF volume-type 6 GHz ECR ion source has been significantly enhanced with a new RF injection and cavity configuration: measured charge-state distributions now peak at higher charge states and the intensities of the charge-states $>Ar^{7+}$ were increased more than a factor of two. Details on the source improvement and comparison of the source performances are given in this report.

INTRODUCTION

The all-permanent-magnet, 6 GHz ECR ion source at the Oak Ridge National Laboratory (ORNL) Holifield Radioactive Ion Beam Facility (HRIBF) was originally designed with an RF injection system consisting of a smoothly tapered rectangular-to-circular transition, starting from a WR137 waveguide and terminating with a cylinder whose diameter matched the plasma chamber dimension [1]. High RF power reflection was often observed with this traveling wave structure. The axial magnetic profile of the source features an extended central flat region which is tuned to be in resonance with single-frequency microwave radiation, as shown in Fig. 1, resulting in a large ECR volume [2,3]. However, there exists a confined parasitic ECR zone in the RF injection region as also shown in Fig. 1. This parasitic ECR zone can absorb a significant fraction of the input RF power and prevent efficient RF coupling into the plasma chamber.

A major modification has been made to the source by adding a small iron plug on axis close to the peak of the injection side axial magnetic field and shifting the RF injection to an off-axis position to avoid the parasitic ECR resonance inside the waveguide. The plasma chamber thus became a cavity structure instead of a traveling wave structure for the injected RF radiation. The performance of the source has been significantly enhanced with the new RF injection system. Details of the source improvement are presented below.

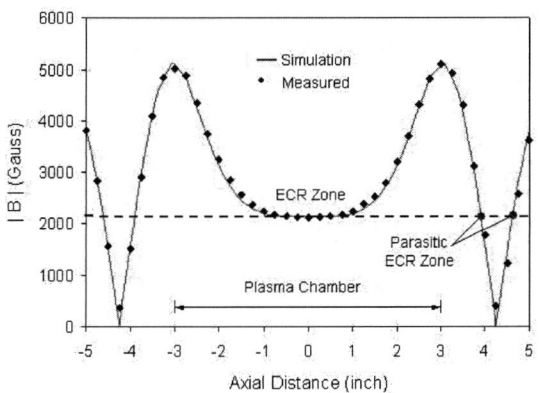

FIGURE 1. Axial magnetic field profile of the volume-type ECR ion source.

SOURCE MODIFICATIONS

A schematic view of source with the new RF injection system is shown in Fig. 2. The tapered waveguide transition section after the plasma chamber was completely blocked with a tapered aluminum cylinder. Thus, the plasma chamber became a cavity structure for the injected RF radiation. An iron rod, 0.5 inches in diameter and 2.5 inch long, was placed inside the aluminum plug on the axis, near the apex of the mirror magnetic field. The iron plug increased the peak mirror magnetic field at the RF injection side to almost one Tesla. The maximum mirror ratio was increased to nearly 5 at the RF injection side. Stronger magnetic fields could provide better plasma confinement and thereby enhance source performance. At a radial distance sufficiently off-axis, the magnetic field was also raised above the ECR value for the 6.4 GHz microwave radiation. Shown in Fig. 3 is the calculated new magnetic field profile at a radial distance r=0.4 inch from the axis. As noted, there is no confined ECR zone at this radial distance. Thus, the parasitic ECR zone in the RF injection region could be eliminated by shifting the RF waveguide off-axis. The new RF injection system consisted of a transition from the WR137 waveguide to a WRD580 double-ridge waveguide. The double-ridge waveguide was tilted at a small angle such that it penetrated the aluminum cylinder off-axis to avoid parasitic ECR resonance.

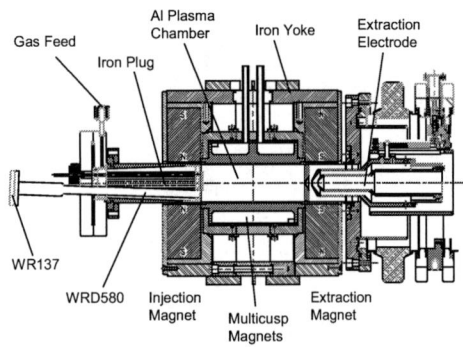

FIGURE 2. Schematic view of the source modified with the off-axis RF injection system.

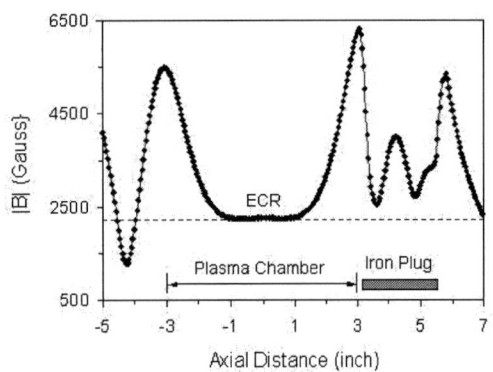

FIGURE 3. The magnetic field profile with the added iron plug at a radial distance r=0.4 inch off-axis.

SOURCE PERFORMANCE

The new RF injection was evaluated using Ar as the operating gas. Ions were extracted from the source at a voltage of 20 kV and mass analyzed with a 45° dipole magnet. The performances of the source in the traveling-wave and cavity configurations were compared based on the observed Ar charge-state distributions, and intensities within a given charge-state with the source optimized individually for optimal high-charge-state ion production.

Traveling-Wave Configuration

Typical Ar charge-state distributions obtained with the traveling-wave RF injection system at different RF powers are displayed in Fig. 4. The charge-state distributions showed an apparent superposition of two distributions as displayed: one peaking at charge q=2+ and the other peaking at q=8+. At lower RF powers (Fig. 4, upper), the first distribution peaked at q=2+ was dominating. With increasing RF power, the charge-states shifted towards higher charges and the second distribution peaking at q=8+ became prominent. At very high RF power, ~1.1 kW forward power, the data clearly showed two distributions in the Ar charge states (Fig.4, lower). About 65% of the 1.1 kW forward RF power was reflected. Such high RF power reflection was often observed with the traveling wave configuration.

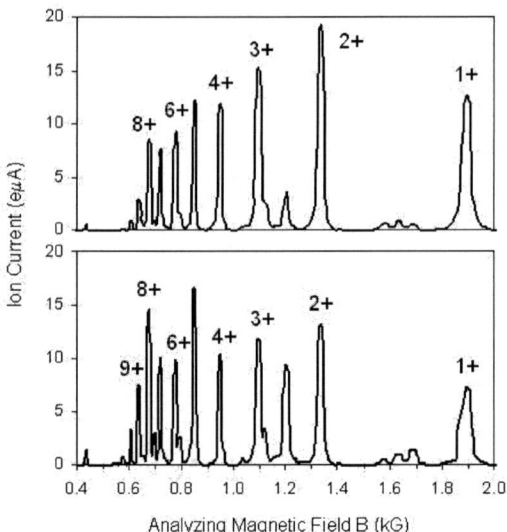

FIGURE 4. Ar charge-state distributions observed with the traveling-wave RF injection system. Upper: forward RF power =210 W, P ~2.7x10^{-7} Torr. Lower: RF power =1.1 kW, P ~ 2.5x10^{-7} Torr.

In general, the charge-state distributions were mostly peaked at Ar^{2+} and Ar^{11+} was the highest charge states observed with intensities > 0.1 eµA. It is believed that contributions to the q=2+ distribution come from the confined parasitic ECR zone located in the tapered RF injection region of the source which injected q=1+ and a small amount of q=2+ into the discharge [4].

Cavity Configuration

The performance of the source has been significantly enhanced after changing to the cavity configuration and off-axis RF injection. Fig. 5 shows an Ar charge-state distribution obtained with the new source configuration. The data were taken under similar operating conditions and with similar RF power as those used for the data shown in Fig. 4 (upper). As noted, the double-peaked feature disappeared and the charge-state distribution was characterized by a single prominent peak at Ar^{8+}. The new source configuration clearly favored the production of higher charge state: the intensities of Ar^{8+} and Ar^{9+} were increased by a factor of 2 to 3 and the highest observed charge-state was increased to Ar^{13+}. Moreover, much less RF power was needed to produce the same high charge intensities than that required for the traveling-wave configuration and reflected RF power was noticeably reduced.

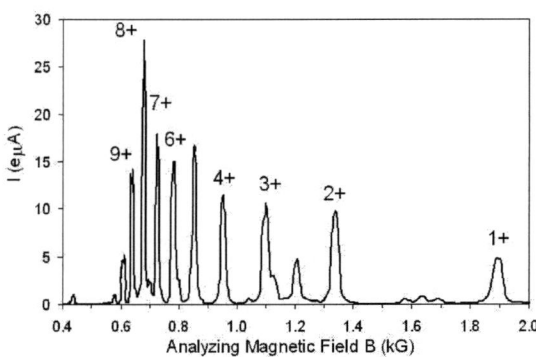

FIGURE 5. Ar charge-state distribution observed with the cavity RF injection system. RF power =110 W, P ~3.8x10^{-7} Torr in the extraction region.

"Surface" ECR Zone Configuration

The magnetic fields of the flat-B source can be reconfigured to that of a conventional minimum-B field configuration that has a parabolic profile in the central region of the source producing an ECR zone that is a closed surface. The performances of the source in the conventional minimum-B mirror magnetic field configuration were also measured for both the traveling-wave and the cavity RF injection systems. Fig. 6 presents the best charge-state distributions for the minimum-B configuration of the source optimized for high charge states with the traveling-wave (Fig. 6, upper) and the cavity RF injection systems (Fig. 6, lower). As can be seen, even more remarkable enhancement in source performance was obtained with the cavity configuration: the intensities of Ar^{7+} and higher charge-state ions were increased by more than one order of magnitude.

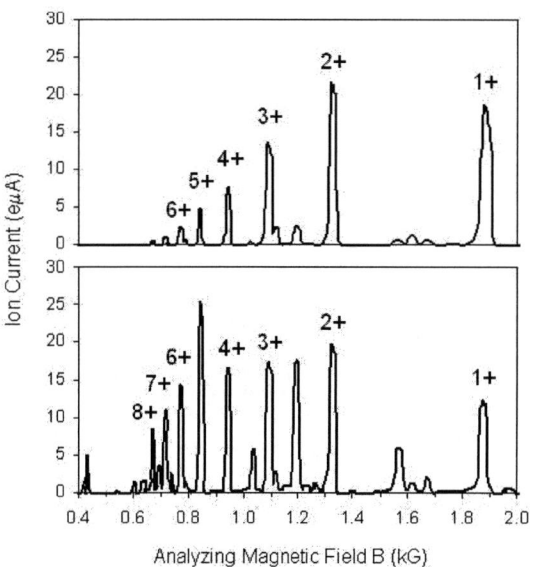

FIGURE 6. Comparison of Ar charge-state-distributions for the conventional mini–B configuration. Upper: traveling-wave RF injection, RF power =70 W, P~ 1.7x10^{-7} Torr. Lower: cavity RF injection, RF power = 160 W, P~ 2.8x10^{-7} Torr.

DISCUSSION

The new RF injection has significantly enhanced the performance of our 6 GHz ECR ion source in both "volume" and "surface" ECR configurations. It is believed that the enhancement was a result of three effects. First, the on-axis iron plug increased the maximum mirror ratio from 2 to about 5 at the RF injection side. Higher magnetic fields improve plasma confinement and thereby enhance the production of higher charge states. Second, the off-axis RF coupling eliminated a confined parasitic ECR zone in the RF injection region, reducing RF losses and improving RF coupling into the plasma chamber. Third, the plasma chamber became a cavity structure, instead of a traveling wave structure for the microwave radiation. Thus, more RF power could be stored in the plasma chamber due to the resonator structure, and the ECR plasma could also be shifted toward the extraction region. This cavity configuration will permit the use of a biased disk at the rear of the plasma chamber that has proven to be an effective method for enhancing the charge-state distributions in these sources. Future studies include a biased disk added to the source.

ACKNOWLEDGMENTS

Research sponsored by the Office of Science, U. S. Department of Energy, under contract DE-AC05-00OR22725.

REFERENCES

1. Y. Liu, G. D. Alton, G. D. Mills, C. A. Reed, and D. L. Haynes, *Rev. Sci. Instrum.* **69** (1998) 1311.
2. G. D. Alton, United States Patent number: 5,506,475.
3. G. D. Alton, and D. N. Smithe, *Rev. Sci. Instrum.* **65** (1994) 775.
4. Y. Liu, et. al, unpublished.

Performances of Volume Versus Surface ECR Ion Sources

Y. Liu, G.D. Alton, H.Z. Bilheux, F.M. Meyer

Oak Ridge National Laboratory, P.O. Box 2008, Oak Ridge, TN 37831-6368, USA

Abstract. An all-permanent, 6 GHz ECR ion source has been constructed at the Holifield Radioactive Ion Beam Facility (HRIBF), Oak Ridge National Laboratory (ORNL), that permits configuration of the central magnetic field in either conventional parabolic or flat minimum-B profiles. The magnitude of the central flat field configuration extends over an axial region of ~ 2 cm to form a large and uniformly distributed ECR volume. The capability of operating the source in either volume or surface modes permits direct comparison of the performances of each source type. The studies show that the volume ECR source produces higher charge-states and higher intensities within a particular charge-state than does the surface form of the source. The X-ray spectra derived during operation of the source also suggest that the enhanced performance of volume ECR source is attributable to its ability to accelerate a larger population of electrons to higher energies than its conventional counterpart.

INTRODUCTION

It has been suggested [1-2] and subsequently demonstrated [3-6] that the performances of ECR ion sources improve whenever the physical sizes of the embedded ECR zones are increased. This can be done by flattening the central magnetic field such that a large ECR volume is created on axis that is in resonance with single frequency microwave radiation [1,2]. Heinen, et al., [3,4] and Liu et al. [5] have demonstrated the performance advantages of a volume ECR ion source that incorporated the flat central-field (flat-B) concept. Alternatively, the number of ECR zones can be increased by injecting multiple frequency [7-11] or broadband [12] microwave radiation.

An all-permanent-magnet, 6 GHz ECR ion source, based on the original flat-field concept [1,2], has been constructed at the Holifield Radioactive Ion Beam Facility, Oak Ridge National Laboratory [6]. The source is also designed so that it can be converted from a flat-B magnetic field configuration to a conventional parabolic central field configuration and vice verse. This design flexibility enables comparisons of the performances of the two source configurations. The results of comparative studies of the performances of the volume and surface ECR ion sources are presented in this report.

SOURCE DESCRIPTION

A schematic representation of the 6 GHz, all-permanent-magnet ECR source is shown in Fig. 1. The magnetic field profile of this source is specially designed to have an central flat field that is in resonance with single-frequency microwave radiation, as shown in Fig. 2. A N= 12 multi-cusp magnetic field is used to confine particles and to extend the ECR zone in the radial direction. The source can also be converted to a conventional parabolic central field profile (Fig. 2) which utilizes a sextupole multi-cusp magnetic field for plasma confinement in the radial direction. In combination with the axial mirror field, a magnetic field strength of 5 kG, approximately equal to that of the axial mirror field, is generated at the inner wall of the plasma chamber. The plasma chamber is made of *Al* and is 15.6 cm in length and 5.4 cm in diameter. The source was originally designed with a traveling-wave RF injection system consisting of a smoothly tapered rectangular-to-circular transition, starting from a WR137 waveguide and terminating with a cylinder whose diameter matched the plasma chamber dimension, as shown in Fig. 1.

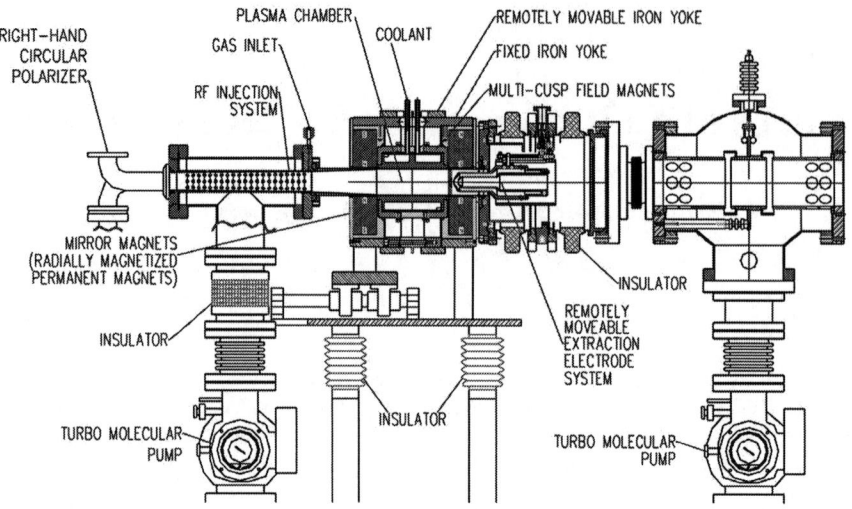

FIGURE 1. Schematic view of the all-permanent magnet, 6 GHz ECR ion source.

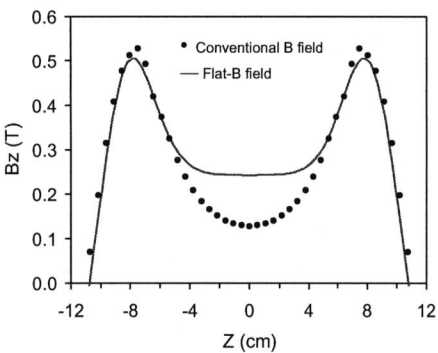

FIGURE 2. Axial magnetic field profiles for the flat-B (solid) and conventional-B (dotted) configurations.

COMPARISON OF SOURCE PERFORMANCES

During all evaluative studies Ar was used as the operating gas for both source configurations. In general, the results derived from each of the source geometries were very reproducible. Our studies emphasized the production of high-charge-state ions (> 6+) based on observed charge-state distributions and ion beam intensities of a given charge-state. Two detectors were used to measure X-ray generation as functions of pressure and RF power: a $Si(Li)$ detector and a $Ge(Be)$ detector for measuring the energies and intensities of low and high energy X-rays, respectively.

Charge State Distributions

The charge-state distributions and intensities of Ar ions were compared under optimized operating conditions for each source configuration, including gas pressure and RF power. In all cases, the volume configuration significantly out performed the surface source. Fig. 3 compares Ar charge-states obtained with the flat-B (lower plot) and conventional geometry (upper plot) configuration without gas mixing.

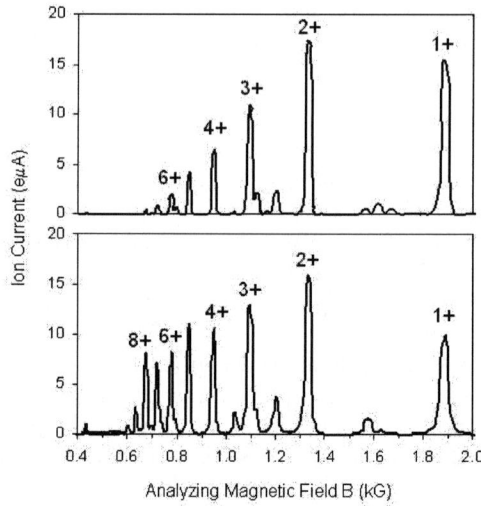

FIGURE 3. Ar charge-state-distributions observed with the conventional-B geometry source (upper) and the flat-B geometry source (lower). RF power: ~40 W for both cases.

The sources were operated with similar RF power while the gas pressures were adjusted for optimal production of high-charge-state ions. As notes, from the data, the volume ECR source produced higher charge-states and much higher intensities within a particular high-harge-state. The total extracted *Ar* ion beam currents and the intensities of individual charge state ions, in general, increased with increasing RF power until an optimal RF power was attained. For the surface source, the optimal RF power was typically less than 100 W when optimized for high- charge-state ions. In comparison, the optimal RF power for the volume source was much higher, as illustrated in Fig. 4, because of the much larger ECR zones within the flat-*B* configuration which should adsorb more microwave power before saturating.

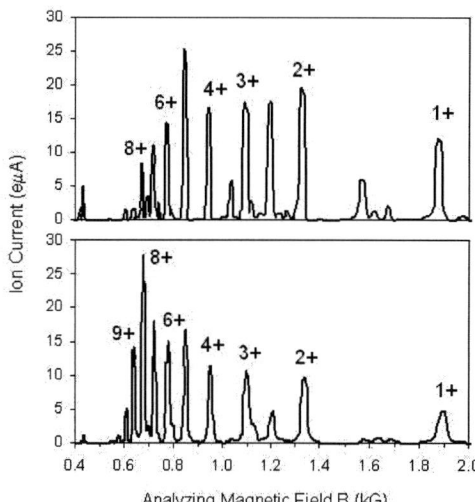

FIGURE 5. *Ar* charge-state-distributions observed with the conventional-*B* (upper) and flat-*B* (lower) configuration, with the new off-axis RF injection system.

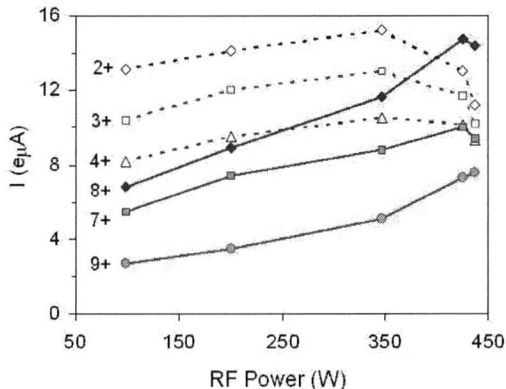

FIGURE 4. Ion beam intensity as a function of RF power for the volume source.

The source was originally designed with a traveling-wave RF injection system consisting of a smoothly tapered rectangular-to-circular transition, starting from a WR137 waveguide and terminating with a cylinder whose diameter matched the plasma chamber dimension, as shown in Fig. 1. The traveling wave RF system was changed to an off-axis cavity RF injection system to avoid parasitic ECR resonances in the RF injection region. The performance of the source was significantly enhanced with the new RF injection system for both volume and surface ECR configurations. More detailed information on the new RF injection system can be found in Ref. 13. A comparison of *Ar* charge states obtained with the new RF injection system is displayed in Fig. 5. The data were taken with RF power and gas pressures adjusted for optimal production of high-charge-state ions for each source configuration. As illustrated, the volume source out performs the surface configuration in terms of high-charge-state production, when operated with the new RF system.

X-ray Spectra

Fig. 6 displays typical *X*-ray spectra measured with the low- and high-energy *X*-ray detectors for the conventional-*B* (upper plot) and the flat-*B* (lower plot) magnetic field configurations. The sources were configured with the traveling wave RF system and operated with ~190 W RF power under similar gas pressures. As illustrated, the high-energy *X*-ray populations are dramatically different for the two source configurations. The flat-*B* configuration produced much higher *X*-ray intensities and energies that correlate to larger high- energy electron populations with maximum electron energy of $E_{electron}$ ~300 keV. These data clearly suggest that more RF power is coupled into the plasma by the presence of larger ECR zones within the volume source than that coupled into the conventional-*B* geometry configuration.

Gas Mixing Effect

Gas mixing is well known to enhance the production of high-charge-state ions in ECR ion sources. We have also investigated this technique with our source. When the source was in the traveling wave structure, the intensities of high-charge-state ions were increased by introducing light gases into the plasma chamber while the low charge states decreased, i.e., the charges state distributions were shifted to higher

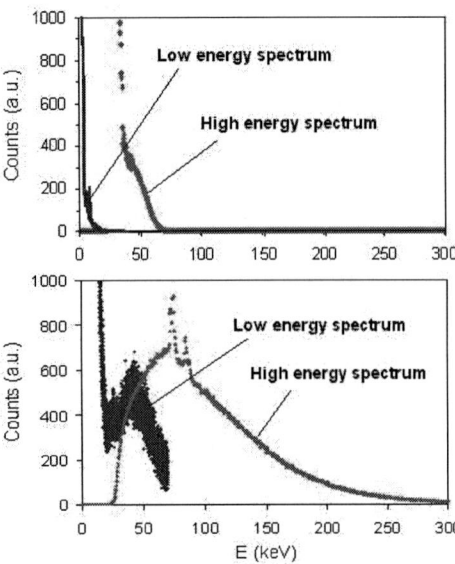

FIGURE 6. Measured X-ray spectra for the conventional-B (upper) and flat-B (lower) configurations under similar operating conditions: RF power ~ 190 W, P ~ 10^{-6} Torr.

charges with gas mixing. Lighter mixing gases were more effective than heavier gases. Figs. 7 displays the high-charge-state distributions observed with different mixing gases for the conventional and the flat-B magnetic configurations. As noted, the largest improvements in source performances, especially for the flat-B configuration was obtained with He mixing gas. Even with the addition of the gas mixing effect to enhance source performance, the flat-B configuration still outperformed that of the conventional-B configuration. With the new cavity RF injection system the effect of gas mixing was much smaller for the flat-B configuration.

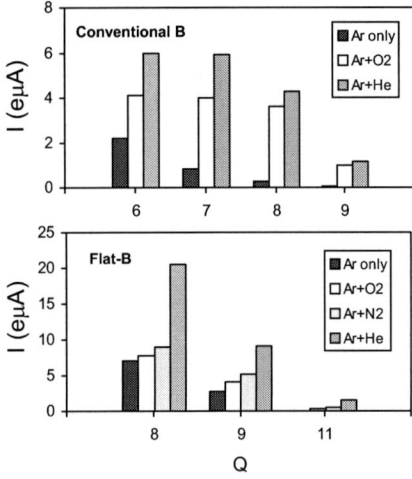

FIGURE 7. Ar charge states obtained with different mixing gases for the conventional and flat-B configurations.

CONCLUSION

We have evaluated the performances of the all-permanent, 6 GHz ECR ion source with two different magnetic configurations. Under all operating conditions optimized for high-charge-state ion production, the volume ECR source produced higher charge states and higher intensities for high charge states than the conventional minimum-B configuration. Measured X-ray spectra also supported the fact that there were more hot electrons generated in the volume ECR source. Much work remains to bring the source performance to levels competitive with existing sources. Further improving the source performances will be the focus of our future studies.

ACKNOWLEDGMENTS

Research sponsored by the Office of Science, U. S. Department of Energy, under contract DE-AC05-00OR22725.

REFERENCES

1. G. D. Alton, United States Patent number: 5,506,475.
2. G. D. Alton, and D. N. Smithe, *Rev. Sci. Instrum.* **65** (1994) 775.
3. A. Heinen, et al., *Rev. Sci. Instrum.* **69** (1998) 729.
4. L. Müller, et al., *Proc. Of the 15th Int. Int. Workshop on ECR Ion Sources* (University of Jyvaskyla, Finland, June 12-14, 2002).
5. Y. Liu, G. D. Alton, H. Bilheux, J. M. Cole, F. W. Meyer, *Rev. Sci. Instrum.* (2004) (in press).
6. Y. Liu, G. D. Alton, C. A. Reed and D. L. Haynes, *Rev. of Scien. Intrum.* **69** (1998) 1311.
7. G. D. Alton, *Production of large resonant volumes in microwave electron cyclotron resonance ion sources*, United States Patent number: 5,841,237.
8. G. D. Alton, *Nucl. Intrum. And Meth. In Physics Research* **A382** (1996) 276.
9. Z.Q. Xie and C.M. Lyneis, *Rev. Sci. Instrum.* **66** (1995) 4218.
10. G. D. Alton, F. W. Meyer, Y. Liu, J. R. Beene, and D. Tucker, *Rev. Sci. Instrum.* **69** (1998) 2305.
11. R.C. Vondrasek and R. C. Pardo, *Proc. of the 15th Int. Workshop on ECR Ion Sources* (University of Jyväskylä, Finland, June 12-14, 2002), 35.
12. G. D. Alton, these proceedings.
13. Y. Liu, et al., these proceedings.

Testing of the "Flat-B" 6-Ghz ECR Ion Source Equipped with a RF Polarizer (Abstract)

H. Z. Bilheux, Y. Liu, J. M. Cole, G.D. Alton

Oak Ridge National Laboratory, Oak Ridge, Tennessee, USA

Abstract. The all-permanent magnet, 6-GHz "flat-B" (or "volume"-type) ECR ion source at the Holifield Radioactive Ion Beam Facility of the Oak Ridge National Laboratory can be equipped with a RF polarizer for injection of right-hand circularly polarized (RHCP) or left-hand circularly polarized (LHCP) microwave radiation into the plasma chamber. Testing of the source with the polarizer is made using Ar as the operating gas. Comparison of the source performance with and without the polarizer will be given in this report.

Managed by UT-Battelle, LLC, for the U.S. Department of Energy under contract DE-AC05-00OR22725.

The Texas A&M ECR Ion Sources: a Status Report

D. P. May, G. J. Derrig, F. P. Abegglen, and H. Peeler

Cyclotron Institute, Texas A&M University, College Station, TX 77843, USA

Abstract. The Texas A&M Cyclotron Institute has two ECR ion sources, ECR1 and ECR2. ECR1 operates at 6.4 GHz and has high-B containment fields. ECR2 operates at 14.5 GHz. Both plasma chambers are aluminum, are large with similar dimensions and incorporate biased disks on the injection ends. The large chamber size has been convenient in allowing a number of sputtering feeds to be used. In ECR1 an eight-lead, high-voltage feedthrough is used for sputtering, enabling the beam to be switched rapidly and easily between eight different metallic elements, some of which are refractory. Many of the metallic elements up through uranium can be sputtered. Sputtering as well as other methods used for generating beams will be described, and the operation of the two ion sources will be described and compared.

INTRODUCTION

The high-B ECR1 ion source was constructed in its present form in 1995, while the ECR2 ion source was constructed in its present form in 2002. Each has been previously described [1], [2]. ECR1 has a hexapole strength at the wall of about 4.9 kilogauss and uses a microwave frequency of 6.4 GHz while ECR2 has a hexapole strength at the wall of 7.7 kilogauss and uses a microwave frequency of 14.5 GHz. The interiors of the aluminum plasma chambers of both ECR1 and ECR2 are approximately 12.7 cm in diameter and are respectively 58 cm and 56 cm in length.

The ECR1 ion source has been remarkably reliable and has been the workhorse source for cyclotron operation, helping to meet a tight schedule that requires several hundred beam changes per year with beams ranging from lithium to uranium. For this purpose it has been outfitted with a gas manifold and with an array of solid feed devices that can be conveniently mounted on the source.

The ECR2 ion source has proved to be more challenging in its operation. The higher microwave frequency and the tighter constraints on cooling of the hexapole have caused problems in its development that are slowly being overcome. However, ECR2 has been used for cyclotron operation, and a sputtering feed has been tested in it.

OPERATIONS

Cyclotron operations at Texas A&M requires that a variety of beams be available from the ion source. For some experiments a single species may be required from the source for a week or more. For other experiments the species may change two or three times a day. For radiation effects testing on materials and on semiconductor devices especially, the ion source may need to be switched between as many as ten different ion species, having a large range of Z, each day.

Solids are introduced into both sources through a radial port. On ECR1 this port has a clearance of 19 mm by 38 mm. ECR2 has three slots on this radial port that measure 10 mm by 40 mm. Both sources are horizontal, and the line through this port is at an angle of 30° below the horizontal.

Ovens

A low-temperature oven modeled after one designed at LBL [3] is used to produce beams of lithium, sodium, phosphorus, and calcium. The oven is temperature stabilized via a thermocouple. Due to its high thermal mass and to the contamination of the feed materials, it takes more than 12 hours for the oven to be mounted, pumped out, and brought to temperature, and then another 24 hours before source operation stabilizes. For lithium a passively heated copper liner is inserted into the plasma chamber as well, so the process takes even longer.

The high-temperature oven used on ECR1 was constructed at NSCL after a design for one used on the RTECR [4]. It is capable of operating up to 1400° C. It is typically used to produce beams from solids that cannot easily be sputtered. Aluminum, chromium and small samples of various separated isotopes fit into this category. The temperature of the oven is usually raised to the point where the vapor pressure of the sample is about 10^{-6} torr. For zinc samples in the oven, energy from the source plasma, probably microwave energy, is sufficient for the sample to reach this point. It takes approximately 6 hours for the oven to be mounted, pumped out and brought to temperature, and it takes another 6 to 12 hours before source operation stabilizes.

Sputter Fixtures

Sputtering as a technique for introducing solids in ECR ion sources was developed at Argonne ATLAS [5]. The first trials with sputtering in ECR1 and ECR2 at Texas A&M used a bellows-sealed positioner mounted on an insulator. With this fixture the ideal radial position for the sputter target was found to be at the radius of the wall, though this position is not critical. The positioner is used for the insertion of larger samples of zinc, bismuth, and uranium. Figure 1 shows an ECR1 charge-state spectrum of uranium using this sputter feed. Small samples of zinc and bismuth melt on the multiple-feed fixtures, and smaller targets of uranium are hard to fabricate. The positioner is also used to insert small samples of iron and nickel. Due to the magnetic force on these metals, these cannot be mounted on the multiple feed fixtures. Feeds of silver, gold, and uranium have been tested in ECR2 with this positioner.

The demands for rapid switching between cyclotron beams for radiation testing led to the investigation of multiple sputter targets mounted in the source. At first a high-voltage (3.5 kV) vacuum feedthrough with three separate leads was tried in ECR1. Each of the three leads on the vacuum side was insulated with a glass tube. Sputter targets of different metals, usually measuring roughly 0.5 cm, were mounted on stainless steel wires by spot-welding and these wires were attached to the leads. The solid feed could be rapidly changed to one of the three different targets simply by switching its lead to the high-voltage supply. Figure 2 shows an ECR1 charge-state

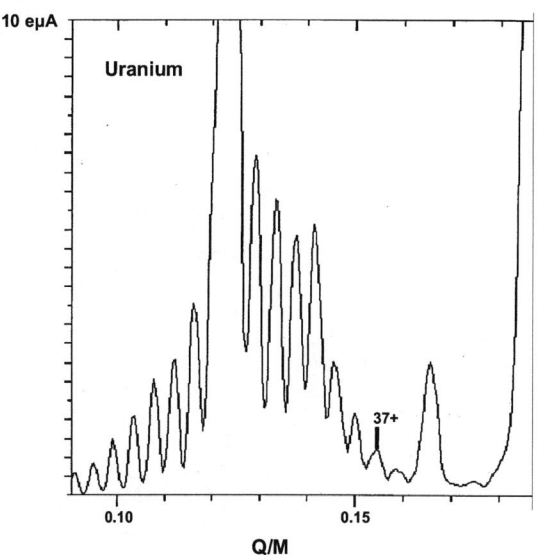

FIGURE 1. ECR1 charge-state spectrum of uranium. The sputtering voltage was 2.0 kV and the microwave power was 1.2 kW.

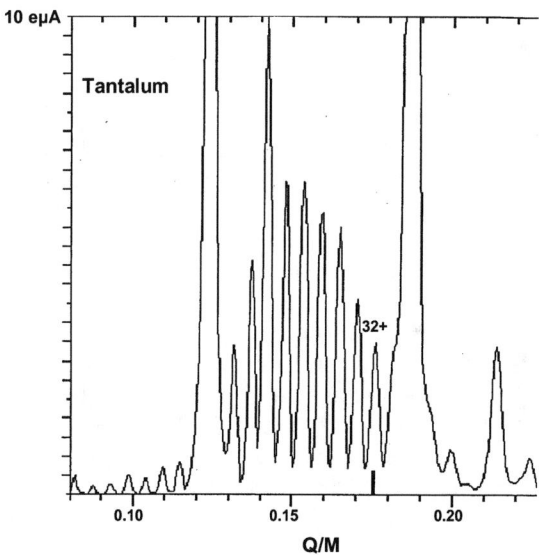

FIGURE 2. ECR1 charge-state spectrum of tantalum. The sputtering voltage was 2.3 kV and the microwave power was 1.3 kW.

spectrum of tantalum using the three-lead feedthrough. Eventually a high-voltage (3.5 kV) vacuum feedthrough with eight leads was tried successfully (Fig. 3). This fixture is now typically in ECR1 for most radiation testing runs.

The metals typically used on these multiple sputtering fixtures include copper, niobium, molybdenum, silver, praseodymium, holmium, ytterbium, tantalum, platinum, and gold. The glass tubes eventually become coated with metal, but the cross-contamination is minimal and disappears after about an hour.

FIGURE 3. Eight-lead sputtering fixture. The sputter targets are on the left.

Problems and Repairs

Recently a gradual degradation in the performance of the ECR1 on source was correlated with a damaged area near the middle of one of the hexapole bars. First noticed in January of 2002 as a melting in the aluminum wall of the plasma chamber, this area corresponded to a dip in the nominal 4.9 kilogauss hexapolar field to 3.9 kilogauss. In the original measurements on the hexapole, made in 1995, this area showed the lowest dip in the field, 4.3 kilogauss. By January of 2004 the field in this area had declined to close to the ECR field of 2.28 kilogauss, so the bar was replaced. Upon close examination, this area was seen to be at the joint between two of the six rectangular blocks making up the bar. Since the repair the source performance has improved considerably.

The ECR2 ion source exhibits a much higher flux of x-rays than ECR1, so a considerable amount of extra lead shielding has been constructed. For the microwave injection into each source, the high-voltage break and vacuum window have been combined into one high-density, polyethylene window, but ECR2 exhibited microwave burning of the Viton o-ring that had been successfully used in ECR1 to seal the window. The o-ring was abandoned for ECR2, and the polyethylene was put under high mechanical pressure to make the vacuum seal.

ECR2 has failed to achieve the low vacuum exhibited by ECR1, and it has been decided that a water-to-vacuum seal is opening with some stress on the plasma chamber. This stress seems to occur with high axial magnetic field and is most likely torque on the permanent magnets of the hexapole caused by interaction with the axial field. Even with this leak ECR2 has achieved 139 eμA of $^{16}O^{7+}$ at 1.3 kW of microwave power, but to achieve comparable results with higher Z the vacuum must be improved.

REFERENCES

1. D. P. May, G. J. Derrig, F. P. Abegglen, and G. J. Kim, *Proc. of the 13th Int. Workshop on ECR Ion Sources*, College Station, Texas, 1997, p. 43.
2. D. P. May, G. J. Derrig, and F. P. Abegglen,, *Proc. of the 15th Int. Workshop on ECR Ion Sources,* Jyväskylä, Finland, 2001, p. 56.
3. C. M. Lyneis, *Proc. of the 6th Int. Workshop on ECR Ion Sources,* Berkeley, California, 1985, p. 51.
4. R. Harkewicz, *Rev. Sci. Instrum.* **67,** 1996, p. 2176.
5. R. Harkewicz, *Rev. Sci. Instrum.* **66,** 1995, p. 2883.

Design of an All-Permanent-Magnet, Volume-Type, 10GHz ECR Ion Source With Field-Forming Iron Yoke

M.Stalder, C.T.Steigies, R.F.Wimmer-Schweingruber

Extraterrestrial Physics, University of Kiel, 24098 Kiel, Germany

Abstract. A new all-permanent-magnet ECR ion source has been designed and assembled to produce highly charged ions. Using four iron rings the central B-field is formed to a plateau, which leads to a "volume" type ECR zone. The shape of the B-field plateau and the minimum field strength can be tuned by moving the iron rings. The source can be operated in the frequency range of 10GHz to 14GHz with axial mirror ratios of $B_{max}/B_{ECR} = 2.8$ and $B_{max}/B_{ECR} = 2$ respectively. To achieve a minimum-B configuration we can either use a hexapole or a dodekapole (12fold multipole). Experimental tests will show whether the highest charge states of the ions produced in the source can be increased using a dodekapole.

1 INTRODUCTION

We are building a new laboratory for solar wind instrument calibration at the University of Kiel. To provide typical solar wind ions such as O^{6+}, Ne^{8+}, Fe^{10+}, we built a new ECR ion source optimized to produce highly charged ions. The ECR ion source lies on a 450kV high-voltage platform to accelerate the ions across the potential drop to ground. Because space and electrical power are very limited on the HV platform, we decided to build an all-permanent-magnet system. In this work we describe the design characteristics and the B-field measurement results of the new ECR ion source.

2 OUTLINE OF THE SOURCE

A cross-section view of the ECR ion source is shown in Fig. 1. The source consists of two permanent mirror magnets, an additional axial magnet ring, alternatively a hexapole or a dodekapole and four iron rings to form the magnetic field. The minimum B-field forms a plateau so that the ECR zone is of "volume" type. The shape and the |B| value of the plateau can be changed by moving the iron rings in the axial direction.

The plasma chamber is water cooled, has an inner diameter of 45mm and is made of aluminum. The magnetic mirror points are 140mm apart and lie at the ends of the hexa-/dodekapole. The magnet system can be moved relativly to the plasma electrode which is fixed in the plasma chamber. The extraction B-field can be lowered by moving the mirror field point axially behind the plasma electrode.

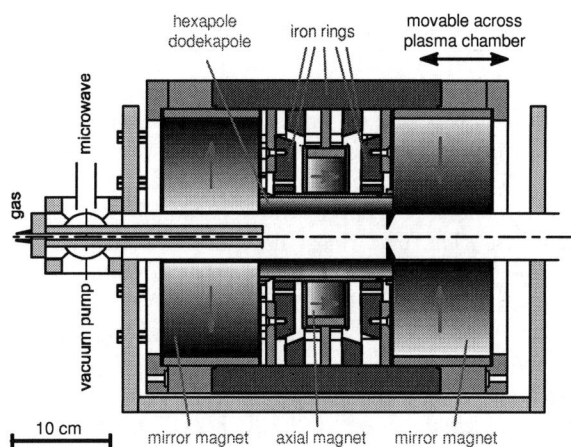

FIGURE 1. Schematics of the ECRIS. In the center the four movable iron rings are shown.

The microwave is coupled into the plasma chamber with a coaxial waveguide. A TWT microwave amplifier provides 100 W in the frequency range from

10 GHz to 14 GHz. The total weight of the source is 150 kg of which the permanent magnets contribute 75 kg. The vacuum turbo pump is placed behind the extraction from the ECR ion source.

3 SOURCE DESIGN

3.1 Concept

The source design is driven by the principle goal to produce highly charged ions. The promising results of the Münster plateau ECR source [1] encouraged us to work out a plateau ECR concept [2]. A long confinement time is necessary for the production of highly charged ions [3]. Increasing the mirror ratio B_{max}/B_{ECR} is one of the possibilities to achieve long confinement.

As it is very difficult to build mirror field magnets with field strength larger than 1T we decided to operate the new ECR ion source at 10 GHz which allows a very good mirror ratio. Extensive simulations of permanent magnet systems with the 2D Finite Element Method Magnetics code femm3.3 [4] showed that it is difficult to build a plateau ECR ion source with high mirror ratio and permanent magnets only. The main difficulty is the fast decline in the |B| from the mirror points towards the plateau. We developed a permanent magnet optimization procedure and designed a magnet system without iron. Unfortunately it turned out that the complex mirror magnets could not be manufactured.

Hence, we decided to build an ECR ion source with field-forming iron yokes to separate the mirror B-field from the plateau region. With the iron rings the plateau concept could be realized with very simple permanent magnets. The iron rings limit the diameter for the hexapole and the dodekapole. The hexapole field strength is only 0.8T at the plasma chamber wall. Probably we will have to lower the extraction B-field below the hexapole field strength for good performance of the source. In this case we are not able to take full advantage of the high axial mirror field. Using a dodekapole could be a solution to this problem as the confinement of ions and electrons in this case is mainly axial. With an additional iron pluge at the microwave side the mirror ratio can be made very big.

The advantage of our design is that we can tune the plateau B-field in a wide range (see Figure 2), and in particular compensate the fabrication tolerances of the permanent magnets. The disadvantages are that the hexapole size and for this the B-field at the plasma chamber wall is limited and the mechanical design to hold the iron rings is complex as the acting forces are large (up to 10^4 N).

3.2 Magnet System

The axial mirror field of 1T is produced by two radially magnetized mirror magnet rings. The rings have an exterior diameter of 240mm, a length of 100mm and a hole with a diameter of 50mm for the plasma chamber. The iron rings enhance the mirror field strength by 0.1T and lower the contribution of the mirror field magnets to the plateau |B| by 0.3T (see Figure 2). The axial magnet has a exterior diameter of 160mm, a length of 40mm and a hole with a diameter of 90mm for the hexa-/dodekapole. The mirror magnets and the axial magnet are built with 44 MGOe NdFeB material.

The hexapole and the dodekapole are built from 24 segments with 40 MGOe NdFeB material. The inner diameter is 50mm, the exterior diameter is 80mm and the length is 140mm. For the hexapole $|B| \sim r^2$ and reaches $|B| = 0.83$ T at the plasma chamber wall (d = 45 mm). For the dodekapole $|B| \sim r^5$ and reaches $|B| = 0.64$ T at the plasma chamber wall.

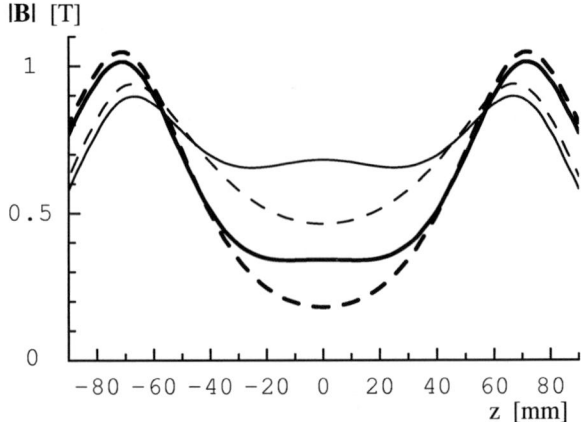

FIGURE 2. Simulated axial |B|. Mirror magnets (dashed fine line), mirror and axial magnets (fine line), mirror magnets with iron rings (dashed bold line), mirror and axial magnets with iron rings (bold line),

Special care has to be taken choosing the material for the hexapole and the dodekapole. The axial mirror field magnets in combination with the iron rings produce radial magnet fields of up to 1T inside the hexapole or dodekapole. The magnet material for the hexapole and dodekapole has to withstand the field strength of 1T which is in the worst case in opposite direction to the local magnetization.

3.2 Calculated and measured magnetic field distributions

In this section we compare the results of the femm simulations with the measured B-field. Before assembling, we measured each magnet ring. The mirror magnet rings are 5% weaker than calculated. This is within the tolerance of the magnet material specification. For all the other magnets the simulation results and the measurements agree within 1%.

For the assembled system a comparison of the simulated and measured B-field is shown in figures 3, 4.

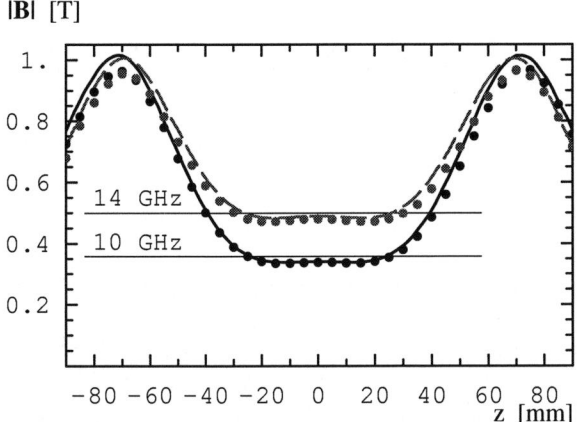

FIGURE 3. Measured and simulated B-field. Simulated B-field (lines) for different iron ring positions with measured B-field (points).

The measured mirror B-field is 6% lower than the simulated one. Simulations with the weaker mirror-field magnets reproduce the measurements. The

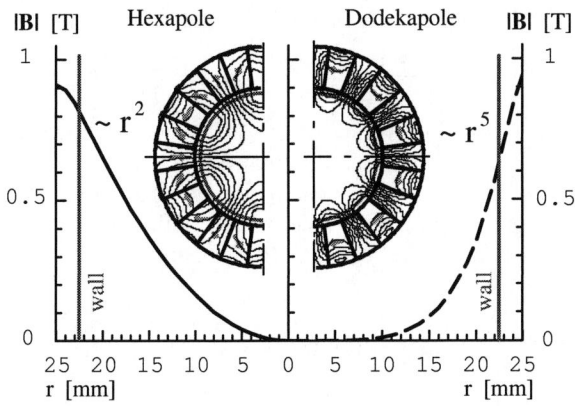

FIGURE 4. Hexapole and Dodekapole. The left-hand side shows the hexapole, the right-hand side the dodekapole configuration with the magnet-field lines and radial |B|.

simulations for the plateau B-field region were very successful.

Figure 5 shows the measured B-field for the plateau region. Full control of the plateau form is possible in the range of 10 GHz to 13 GHz. Moving the iron rings the plateau B-field can be shaped to form either a local maxima (2% over the minimum B-field, long dashes), be flat (solid) or have a parabolic shape (short dashes).

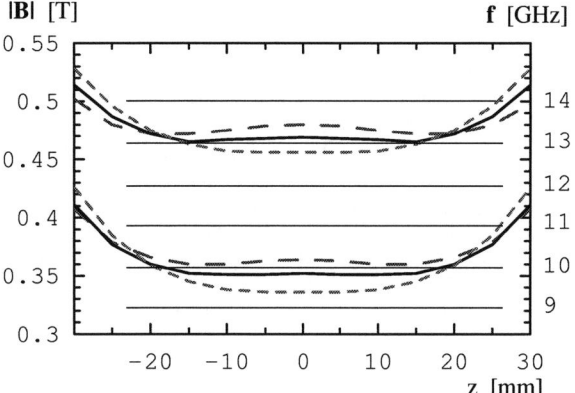

FIGURE 5. Plateau B-field. Measured B-field is shown for different iron ring positions.

The iron rings are manufactured from steel 52 and put into the simulations with the Steel1010 property. As described in the manual of femm [4] we used Kelvin transformation to solve the open boundary problem in the magneto-static simulations.

4 CONCLUSION

We successfully built and assembled a new ECR ion source with field-forming iron rings. By moving the iron rings we can change the resonance frequency in a wide range and vary the plateau shape. The measured an simulated B-fields are in good agreement.

5 REFERENCES

1. Müller, L. et al, *Proceedings of the ECRIS02*, Jyvaskylä Finland, 2002
2. Alton, G. D., and Smith, D. N., *Rev. Sci. Instrum.* **65**, 775 (1994).
3. Geller, R., "Confinement in magnetic mirror fields," in *Electron Cyclotron Resonance Ion Sources and ECR Plasmas,* Bristol, Institute of Physical Printing, 1996
4. Meeker, D., www.femm.foster-miller.com, Foster-Miller Inc.

Improved Spindle Cusp Magnetic Field for ECRIS

M. H. Rashid[*], C. Mallik and R. K. Bhandari

Variable Energy Cyclotron Centre, Sector-1, Block-AF, Bidhan Nagar, Kolkata- 700 064, India

Abstract. Magnetic field of minimum-B configuration is very important for achieving more plasma confinement and closed electron cyclotron resonance (ECR) surface for electron heating and plasma discharge. The spindle cusp magnetic field configuration forms the modified minimum-B configuration. The absolute magnetic field at the chamber surface on mid-plane has been optimized and improved sufficiently and symmetrized to the field at the point cusp positions on the central axis. With enhancement of electrostatic and magnetic mirror action at the cusp positions the density of the plasma as well as confinement is boosted. The system becomes simpler, more compact and cost-effective compared to the conventional one to generate and extract highly charged heavy ions (HCHI). A co-operative and collaborative effort is essential to develop and test such conceived new ECRIS.

INTRODUCTION

Earlier ECR ion sources (ECRIS) using the configuration were little successful to be dismantled and it was attributed to the huge constant loss of plasma on the cusp positions because of insufficient magnetic field for providing mirror reflections of charged particles [1,2]. Plasma was lost along the length of cusp lines of the multipole because of loss of electrons. We have to reduce the total length of cusp loss line for better confinement of plasma. So, it is very essential to develop some method of generating more confining ring cusp field of sufficient strength employing either a coil system (normal temperature or low temperature) or a powerful permanent magnet ring pole. The features and designs of such magnet for ECRIS are described in this paper.

People now want simple, compact and cost effective source as far as possible without compromising much and sacrificing the extracted beam using either high quality and powerful permanent magnets or coil system. A conventional ECRIS uses the principle of adiabatic invariance for mirror reflection and HBM operation to successfully generate and confine plasma. This principle can be utilized in a properly configured cusp field also. The main objective is to achieve closed ECR magnetic field surface far off the plasma chamber surface. When microwave (μ-wave) power at the resonance frequency is injected into the box, plasma electrons crossing this closed surface will, in general, be heated to hundreds of electron volt due to transfer of energy from the EM wave to electron at ECR resonance ($\omega_\mu = \omega_{c(e)}$). The heated electrons strike the atoms and ionize them to high charge states by stripping them of electrons in stepwise manner and generate plasma. It can be now possible to make more powerful and bigger ion device using only superconducting coils and millimeter wave high frequency gyrotrons if sufficient field at the cusp regions is generated.

CUSP FIELD CONFIGURATION

Centres of curvature of the field lines are situated outside the plasma at the centre when the lines of forces are convex as in the case of cusped field. The cusp geometry, indeed, has more confining property and can be used to designing ECR ion device.

Feature of the Cusp Field

When two coaxial current loops kept apart are energised oppositely, cusp magnetic field is produced

---

[*] Mail to: haroon@veccal.ernet.in

having azimuthal component of the vector potential, radial and axial component of magnetic field expressed as in equations $A_\theta = G\,r\,z$, $B_r = G\,r$ and $B_z = 2G\,z$ respectively, where G is a constant. If the current carrying loops of diameter D are D distance apart then the axial field gradient is twice the radial field gradient at certain distance from the centre. If the field gradients in both the directions are to be made equal, then these loops should be just $D/2$ distance apart. Moreover, the magnitude of the magnetic fields at the mid-plane ring of certain radius is never sufficient and equal to the field at same distance from centre on the axis of rotation (z-axis).

We assume an off-axis flux tube of elementary length, dl, formed by a thin bunch of field lines (ϕ) having area of cross section dS (Fig. 1). There are more flux tubes at its vicinity. As a result of interchange of neighbouring tubes of magnetic flux we get the variation in magnetic energy in Eq. (1), which imply the necessary condition for stability of plasma $\delta E_{mag} > 0$. The length of magnetic lines of forces (MLF) should decrease with the increase of magnetic field.

$$\delta E_{mag} \cong -\delta\left[\frac{\phi^2}{2\mu_0}\right]\cdot\delta\left[\int\frac{dl}{dS}\right] \quad (1)$$

The variation in thermal energy of plasma in the tube due to the exchange of the flux tube of shaded region with tube at its vicinity can be expressed by δE_{pla} given by Eq. (2) in adiabatic condition in terms of variation in pressure (P) and volume (V).

$$\delta E_{pla} \cong \delta P \cdot \delta V \quad (2)$$

The sufficient condition for plasma stability, $\delta E_{pla} > 0$, implies that with the decrease of pressure P the volume V also must decrease. If the cusp geometry is filled with plasma, the shaded flux tube is also filled with plasma, which tends to migrate towards the side where the flux tube volume V given by Eq. (3) will increase.

$$V = d\phi \int (dl/B) \quad (3)$$

We get from inequality (2) and (3) the following inequality for stability in plasma containment in the (cusp) magnetic field.

$$\delta \int (dl/B) < 0 \quad (4)$$

The variation of integral is along the perpendicular direction to the plasma surface between two infinitesimally close field lines. It means physically that the necessary and sufficient condition of plasma stability is the increase of the average field along the field lines outwards from the plasma boundary. This is called the *modified minimum-B* concept [3]. In this situation the field lines are convex towards the contained plasma at the centre.

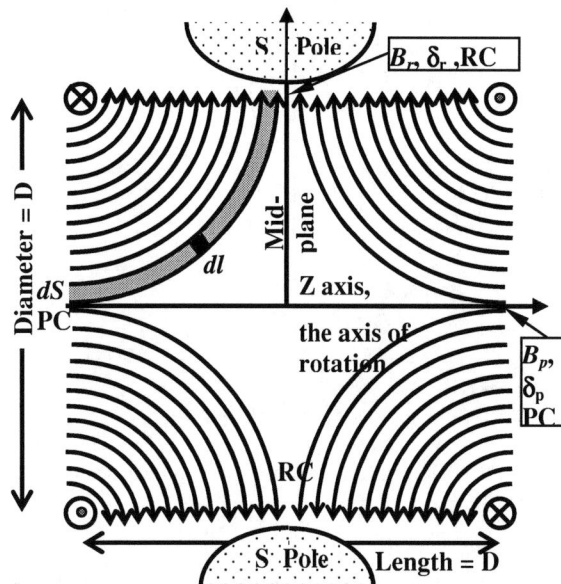

FIGURE 1. Cusp field configuration produced by the coil system or permanent magnet ring pole (S Pole).

Plasma in the Cusp Field

The magnetic mirror ratio R_m is defined as $R_m = B_{max}/B_{min}$ on the MLF's. An electron have parallel ($P_{\bullet\bullet}$) and perpendicular (P_γ) component of its total momentum (P) with respect to vector magnetic field **B**, which makes it move along and gyrate around the MLF respectively at an angle given by Eq. $\alpha = \sin^{-1}(P_\gamma/P)$. The apex angle of the electron is defined as $\alpha_{apex} = \sin^{-1}(1/\sqrt{R_m})$.

Instead of providing an idealised cusp field with sharp interface separating vacuum magnetic field from field free plasma, this configuration have open cusps. It has unconfined plasma expanding to infinity on the axis of rotation (two holes at PCs of size A_p area) and at the middle position of the current loops (ring gap at RC of A_r area) shown in Fig. 2. The opening areas are directly related to the geometric mean of gyro-radii $\rho_{c(e)}$ and $\rho_{c(i)}$ of electron and ion $\delta_j \approx k_j(\rho_{j,c(e)}\rho_{j,c(i)})^{1/2}$, where k_j is a constant and subscript j indicates point or ring cusp for $j=r$ or $j=p$ respectively. It is written as k (~2, which is imprical) for symmetric field subsequently. Rough idea of cusp hole sizes can be obtained from $2\delta_p$ and $\delta_r L$ at PCs and RC respectively, where L is the length of the RC. More accurate opening areas at PCs, RC and particle fluxes are discussed in [4]. The cusp-hole size has been estimated to be the gyro-radius in some two dimensional cases. The loss rate through the cusps is proportional to geometrical surface area at the cusps.

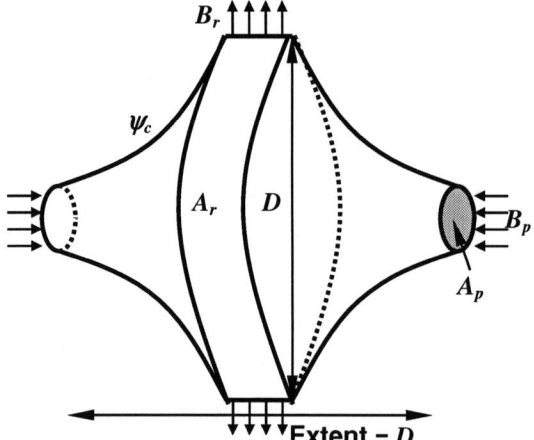

FIGURE 2. View of loss area at the PC and RC positions in the cusped field.

Magnetic field B_r at the RC is insufficient and less than the field B_p at the PC. The loss cone is large due to small R_m at the RC and the charged particles (electrons) fall easily into the loss cone defined in the momentum space. Thus, in absence of collisions, particles for which $\alpha < \alpha_{apex}$ leak through the magnetic mirrors otherwise they will be reflected back towards the plasma centre. So, the radial loss rate of charged particles is much larger than the axial loss rate, no enhancement of high charge state components results in the extracted beam. It is further enhanced by the inappropriate radial plasma potential generated. The plasma around the centre is extremely starved of electrons and the production of ions particularly highly charged heavy ion (HCHI) becomes very difficult and whatever ions produced prefer to migrate radially towards the (ring) cusp positions. The radial loss is further enhanced by the negative radial density gradient of plasma; however, it is restricted by stronger electric and magnetic mirror reflection. The low charge state ions can be trapped by the potential but experience more end loss then high charge state ions at the cusp positions. So, extraction of high charge state ions becomes difficult. If the B_p and B_r fields are symmetric and obtained sufficiently high, then it may be referred to a theory of equilibrium, which takes into account mirror reflection of particles and therefore permits the existence of finite contained plasma without flow. The large angle scattering is less efficient in causing diffusion in the velocity space than the more frequent small angle scattering. However as an ambipolar diffusion, the actual loss of plasma is determined by the slowly diffusing component i.e. the positive ions and this loss also decreases with increasing electron reflection i.e. increasing R_m.

DESIGNING IMPROVED CUSP FIELD

The first attempt here is to achieve sufficiently strong magnetic field of equal strength (for symmetry) at the positions of PC and RC on the chamber surface for HBM operation i.e. it should follow the inequality $B_p = B_r \geq 2B_{ECR}$. The double walled cylindrical plasma chamber 'CH' is kept large enough compared to the wave-length, 60 mm corresponding to 10 GHz µ-wave. Proper cooling arrangement of chamber can be provided. The inner diameter and length of the plasma chamber (D) are 120 mm. This is to realise a so-called *multi-mode cavity*. It contains a large volume of dense plasma corresponding to large volume oblate spheroid of the ECR resonance surface around its centre. A design of improved cusp field configuration was reported in ref. [5] using the normal temperature coils for 14.4 GHz frequency and in ref. [6] using a permanent magnet ring pole for 10.0 GHz frequency. The designed parameters of an ion source of frequency 18.0 GHz are described herein for example. The magnetic field at the ring cusp is made stronger by placing a specially shaped disk (MID) or permanent magnet ring in the $z=0$ plane (mid-plane). Yokes and plugs made of highly permeable material like MS (magnetic steel) are also properly placed. These techniques are meant to improve the cusp field configuration to increase the density of plasma [7] consequently high current extraction of HCHI's. The geometry for field computation is shown in Fig. 3 and used the well-known POISSON code. The coils can be normal or low temperature allowing ~200 kA-turn/coil magneto-motive force to generate 13.2 kG magnetic field at the chamber surface on the central axis and the mid-plane radius both. The magnetic field has been plotted in Fig. 4 when the iron disk on the mid-plane was not present. The remarkable improvement in magnetic field at those cusp positions is depicted in Fig. 5, which corresponds to the HBM operation of the field configuration for 18.0 GHz µ-wave frequency. The magnetic field is proper and the ECR resonance surface is shown in Fig. 3 inside the chamber CH. The curved line is a part of the surface of an oblate spheroid. The resonance surface area and the volume enclosed corresponding to hot plasma are large enough to sustain generation and extraction of HCHI's.

CONCLUSIONS

The sufficient improved spindle cusp field was achieved. The shaped plugs made of highly permeable material are also used. The property of mirror reflection in the improved cusp field for 14.4 GHz ECR cusp field was investigated also by electron simulation [8] and found to be robust. The improved spindle cusp geometry does not need any multipole magnet. The length of the loss-line in this configuration is much less while the ECR surface area and hot plasma volume are much more than in a corresponding conventional ECRIS. It is possible to develop a compact and more efficient ECRIS now. The application of biased electrodes at the cusp positions would further improve its performance.

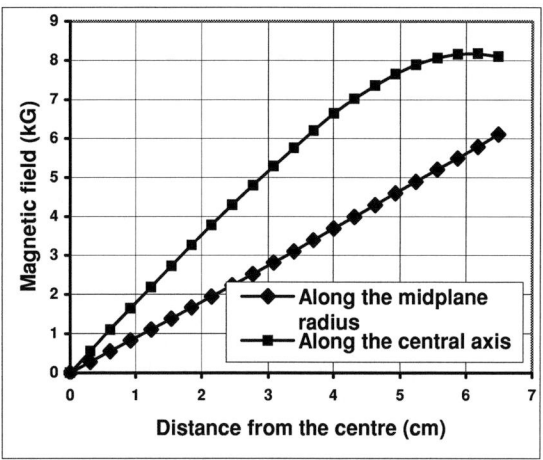

FIGURE 4. Magnetic field plot without the mid iron disk.

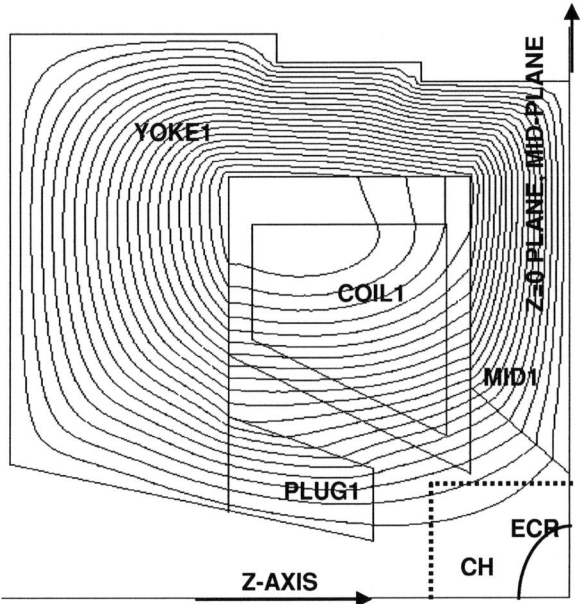

FIGURE 3. The magnet geometry is in midplane cylindrical symmetry. The position of the quarter plasma chamber is shown in dotted lines at the distance of 6 cm from the center.

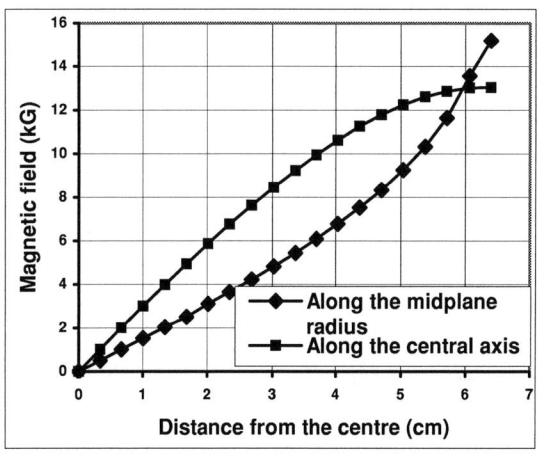

FIGURE 5. Magnetic field plot with the mid iron disk.

REFERENCES

1. K. Sudlitz *et al.*, Proc. 12 th. Int. Workshop on ECR Ion Sources, RIKEN Rep. **INS-J-182**, Saitama, 217 (1995).
2. K. Sudlitz and E. Kulczycka, Proc. 14 th. Intl. Workshop on ECR Ion Sources, **ECRIS99**, CERN, (1999).
3. L. A Artsimovich, *A Physicist's ABC on Plasma (English translation by Oleg Glebov)*, Mir Publishers, Moscow, 101 (1985).
4. M. H. Rashid, Thesis submitted (2003); Summary in DAE-BRNS Symp. Nucl. Phys. at BARC, Mumbai, **46B**, 596 (2003).
5. M. H. Rashid and R. K. Bhandari, Rev. Sci. Intrum., **74**(9), 4216 (2003).
6. M. H. Rashid and R. K. Bhandari, Indian. J. Phys., **78**(9), 727 (2004).
7. M. Leitner, D. Wutte, J. Brandstotter, F. Aumayr and H. P. Winter, Rev. Sci. Intrum., **65**(4), 1091 (1994).
8. M. H. Rashid, C. Mallik and R. K. Bhandari, DAE-BRNS Symp. Nucl. Phys. at BARC, Mumbai, **46B**, 550 (2003).

Preliminary Bremstrahlung Measurements on VENUS at 18 and 28 GHz [1]

C.M. Lyneis and D. Leitner

Lawrence Berkeley National Laboratory, 1 Cyclotron Road, Berkeley, CA 94720, USA

Abstract. The bremstrahlung produced by the VENUS ECR ion source at 18 GHz and 28 GHz in the axial direction has been measured with a germanium detector. The bremstrahlung spectrum goes out beyond 1 MeV at 28 GHz and this complicates analysis of the data and the design of the collimators and detection system. Preliminary spectra and the geometry of the detection system will be described.

---

[1] Paper included in
D. Leitner, C.M. Lyneis, S.R. Abbott, R.D. Dwinell, D. Collins, and M. Leitner: *"First Results of the Superconducting ECR Ion Source VENUS with 28 GHz"*, Proceedings of the 16th International Workshop on ECR Ion Sources (ECRIS 04), edited by M. Leitner, AIP Conference Proceedings 749, Melville, New York, 2005, pp. 3-9.

THE 28 GHZ, 10 KW, CW GYROTRON GENERATOR FOR THE VENUS ECR ION SOURCE AT LBNL

M. Marks, S. Evans, H. Jory, D. Holstein, R. Rizzo, P. Beck, B. Cisto *
D. Leitner, C.M. Lyneis, D. Collins, R.D. Dwinell †

CPI, 150 Sohier Rd., Beverly, MA 01915, USA
† Lawrence Berkeley National Laboratory, 88 Inch Cyclotron, 1 Cyclotron Road, Berkeley, CA 94720, USA

Abstract. The VIA-301 Heatwave™ gyrotron generator was specifically designed to meet the requirements of the Venus ECR Ion Source at the Lawrence Berkeley National Laboratory (LBNL). VENUS (Versatile ECR ion source for NUclear Science) is a next generation superconducting ECR ion source, designed to produce high current, high charge state ions for the 88-Inch Cyclotron at the Lawrence Berkeley National Laboratory. VENUS also serves as the prototype ion source for the RIA (Rare Isotope Accelerator) front end [1].

This VIA-301 Heatwave™ gyrotron system provides 100 watts to 10 kW continuous wave (CW) RF output at 28 GHz. The RF output level is smoothly controllable throughout this entire range. The power can be set and maintained to within 10 watts at the higher power end of the power range and to within 30 watts at the lower power end of the power range. A dual directional coupler, analog conditioning circuitry, and a 12-bit analog input to the embedded controller are used to provide a power measurement accurate to within 2%. The embedded controller completes a feedback loop using an external command set point for desired power output. Typical control-loop-time is on the order of 500 mS. Hard-wired interlocks are provided for personnel safety and for protection of the generator system. In addition, there are software controlled interlocks for protection of the generator from high ambient temperature, high water temperature, and other conditions that would affect the performance of the generator or reduce the lifetime of the gyrotron. Cooling of the gyrotron and power supply is achieved using both water and forced circulation of ambient air. Water-cooling provides about 80% of the cooling requirement. Input power to the generator from the prime power line is less than 60 kW at full power. The Heatwave™ may be operated locally via its front panel or remotely via either RS-232 and/or Ethernet connections. Through the RS-232 the forward power, the reflected power, the interlock status and crucial operating parameters are transmitted and tied into the VENUS PLC control system.

The paper describes the gyrotron system, control software, the user interface, the main system parameter, and performance in respect to output power stability.

THE LBNL REQUIREMENT

In brief, the VENUS microwave generator must provide controlled microwave output power at 28 GHz that is up to 10 KW in amplitude. In addition, the generator must interface with the ion source PLC control system for remote operation. The generator must also have personnel safety and self-protection interlocking sub-systems.

CPI's BASIC APPROACH

A generator based around a gyrotron vacuum tube type seemed to be the most reasonable approach to meet this requirement. CPI manufactures gyrotrons of many frequencies and power outputs at its facility in Palo Alto, CA. The VGA8028, has a proven reliability record in several fielded applications, so it was selected as the microwave device around which the generator would be built. In addition, CPI had previously manufactured similar generators using this specific gyrotron type for industrial and research uses, so the basis for the VENUS microwave generator had been well established within the company.

Primary Gyrotron Operating Parameters

Figure 1 shows the gyrotron power output as a function of current in the main magnet with all other parameters held constant. The beam voltage, beam current, and gun magnet current were held to constant values.

Figure 2 shows power output as a function of beam voltage. The 1.08 to 1.16 amp change in beam current results from normal gyrotron beam dynamics with the heater power being held constant. Changing

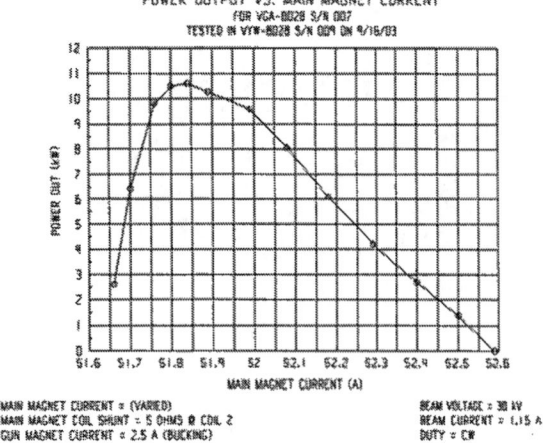

Figure 1.

beam voltage is a very convenient and fast way to change power output over a power range of as much as 20 to 1 with other gyrotron operating parameters either fixed or very slowly changing. In the Heatwave system, primary power control is achieved via beam voltage inputs. Secondary power control is accomplished via main magnet current inputs. The Heatwave controller uses heater power inputs to maintain a relatively fixed beam current from the gyrotron's temperature limited cathode. The beam current setpoint is user setable if operation at other than the factory setpoint is desired. The gun magnet current is fixed over all operating power ranges but it may also be modified via a command input from the user if that is desired.

In gyrotrons, the cathode temperature is such that "temperature-limited" operation results. That is, the temperature of the cathode has a first order and direct effect in determining the gyrotron beam current. The beam-voltage (E_b) has a secondary effect on the value of beam current. In the Heatwave generator, the embedded controller reads the gyrotron beam current on each control loop cycle and it makes appropriate adjustment to the gyrotron heater power in order to maintain a relatively constant gyrotron beam current.

Heatwave Microwave Generator Topology

The power supplies, the embedded system controller, and the controller software combine to provide the inputs necessary for the gyrotron to maintain the desired microwave output level from the Heatwave generator. A block/schematic diagram of the VGA8028 Gyrotron with its output components is shown in figure 3. A block/schematic diagram of the VIA-301 Heatwave generator is shown in figure 4.

In the Heatwave generator, fig. 4, the 480Y277 volt prime power is filtered for EMI and then it is controlled by a 125 amp shunt-trip circuit breaker. The shunt-trip is activated by any input to it from an "emergency off" button-switch, Heatwave enclosure cover/door interlock switch, externally wired emergency off or other switches, and the Heatwave embedded controller. The power is delivered to the beam power supply, main magnet power supply, and the distribution transformer that provides electrical power to the remainder of the system components.

The 500 mS loop-time of the embedded controller is too slow to adequately protect the gyrotron from damage that would be caused by a waveguide arc event, or a gyrotron internal arc event, or a main magnet field that is substantially higher or lower than the window for acceptable operation. There is hard-wired logic in the system for detection of those events

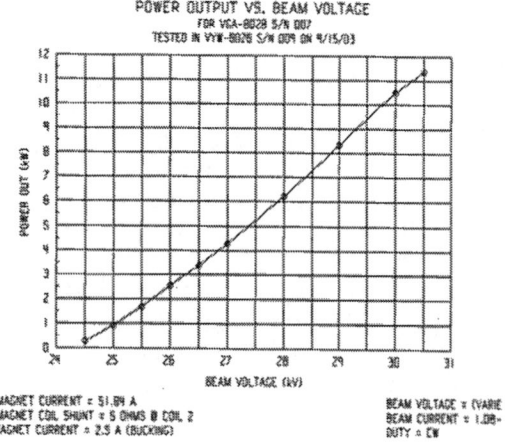

Figure 2.

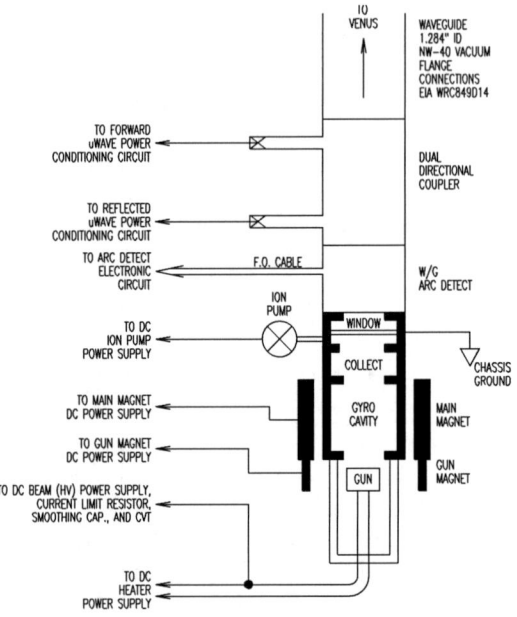

Figure 3. Schematic/Block Diagram Gyrotron and output components.

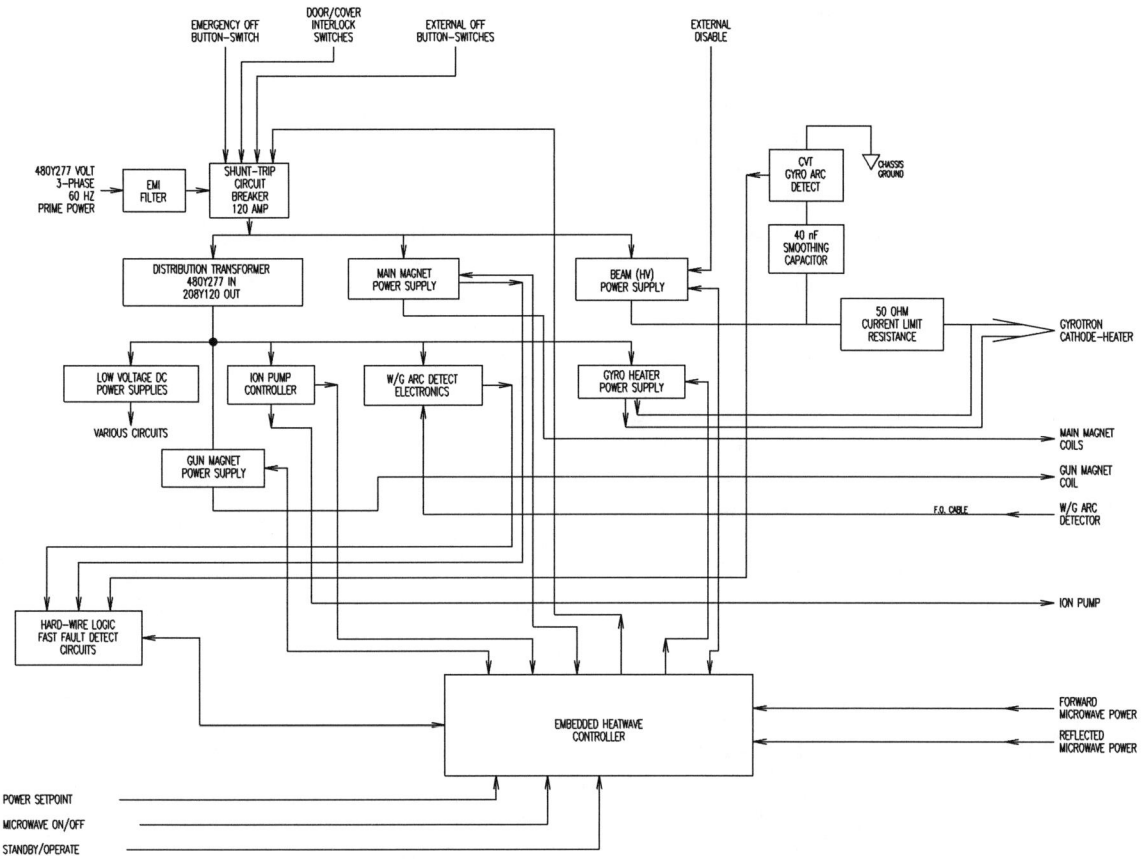

Figure 4. Heatwave Generator Block Diagram

and the beam power supply is inhibited quickly when a fault is detected. The inhibit time is less than 50 µS. If the event is a gyrotron internal arc, the charge stored in the 40 nF smoothing capacitor and about 2 amps of beam power supply follow-on current will flow through that internal arc for the 50 (or less) µS it takes to remove the cathode voltage from the gyrotron. Most of the energy delivered in this time is dissipated in the 50 ohm current limit resistor. Less than 3 joules are dissipated in the gyrotron itself.

The remainder of the interlocking events do not require a fast response. The 500 mS response of the embedded controller is an adequate duration for removing the beam voltage from the gyrotron. So, when those faults are detected, the embedded controller disables the beam (HV) power supply. The exceptions to this are the safety interlocks on enclosure panels/doors that are direct wired to the shunt-trip of the main Heatwave system circuit breaker. To continue operation after a fault event of any type, the Heatwave system must be reset at the Heatwave system front panel.

There are 3 basic operating commands that are input to the embedded controller. Those commands may be initiated at the Heatwave system front panel through switches and a 10-turn knob, or via a serial RS-232 connection, or via a TCP/IP Ethernet connection. When the Heatwave system is set for "local" operation, the controller will recognize only commands originating at the system front panel. When the Heatwave system is set for "remote" operation, the controller will recognize only commands originating at either the RS-232 or Ethernet ports. The basic command set is as follows.

- Standby/Operate
- Microwaves On/Off
- Microwave power setpoint (0 to 10 KW)

The primary supply operating in "Standby" is the ion pump supply. The other supplies are either off or set to very low output levels. The beam (HV) supply is disabled during "Standby."

The controller ramps up the heater supply, magnet supplies, and then the beam (HV) supply during "Operate" to the point where the gyrotron beam is stable at the desired level but the voltage is below the level required for the gyrotron to produce a microwave output.

When the "Microwave On" command is issued to the controller, and the gyrotron beam is stable at the proper current, the controller will make the changes necessary to achieve a microwave power output that is the same as the setpoint level being commanded.

Even for significant changes in the commanded power setpoint, the new desired microwave power output is usually achieved in less than 15 seconds. This result happens because the beam voltage is the first and most quickly acting way to achieve the desired power. The controller also makes appropriate changes in the heater power to achieve the desired beam current. But, since the resulting beam current has a very slow drift in the desired direction, the microwave power output is kept stable by appropriate modification of the beam voltage on each control loop cycle. The beam voltage and current become stable over a period of several minutes even though the power output has long since been stable. The main magnet current and gun magnet current are usually not changed in this process.

The user may elect to change the magnet currents to improve operation at lower power levels if that is desired. The default levels for main and gun magnet currents are aimed at providing better performance in the range from about 1 – 10 KW. The magnet current setpoint changes are accomplished by a command via either the serial or Ethernet port to the embedded Heatwave controller. So, within a few seconds, good control of microwave power output is achieved for levels of power that can be less than 100 watts of microwave power output.

The Heatwave System Embedded Controller

The Heatwave system controller has a configuration file that is maintained in non-volatile flash memory. Constants such as the limits for declaring faults and operating parameters are stored in the configuration file. The complied LabView generated operating software is also stored in the flash memory. The controller is booted upon the application of prime power to the Heatwave system. The operating software is automatically loaded in the processor and the controller first checks the various sensors and components of the Heatwave system before it launches into the normal control loop. The configuration file may be restored to the factory default values via a command through either the RS-232 or Ethernet ports.

The command set for the controller allows for reading the current value of each of the interlocks as well as the raw voltage levels being input to each ADC and the raw voltage or current levels being output at each DAC of the controller. In addition, the current microwave power level in KW is available by using a command query. The controller either responds with the value being requested or "OK" as appropriate. If the command is not proper, the controller responds with a reply that the command was not properly formed. So, when operating the Heatwave system from a remote location, there is always a response to a command/query.

Pulsed output capability

As implemented at LBNL, the Heatwave system can be operated only in the CW mode. However, with appropriate changes to the software used in the Heatwave embedded controller, pulse mode operation may be achieved.

The pulse mode operation would be achieved by inhibit/un-inhibit commands to the beam power supply. That power supply is capable of providing sub-millisecond rise and fall times for the microwave output pulse. Operation at any duty cycle up to CW is acceptable for microwave power outputs of 10 KW or less. Algorithms in the embedded controller and related hard-wired logic circuits would control the flatness of the top of the microwave pulse. Microwave power output in excess of 10 KW could be achieved under some pulse conditions.

For further details about pulsed operation, please contact CPI.

CONCLUSIONS

- Smooth control of microwave output power from a few hundred watts to 10 KW.
- Gyrotron is protected from potentially damaging events such as waveguide arcs, high VSWR, high temperature, overcurrent, overvoltage, and other conditions and events.
- Operating personnel are protected from potential hazards such as excessive leakage of microwave fields, X-rays, magnetic fields.
- The use of a basic command set and a desired microwave power setpoint relieve the microwave generator's human operator from close monitoring and adjustment of the various power supplies that would otherwise be required in a more basic manually operated gyrotron-based microwave generator system. Because of the protections built into the Heatwave system, gyrotron life and maintenance will be positively impacted when compared to a more manually operated gyrotron-based microwave generator.

Influence of Wall-Current-Compensation and Secondary-Electron-Emission on the Plasma Parameters and on the Performance of Electron Cyclotron Resonance Ion Sources

L. Schachter[1], S. Dobrescu[1] and K. E. Stiebing[2]

[1]*National Institute for Physics and Nuclear Engineering, Bucharest, Romania*
[2]*Institut für Kernphysik der Johann Wolfgang Goethe-Universität, Frankfurt/Main, Germany*

Abstract. Axial and radial diffusion processes determine the confinement time in an ECRIS. It has been demonstrated that a biased disk redirects the ion- and electron currents in the source in such a way that the source performance is improved. This effect is due to a partial cancellation of the compensating currents in the conductive walls of the plasma chamber.

In this contribution we present an experiment, where these currents were effectively suppressed by using a metal-dielectric (MD) disk instead of the standard metallic disk in the Frankfurt 14-GHz-ECRIS. Lower values of the plasma potential and higher average charge states in the presence of the MD disk as compared to the case of the standard disk indicate that, due to the insulating properties of its dielectric layer the MD disk obviously blocks compensating wall currents better than applying bias to the metallic standard disk.

A comparison with results from experiments with a MD liner in the source, covering essentially the complete radial walls of the plasma chamber, clearly demonstrates that the beneficial effect of the liner on the performance of the ECRIS is much stronger than that observed with the MD-disk. In accord with our earlier interpretation, it has to be concluded that the "liner-effect" is not just the effect of blocking the compensating wall currents but rather has to be ascribed to the unique property of the thin MD liner as a strong secondary electron emitter under bombardment by charged particles.

INTRODUCTION

One generally accepted method to increase the ion-beam output and to improve the charge state distribution (CSD) from an ECRIS is to install a negatively biased metallic disk at the injection side of the source. In a previous work [1] it has been demonstrated that his effect is an essential redirection of the axial and radial components of the electron and ion diffusion currents by influencing the compensating wall currents (Simon currents) in the plasma chamber walls of the source, which are required to allow nonambipolar diffusion [2]. Therefore, the occurrence of the biased disk effect may be taken as evidence for the quite strong nonambipolarity of the magnetically confined ECRIS plasma, which favors radial losses of the magnetically weakly bound ions and axial losses of strongly bound electrons. The biased disk effect is therefore an effective hindrance (partial blocking) of the compensating wall currents.

Previous good experience with a metal-dielectric (MD) structure of the type $Al-Al_2O_3$ instead of a biased metallic disk [3, 4], led to the idea of using a MD cylindrical liner as plasma chamber wall. With this setup remarkable improvements of ECRIS performances could be achieved at two basically different 14 GHz ECRIS installations [5, 6], which were much superior to those improvements gained by the use of the biased disk. In dedicated experiments it was shown that the mechanism of the biased disk effect is quite different from the effect of a MD liner. Whereas the use of a liner goes along with an increase of the plasma potential and distinct changes of intrinsic plasma parameters (higher electron- and hence ion density, higher plasma temperature etc), the biased disk effect leaves the basic plasma properties practically unchanged and leads only to a decrease of the plasma potential.

In order to further pursue this essential difference, we have performed a new experiment at the Frankfurt 14 GHz ECRIS, where the compensating wall currents from the disk to the radial walls were completely suppressed by the use of metal-dielectric (MD) disks instead of the standard metallic biased disk. The influence of the MD disk on the plasma parameters, on the average charge state values and on the extracted argon ion currents was measured and compared with the data for the biased metallic disk.

EXPERIMENTAL SETUP AND PROCEDURES

During these experiments, the Frankfurt 14-GHz-ECR ion source was operated with pure argon as working gas. A standard stainless steel disk of 8 mm diameter was mounted on-axis. Its position and bias were optimized by maximizing the intensity of the extracted ion currents measured in a Faraday cup after the 90°-analyzing magnet.

The MD disks were made of 1 mm thick, 17 mm diameter aluminum plates. The dielectric layer consisted of an Al_2O_3 film formed by a special electro-chemical process. Two types of disks were used, one with a thin dielectric layer that was tested to provide a high secondary electron emission coefficient and one with a thick dielectric layer (at least ten times thicker), which had much lower coefficients of secondary electron emission. The MD disks were mounted on top of the stainless steel disk with the MD-layer facing the plasma. In order to test the essentially insulating character of the MD-layer of the disk ($R_{layer} > 10^9 \, \Omega$), we applied bias on the backside of the disk. No changes of the ECRIS performance were observed. Therefore all MD-disk experiments were carried out with 0 V on the disk-rod. Both MD-disks were of good shape after their exposure in the experiments and have kept their electrical properties.

For the purpose of plasma diagnosis, a moveable Langmuir probe (tungsten wire of 0.5 mm diameter with an active length of 4 mm) was inserted at the microwave injection side of the ECRIS [7]. Classical Langmuir plasma diagnosis was applied. The probe crossed the source axis close to the maximum of the axial magnetic field. In order to avoid that the disk was shadowed by the probe, positions well outside this position were used for the measurements reported here.

The voltage-current characteristics of the probe allowed determining the plasma potential U_{Plasma} at the position of the Langmuir probe, well outside the ECR zone. However, they certainly reflect the situation inside the plasma and their relative changes are therefore indicative of the evolution of these parameters inside the plasma.

The CSD spectra were taken after optimizing source and ion transport for the charge states Ar^{8+} and Ar^{11+}. In order to monitor the energies of the hot electrons in the ECR plasma, Bremsstrahlung spectra were recorded. A shift toward higher energies was observed when the MD disk was inserted as compared to the standard configuration.

RESULTS AND COMMENTS

For stable plasma, quasi neutrality demands for equal net loss rates of positive (ions) and negative (electrons) charges. Due to their higher mobility, electrons tend to escape much faster from the plasma than the ions. As a result, a positive plasma potential builds up in the sheath of the plasma in order to retard the escape of electrons and to accelerate the escape of ions from the plasma for maintaining quasi neutrality. Although this process is very similar to the ambipolar diffusion effect in a lowly ionized plasma one has to bear in mind that in an ECRIS the strong magnetic confinement in conjunction with conducting plasma chamber walls leads to a significant nonambipolar diffusion (electrons escaping axially, ions escaping radially). Irrespective of this difference, the average ion confinement time in the plasma is linked to the net loss of the electrons and hence to the average plasma potential.

By applying bias to a metallic disk at the injection side of the ECRIS the diffusion of charges in the plasma is fundamentally changed [1]. The ion diffusion takes place axially and the electron diffusion is hence dominated by radial movement, resulting in a better extraction of ions through the extraction opening. As a consequence the total beam extracted from the source may be drastically improved. This additional positive loss current is compensated by a reduced plasma potential as observed by Mironov et al. [8]. As a secondary effect, by better extracting the highly charged ions, the effective ion confinement for lowly charged ions is enlarged and hence the CSD is shifted towards higher charge states. It has been proven by Bremsstrahlung measurements, that this effect does not coincide with strong changes of the plasma parameters (i.e. higher electron density etc). This is in clear contrast to the case of an MD-liner, where both electron density and temperature are enhanced and are reflected by a higher plasma potential [9].

The aim of the experiment presented here is to study the physics of the ECR plasma source when the disk is non conducting and to compare it to the case of a standard metallic disk and to the case of a MD-liner, covering the radial walls of the plasma chamber [5, 6].

A synthesis of our measurements is presented in Table 1. The main consequence observed when the MD disk is introduced is a shift of the CSD to values higher than in the case of the optimized biased metallic disk. As a clear indication of the fact that this is related rather to the non conducting properties of the MD-disk than to the strength of its secondary-electrons emission, the average charge state q_{mean} for the thick MD-disk is even higher than that of the high emissive thin disk.

Table 1. Results from the measurements			
	Metallic disk (standard)	MD-disk thin	MD-disk thick
Disk Voltage	0 V	- 400 V	0 V	0 V
q_{mean}		7.091	7.785	8.358
U_{plasma}	34.3	29	27.6	27
Ratio 1 for Ar^{11+}			6.3	16
Ratio 1 for Ar^{12+}			10	33
Ratio 2 for Ar^{11+}		2.19		
Ratio 2 for Ar^{12+}		1.11		

Ratio 1 is defined as I_{MD}/I_{BD} of ion currents for the case of the MD-disk (MD) relative to the case of the biased standard disk (BD) (U_{BD}=-400V). No bias was applied to the back of the MD-disk.

Ratio 2 is defined as $I_{(BD = 400V)}/I_{(BD = 0V)}$ of ion currents for the biased standard disk (U_{BD}=-400V) relative to the unbiased standard disk (U_{BD}=0V).

Compared to the unbiased standard disk the plasma potentials U_{plasma} are lower by 4-6V for the optimized biased standard disk and they are lower by 6-10V for the MD disk. The relative enhancement of the high-charge-state components in the CSD is illustrated by the ratios I_{MD}/I_{BD} (Ratio 1 in table 1) of the yields of Ar^{11+} (Ar^{12+}) ions for the unbiased MD disk relative to their yield for the optimized biased disk. This ratio amounts to 6 (10) for the thin MD disk and 16 (33) for the thick MD disk. That this effect is not a generally higher ion production but is rather a shift of the CSD towards higher charge states becomes evident if one analyzes the ratio of relative weights of two high charge states per CSD:

$$I[Ar^{11+}/Ar^{8+}]_{MD} / I[Ar^{11+}/Ar^{8+}]_{BD} = 3.86$$

All data indicate that the effect of the MD disk seems to be similar but more efficient than the biased-disk effect. The higher argon ion beam current for the thick MD disk probably is due to an additional gas mixing effect by the presence of contaminations by light elements (O_2, C, etc), which always is observed in conjunction with MD layers. It cannot be avoided due to the processing of the layers. Its relative amount is proportional to the thickness of the layer and was a significantly higher contribution for the thick layer than for the thin layer. .

The enhancement effect by biasing the disk (ratio 2 in table1) is much lower than the effect by insertion of a MD-disk (ratios 1 in table1) and exhibits the opposite effect (biasing is less effective for the higher charge state)

We summarize that the secondary electron emission obviously plays no role when the MD structure is inserted as a disk on the end plate at the injection part of the source. This may be understood as follows. Due to the bombardment of the disk by ions through the loss cone, the dielectric layer on the surface of the disk is positively charged. As a consequence electric loop between the side walls and the end plates is interrupted. This is in contrast to the standard disk, which, dependent on the imposed bias always allows a certain amount of positive current from the disk to the walls. This current, however, is the dominant contribution to the non ambipolar diffusion. In this way the MD disk restores ambipolarity. It has been pointed out elsewhere [2] that an ambipolar plasma should provide the longest ion lifetimes and hence should favor the creation of more intense high charge states, as observed in the experiments reported here.

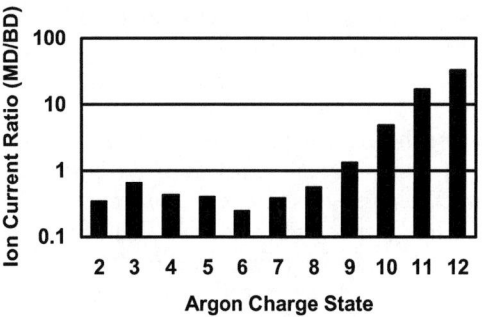

Fig. 1 The current ratios MD disk/ Biased disk for different argon charge states.
(P_{RF} = 600 W, p_{inj} = 3.5 10^{-7} mbar)

In Fig. 1 the influence of the MD disk on the argon charge state distribution relative to the optimized standard disk are presented. It is evident that high charge states are significantly enhanced whereas low charge states are reduced.

The general operation of an emissive MD structure (MD-liner) inside an ECRIS has been demonstrated by our measurements of 1999 and 2000 [5, 6]. The massive bombardment with energetic particles from the plasma creates an extremely high electric field strength inside the dielectric layer, which, in turn, leads to an intense secondary electron emission from the MD structure into the plasma. This mechanism has been verified by experiments. The Langmuir probe measurements indicated that the electron density of the plasma was increased by 250%. The higher density gradient due to a higher electron density from the electron secondary emission results in a higher plasma potential. An increase of the argon plasma potential by about 30 % (from 26 V to 35 V) has been measured when the liner was inserted.

Since in stable plasma the net current of electrons and ions to the walls have to be zero, the enhancement of ion lifetimes has to be accompanied by an enhancement of the electron lifetimes as well, allowing a better heating of the electron component. It was observed that the average electron temperature measured by the Langmuir probe method is increased by about 70 % in the presence of the MD liner [9].

The important prerequisite for this effect is the build up of the strong electric field inside the layer. In order to achieve these high fields, the MD structure has to be sufficiently thin and the bombardment by charged particles has to be intense and energetic enough.

As pointed out, the non ambipolar diffusion in an ECRIS with extraction voltage directs the ions to the end plates and the electrons to the radial walls. Due to the low ion energy ($E_i \approx 10^{-3} \times E_e$), a comparatively small potential at the surface of the MD disk already leads to a complete rejection of these ions. This surface potential is not sufficient to create secondary electron emission by the layer, and hence the disk simply works as a perfect blocker. In contrast to this, the radial walls are bombarded with high energetic loss electrons allowing the build-up of much higher surface potentials as they are needed to create the positive effect, described above.

From our new experiment it is evident that the influence of the MD liner must not be confused with the influence of a normal insulator, which allows no charge transfer over the barrier. For strictly isolating walls it plays no role where the current is interrupted. Hence one would not expect any difference between the experiments with a MD liner and with a MD disk in clear disagreement to our results. This experiment also demonstrates that just cutting the compensating currents is clearly less effective for the source performance than to use a MD-liner.

ACKNOWLEDGMENTS

This research was partly supported by the Center of Excellence IDRANAP under Contract with European Commission Grant No. ICAI-CT-2000-70023. The Frankfurt ECRIS-(ve)RFQ facility is a HBFG project of Hessisches Ministerium für Wissenschaft und Kultur (HMWK) and Deutsch Forschungsgemeinschaft (DFG) project number: III P 2-3772-116-246.

REFERENCES

1. K.E.Stiebing, L. Schmidt, H. Schmidt- Bocking, V. Mironov, G. Shirkov, S. Biri., Proceedings of the 15th International Workshop on ECR Ion Sources, Editors J.Arje, H.Koivisto, P.Suominen, Jyvaskyla, Finland, 2002, p. 146.
2. G. Drentje, U. Wolters, A. Nadzeyka, D. Meyer, and K. Wiesemann, *Rev. Sci. Instrum.* **73**, 516, 2002.
3. L. Schachter, S. Dobrescu, Al.I. Badescu-Singureanu, *Rev. Sci. Instrum.* **69**, 706 (1998).
4. L. Schachter, K. E. Stiebing, S. Dobrescu, Al. I. Badescu – Singureanu, L. Schmidt, O. Hohn, S. Runkel, *Rev. Sci. Instrum.* **70**, 1367 (1999) .
5. L. Schachter, K. E. Stiebing, S. Dobrescu, Al.I. Badescu-Singureanu, S. Runkel, O. Hohn, L. Schmidt, A. Schempp and H. Schmidt-Böcking, *Rev. Sci. Instrum.* **71**, 918 (2000).
6. L. Schachter, S. Dobrescu, G. Rodrigues, A.G. Drentje, *Rev. Sci. Instrum.* **73**, 570 (2002).
7. Homepage of the ECR-RFQ – Group at IKF (http://hsbpcl.ikf.physik.uni-frankfurt.de/ezr/)
8. V. Mironov, K. E. Stiebing, O. Hohn, L. Schmidt, H. Schmidt- Bocking, S. Runkel, A. Schempp, G. Shirkov, S. Biri, and L. Kenez, *Rev. Sci. Instrum.* **73**, 623 (2002)
9. L. Schachter , S. Dobrescu , K. E. Stiebing, *Rev. Sci. Instrum.* **73** ,4172 (2002).

2Q-LEBT Prototype for the RIA Facility

N.E. Vinogradov[†], V.N. Aseev, M.R.L. Kern, P.N. Ostroumov, R.C. Pardo, R. Scott, R.C. Vondrasek

Argonne National Laboratory, 9700 S. Cass Ave., Argonne, IL 60439, USA

Abstract. The Rare Isotope Accelerator (RIA) facility utilizes the concept of simultaneous acceleration of two charge states from the ion source. We are building a prototype two charge-state (2Q) injector of the RIA Driver Linac, which includes an ECR ion source, a LEBT and one-segment of the prototype RFQ. Currently, the 2Q-LEBT Facility consists of Berkeley Ion Equipment Corporation BIE-100 ECR ion source. The rf transmitters, high voltage power supplies, turbo pumps and other related equipment were received with the source. BIE-100 is an all-permanent-magnet source and has the highest magnetic field strengths for an ECR ion source of this type ever built. The magnetic field achieves a maximum strength of 11 kG at the plasma chamber surface and 13 kG on the axis. The source can operate with two-frequency plasma heating of 12.75 and 14.5 GHz. The reassembly of the source has been completed and beam production was achieved in the June 2004. This report includes measured beam current and emittance for ^{16}O from the source along with the beam dynamics simulations. Detailed design of the 2Q-LEBT and the current project status are also presented.

INTRODUCTION

The concept of simultaneous acceleration of two charge states in the RIA Driver Linac was developed earlier [1]. The proposed design of the RIA Driver Front End [2] may require production of multiple-charge-state beam in the cw regime. An ECR ion source is the only type of source that can serve this purpose. We have undertaken to build a prototype 2Q-injector which includes an ECR ion source, a LEBT and one-segment of the prototype RFQ. The project, called the 2Q-LEBT Facility, is being developed at the high bay area of the ANL Physics Division Dynamitron. At present, the Facility consists of BIE-100 ion source [3], which was designed and built by Berkeley Ion Equipment Corporation and transferred to ANL with related equipment as part of termination of a SBIR Phase II project. BIE-100 is an all-permanent-magnet source utilizing NdFeB magnetic material, grade N45H, and is the first all-permanent-magnet source developed in the USA. It has a number of advantages compared to ion sources built with electromagnets such as moderate dimensions, simpler source structure, easier operation and lower cost. An additional important feature is possibility to bias the source up to 30 kV or higher for better ion beam extraction and transport. Reassembly and commissioning of the BIE-100 ion source have been completed at ANL. Accurate measurements of the output beam emittance are important for further optimization of the source parameters as restrictions on the emittance growth and requirements for the beam quality are very stringent in the 2Q-LEBT. A diagnostic station that includes an adjustable slit system, removable Faraday cup and emittance measurement device has been built and added to the Facility. Description of the BIE-100 commissioning and results of the first run of the source are the main subjects of this report.

BASIC DESIGN OF THE 2Q-LEBT

The Front End of the Driver Linac must form high quality dual charge state heavy-ion beams with total transverse normalized emittance ~0.6 π mm-mrad and longitudinal emittance ~2 π kev/u-nsec at 99.5% level

[†] Corresponding author

of total intensity. The front end consists of an ECR ion source located on a high-voltage platform, low energy beam transport (LEBT), RFQ and medium energy beam transport (MEBT) [2]. The LEBT includes two main sections: 1) an achromatic bending system for charge-to-mass analysis that can select one or two charge-state heavy-ion beams; 2) a straight section that forms the longitudinal emittance and provides beam matching to the following RFQ (see Fig. 1). The bending system consists of two 60° bending magnets, six electrostatic quadruple lenses and a solenoid. A high dispersion area is formed by the first magnet where the required one- or two-charge state beams can be defined and transported to the RFQ.

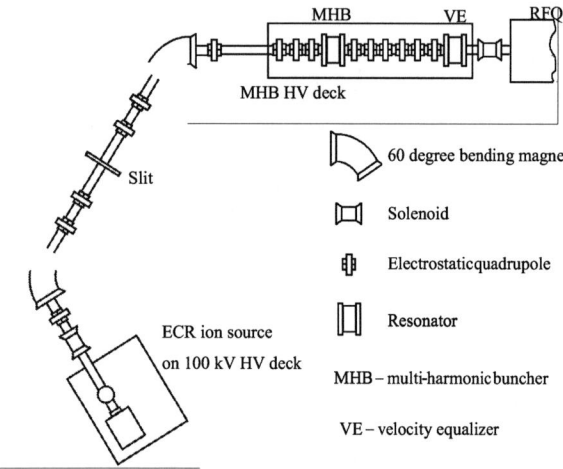

FIGURE 1. Layout of the RIA Driver Front End.

Our plans for development of the 2Q-LEBT prototype comprise of installation of the BIE-100 on a 100 kV high voltage platform, building of the achromatic bending system and transport system including a multi-harmonic buncher (MHB), and a full power 57.5 MHz RFQ segment [4]. The design of the 60° bending magnets and MHB resonators is now in progress. The 57.5 MHz RFQ segment is being fabricated. Full tests of 2Q-LEBT capability are planed during FY05.

NUMERICAL SIMULATION

Beam dynamics simulations in the 2Q-LEBT were carried out using the dedicated code TRACK [5]. TRACK transports a multiple-charge-state beam through three dimensional beam-optics fields including space-charge effects. In our simulations the ions are emitted uniformly from the plane plasma outlet surface. Velocities of the extracted ions are uniformly distributed within a loss cone $\sin\theta \leq B_{center}/B_{extract}$ (θ is angle between normal to outlet plasma surface and the ion velocity; B_{center}=3.7 kG is the axial magnetic field in the ECR center; $B_{extract}$=6.5 kG is the axial magnetic field in plasma outlet hole). TRACK simulations do not consider the ion production and neutralization processes and plasma effects. We plan to perform more accurate simulations using other independent codes. In ECR sources, rotation of the extracted beam due to decreasing magnetic field gives the dominating contribution to the ion beam emittance [6]. The TRACK simulations of the beam extraction produce reasonable parameters of the ion beam for dynamics simulations in the 2Q-LEBT (see Fig.2; M/Q<10 and $B_{extract}$=6.5G). Comparison of calculated and measured transverse emittances is given in the next chapter.

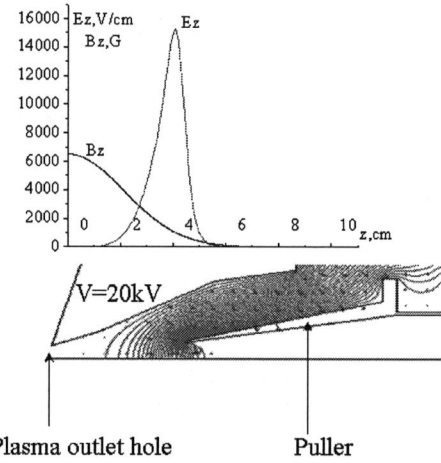

FIGURE 2. Axial magnetic and electrical field distribution along the axis in the extraction region.

COMMISSIONING OF THE SOURCE

Once the BIE-100 was received at ANL, we started installation, assembly and commissioning of the source to integrate it into the 2Q-LEBT Facility. During the commissioning the injection assembly of the source was redesigned and manufactured to increase the beam production performance. The structure of cooling channels was changed to optimize the temperature distribution along with fabrication simplicity. We also extended the heating oven input and combined it with the production gas inlet unlike the original design where the gas input was provided through one of the rf waveguides. A new Einzel lens and puller with its base were fabricated from stainless steel. The ceramic insulator between the ECR source body and beamline was originally designed as one piece and had the shape of a cylinder with two flanges. Such a construction did not allow any flexibility and one of the flanges was broken during installation. To prevent a possibility of breakdown in the future we made this unit consisting of two parts: the ceramic cylinder with one flange on it attached to the source body and aluminum adapter flange with O-ring groove that seals the inside diameter of the ceramic cylinder. The further tests have shown that this design allows enough mechanical flexibility and provides a good vacuum isolation as well. Also, we designed and built a new interlock system and X-ray and high voltage shielding to satisfy ANL safety requirements. The shielding is equipped with sliding lead doors locked by Kirk key, which is incorporated into the interlock system of the Facility. The BIE-100 operates with the possibility of two-frequency plasma heating. The 12.75 GHz rf power is provided by a 650 W traveling wave tube amplifier (TWTA) while the microwaves at 14.5 GHz are produced by a 2 kW klystron amplifier. Both amplifiers were manufactured by Communications & Power Industries (CPI). Two separate rf drivers are used to generate the input signals to the amplifiers. At present the BIE-100 operates at only the primary frequency of 14.5 GHz due to breakdown of the traveling wave tube in the TWTA. Two-frequency heating will be restored as soon as the replacement tube will be available.

consists of a system of adjustable slit, removable Faraday cup and emittance measurement device (EMD), see Fig. 3. The slit system consists of four linear actuators with rectangular plates on each one thus forming one vertical and one horizontal slit of adjustable size. The emittance scanner is based on conventional combination of narrow slit and wire for each transverse direction. Two EMD slits are cut in the same stainless steel plate, which is inserted into the beamline by a linear actuator orientated at 45° with respect to the vertical line. The slits are perpendicular to each other and cut at 45° with respect to the axis of actuator motion. Therefore, the beam is collimated first by the vertical and then by the horizontal slit during one linear movement. The scanning wires work the same way. The remote control of the EMD equipment, data acquisition and data analysis are provided by user-friendly software implemented using LabVIEW 6.1. The code affords transverse beam profiles, emittances and Twiss parameters online. Another LabVIEW code incorporates remote control of the analyzing magnet power supply, gaussmeter and picoammeter connected to the Faraday cup, and provides users with online mass scan analysis. Later on, both codes will be included into integrated software environment enabling full remote control of the whole 2Q-LEBT Facility.

FIGURE 3. Diagnostic station of the BIE-100 including adjustable slit system (a), removable Faraday cup (b) and emittance measurement device (c).

For careful studies of the output beam we have designed and built the diagnostic station, which

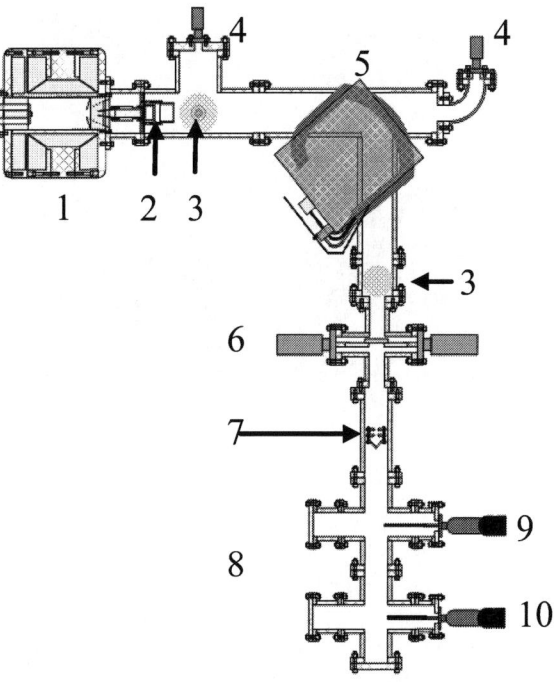

FIGURE 4. Current layout of the 2Q-LEBT Facility: 1-source body; 2-Einzel lens; 3-turbo pump; 4-vacuum gauge; 5-analyzing magnet; 6-slit system; 7-Faraday cup; 8-EMD; 9-EMD slits actuator; 10-EMD target wires actuator

The layout of the final experimental setup is shown on Fig. 4. The first tests of the source after commissioning were run with Oxygen as a production material. The ion beam extraction was carried out at 15÷25 kV of source bias. Ion beam current measurements were performed with the Faraday cup located between slit system and EMD (see Fig. 4). The Faraday cup was biased at -150 V to suppress the secondary electrons coming off it.

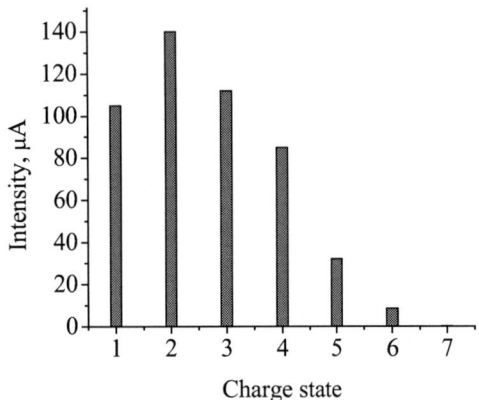

FIGURE 5. Oxygen beam from the BEI-100.

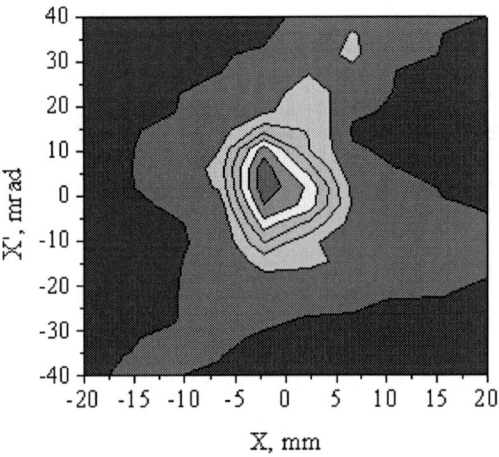

FIGURE 6. View of the $^{16}O^{2+}$ beam measured with EMD. The normalized emittance is 0.11 π·mm·mrad.

The voltage applied to the Einzel lens was limited by sparking due to interaction of the secondary electrons with the lens electrode. The higher level of rf power was input to the source the lower value of the Einzel lens voltage was allowed. At 250 W rf power it was possible to apply about 9 kV to the lens which is not enough for appropriate beam focusing. Therefore, the heating microwave power level was limited by beam dynamics properties in the extraction region.

This early limited source performance is shown in Fig. 5. Figure 6 represents an example of transverse beam emittance measured by EMD. The measured and calculated transverse emittances are 0.11 π·mm·mrad and 0.13 π·mm·mrad respectively and are close to each other. In the near future we plan to improve the geometry of the Einzel lens electrode to eliminate the influence of secondary electrons on its operation.

CONCLUSIONS

Construction of the 2Q-LEBT Facility is in progress. First beam from the BIE-100 ECR ion source has been achieved and characterized. The further development of the 2Q-LEBT prototype will allow us to carry out the experimental investigations of the multiple-charge-state concept and to establish the final design of the RIA Driver Injector.

REFERENCES

1. P.N. Ostroumov et al, Proc. of the XX LINAC Conf., Monterey, CA, August 21-25, 2000, p. 202.
2. A.A. Kolomiets et al, Proc. of the 2003 PAC Conf., Portland, OR, May 12-16, 2003, p. 2876.
3. Dan Z. Xie, Rev. of Scient. Instr, vol. 73 (2), p. 531, 2002.
4. N.E. Vinogradov et al, Proc. of RIA Research and Dev. Workshop, August 26-28, 2003, Bethesda, Maryland, USA. 6. 2.
5. P.N. Ostroumov et al, PRST-AB, vol. 7, 000101 (2004).
6. M.A. Leitner, D.C. Wutte, C.M. Lyneis, Proc. of the 2001 PAC Conf., Chicago, IL, June 18-22, 2001, p. 67.

Formation of Ion Beam from High Density Plasma of ECR Discharge

I. Izotov, S. Razin, A. Sidorov, V. Skalyga, V. Zorin

Institute of Applied Physics, Nizhny Novgorod, Russia

R. Geller, T. Lamy, P. Sortais, T. Thuillier

The Laboratoire de Physique Subatomique et de Cosmologie, Grenoble, France

Abstract. One of the most promising directions of ECR multicharged ion sources evolution is related with increase in frequency of microwave pumping. During last years microwave generators of millimeter wave range - gyrotrons have been used more frequently. Creation of plasma with density 10^{13} cm^{-3} with medium charged ions and ion flux density through a plug of a magnetic trap along magnetic field lines on level of a few A/cm^2 is possible under pumping by powerful millimeter wave radiation and quasigasdynamic (collisional) regime of plasma confinement in the magnetic trap. Such plasma has great prospects for application in plasma based ion implantation systems for processing of surfaces with complicated and petit relief. Use it for ion beam formation seams to be difficult because of too high ion current density. This paper continues investigations described in [1] and shows possibility to arrange ion extraction in zone of plasma expansion from the magnetic trap along axis of system and magnetic field lines.

Plasma was created at ECR gas discharge by means of millimeter wave radiation of a gyrotron with frequency 37.5 GHz, maximum power 100 kW, pulse duration 1.5 ms. Two and three electrode quasi-Pierce extraction systems were used for ion beam formation.

It is demonstrated that there is no changes in ion charge state distribution along expansion routing of plasma under collisional confinement. Also ion flux density decreases with distance from plug of the trap, it allows to control extracting ion current density. Multicharged ion beam of Nitrogen with total current up to 2.5 mA at diameter of extracting hole 1 mm, that corresponds current density 320 mA/cm^2, was obtained. Magnitude of total ion current was limited due to extracting voltage (60 kV). Under such conditions characteristic transversal dimension of plasma equaled 4 cm, magnetic field value in extracting zone was about 0.1 T at axisymmetrical configuration.

EXPERIMENTAL SETUP

Ion beam was formed from plasma created at ECR gas discharge by means of millimeter wave radiation of a pulsed gyrotron with frequency 37.5 GHz, maximum power 100 kW, pulse duration 1.5 ms, confined in a magnetic trap of cusp type. It is demonstrated [2] that quasigasdynamic (collisional) regime of plasma confinement with plasma density 10^{13} cm^{-3}, electron temperature 50 – 100 eV and with ion life time about 10 μs is possible under such conditions. Forming inside the trap, plasma spread along magnetic field lines and in particular along system axes over the trap plug, as it is shown in Fig.1 Extractor was placed on the trap axis and could move along it that enables to regulate ion current density on plasma electrode. Two and three electrode quasi-Pierce extraction systems were used for ion beam formation. Ion current was measured by the first Faraday cup with a diameter exceeding the diameter of puller. The first Faraday cup was located just downstream of the puller. All the elements of the system of electrodes (plasma electrode, puller, Faraday cup) were moving with constant locations relatively to each other. Ion lens was used for parallel ion beam formation with following analysis of ion charge states. Ion lens was placed also downstream of the puller instead of the Faraday cup. Its installation demanded to open vacuum chamber. The second Faraday cup (used for measurement of ion beam current after analyzer) had diameter of aperture 30 mm.

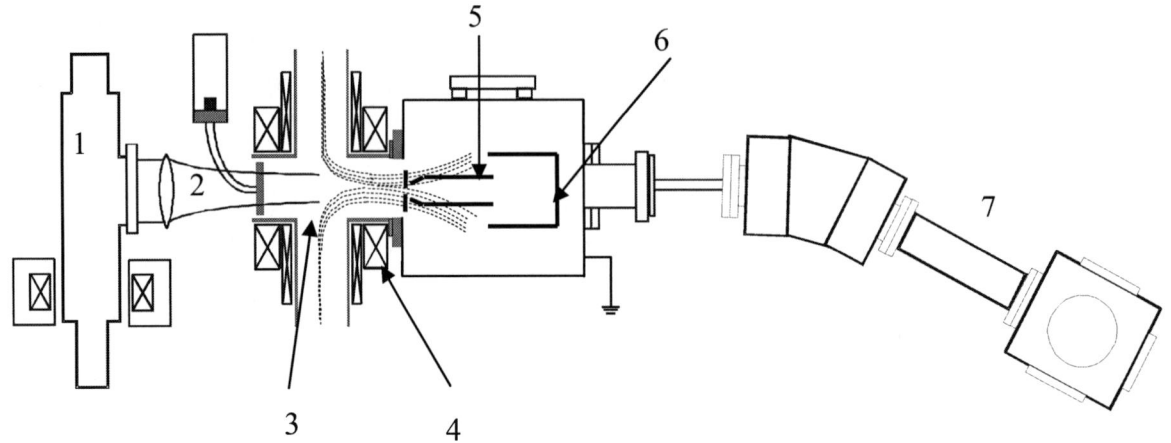

FIGURE 1. Experimental Setup. 1 – gyrotron, 2 – microwave beam, 3 - vacuum chamber, 4 – coils of magnetic trap (some magnetic field lines are shown on the figure), 5 - extractor, 6 – the first Faraday cup or it is possible to put ion lens instead, 7 - ion analyzer.

FORMATION OF ION BEAM

It is known that mean ion beam charge depends on the location of plasma electrode of extractor. Practically all ECR sources of multicharged ions (MCI) work when classical (Pastukhov) regime of plasma confinement is realized. In this case ion blocking in the trap realizes by potential electrostatic dip. Highly charged ions are confined in the dip quite well; therefore introduction of the system of extraction over energy barrier is necessary for their extraction. In the case of quasigasdynamic (collisional) confinement potential distribution along axis has different characteristics [3], potential electrostatic dip is absent. Ambipolar electric field accelerates electrons throwing them out of the trap. Thus ion distribution over charge states has not to depend on extractor location relatively to the trap plug (assume the extractor is always on the axis of the system). For verifying this statement the measurement of ion distributions over charges with different distance between the plug and plasma electrode was conducted. With approximate experimental accuracy 20% (due to necessity to open the chamber) it is demonstrated that there is no changes in ion charge state distribution when extractor is moving and it looks as it is shown in Fig. 2. The distribution is obtained under extracting voltage (20 kV). This voltage is lower than optimum for extraction under such ion current density. Thus distribution has some distortions which overstate hydrogen current value significantly. Nevertheless amplitude of the nearest nitrogen lines are reflected by the distribution with appropriate accuracy.

Creation of intensive ion flux from the trap is possible under quasigasdynamic (collisional) regime of plasma confinement. Magnitude of this flux can

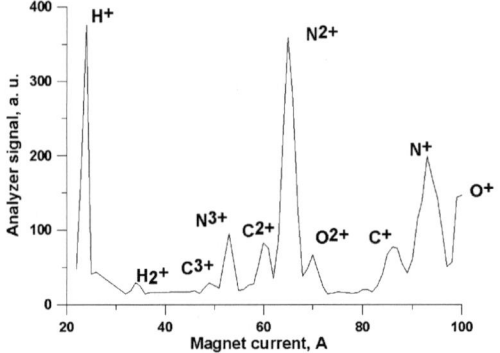

FIGURE 2. Ion spectrum, extracting voltage 20 kV

reach several A/cm^2. Ion flux density decreases with plasma expansion along magnetic field lines. It is given in Fig.3 where curve demonstrates calculation result of ion flux out of the trap with the diameter of extracting hole 1 mm. The calculation was made for plasma characteristics corresponding to the given experimental conditions with the assumption that ion current density is inversely proportional to the magnitude of magnetic field (plasma spread along magnetic field lines) [3]. Experimental points correspond to measured total Faraday cup current (located downstream of the puller) and puller current. At every distance measurements were made for optimum voltage under which Faraday cup current reaches its maximum. When the distances between plasma electrode and the plug were less then 15 cm (current density was too high even if maximal extracting voltage 55 kV applied) magnitude of optimum current was not realized and current of Faraday cup enhanced with the voltage increase up to maximal magnitude. At the voltage higher than 55 kV probability of breakdown between plasma electrode and puller was too high and made measurements impossible.

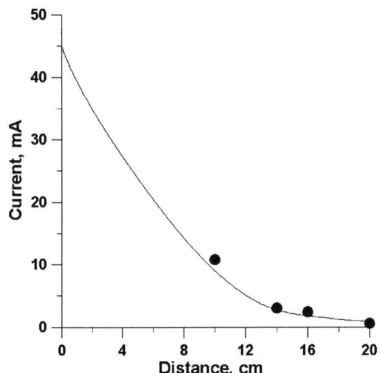

FIGURE 3. Experimental magnitude of total Faraday cup current and puller current (dotes) and results of current calculation (curve).

Proportion of magnitudes of the first Faraday cup current and puller current depended on extracting voltage and density of ion current. Under high ion current density intensive Coulomb ion pushing arises, it leads to ion flux dispersion and to partial spread of ion current on puller. Maximal Faraday cup currents obtained under different distances between plug of the trap and plasma electrode are given in Fig.4 (squares). Magnitudes of puller currents (obtained under the same experimental conditions) are also demonstrated in Fig.4 (crosses). It is clear that under fixed shape of extractor electrodes and their constant locations relatively to each other and fixed maximum extraction voltage, there is an optimum ion flux density with which Faraday cup current has maximum value. In present case (for two electrode extraction system) the magnitude was 250 mA/cm². Using three electrode extraction system enabled to increase total Faraday cup current up to 2.5 mA that corresponds ion current density from plasma 320 mA/cm². It is quite agreed to the magnitude obtained from simplified model. The model assumes that optimal current magnitude (current, under which flux dispersion is minimal) is realized when the thickness of double plasma layer

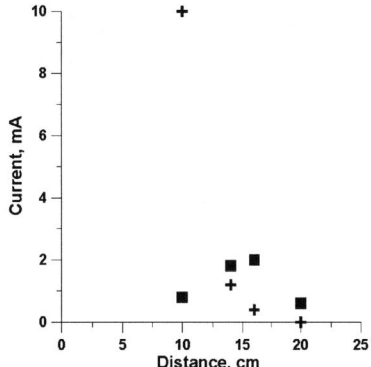

FIGURE 4. Faraday cup (squares) and puller (crosses) currents.

equals the distance between electrodes (meniscus is absent) [4]. Thus optimal current is defined by Child-Langmuir equation

$$j_{opt}(mA/cm^2) = \frac{1.7}{\sqrt{M}} \cdot \frac{U^{3/2}}{L^2},$$

where L - length of accelerating interval (here L is the sum of intervals between electrodes, thickness of plasma electrode and puller hole radius $\approx 0.7 mm$), U-accelerating voltage (50 kV), M equates to 14 for nitrogen, so $j \approx 300 mA/cm^2$.

ION CHARGE DISTRIBUTION

The use of ion lens enabled to focus ion flux and analyzed it by turn magnet and the second Faraday cup located at the distance of 1.5 m from the plug trap. The ion analyzer is a focusing system with focus distance equals to 50 cm. With this scheme optimal ion focusing in the second Faraday cup realized under parallel flux dispersion after the lens. Ion distribution over charge states under extracting voltage 40 kV is shown in Fig. 5. Magnitude of N^+ ion current reconstructed over the spectrum got under voltage 20 kV. It was made this way because under given conditions analyzer current magnitude restricted to 200 A, at that it is impossible to detect N^+ ions by the analyzer when extracting voltage exceeds 20 kV. Errors of N^+ current determination obtained under such extrapolation do not seem to exceed 20%. Maximal achieved current of N^{2+} equals to 0.6 mA. Demonstrated in diagram total ion current is 1.3 mA. Measurements of the first Faraday cup current are necessary for transmission factor of ion line determination. Configuration of experimental setup makes impossible to conduct measurements without vacuum chamber opening and exchange one of the elements thus reducing accuracy. Experience shows that reproducibility of experiments after chamber opening is about 20%. Measured with the aid of the first Faraday cup flux current in the plane downstream of extractor was 1.7 mA. More over it is necessary to take impurities into considerations. Main of them are oxygen and carbon which are not marked on diagram. Their total magnitude reaches up to 50% of N^{2+} current (estimated over the spectrum given in Fig.2). Finally, transmission factor of ion line equals to (75 – 95) % with all errors took into consideration.

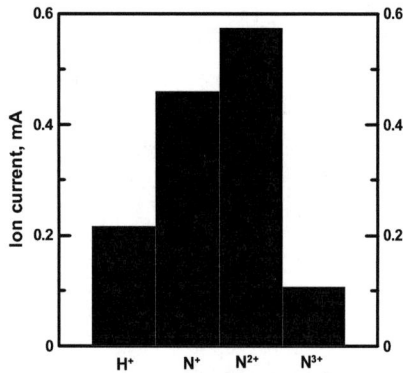

FIGURE 5. Ion charge distribution.

EMITTANCE ESTIMATION

Emittance measurement was not executed under given experimental setup. Here we bring the emittance magnitude estimation. One can assume that ion flux is neutralized downstream of the puller by electron cloud, and further expansion of the beam occurs without emittance change. Dispersion angle of that part of ion flux taking up by Faraday cup and possible for utilization does not exceed 4 degrees (70 mrad), diameter of puller hole equals to 2 mm. Under these assumptions emittance does not exceed 70 π mm mrad, normalized emittance equals 0.2 π mm mrad.

FUTURE PLANS

It is clear from Fig. 4 that optimal distance between the plug trap and plasma electrode under which optimal Faraday cup current is achieved equals to 16 cm. At this distance magnetic field value significantly decreases and equals to 0.07 T. This value far less than magnetic field values in the extraction zone of traditional ECR sources of multicharged ions with high current. One should take into consideration axisymmetrical configuration of magnetic field that is important for formation of ion beams with low emittance. In the observed experiment typical plasma transversal dimension equals 4 cm that is significantly exceeds the hole diameter of extractor and puller. Thus significant increasing of ion current from the source can be expected under the use of multi aperture extraction system.

As it has been demonstrated at the work [3] the use of the cusp trap optimized for maximal ion life time achieving enables to increase mean ion charge at their high flux density from the trap.

ACKNOWLEDGMENTS

The study received support of ISTC (project № 1496). The authors are grateful to A. Bokhanov and M. Kazakov (IAP RAS) for their help with the experiments and technical assistance.

REFERENCES

1. V. Zorin S. Golubev, S. Razin, A. Sidorov, V. Skalyga, A. Vodopyanov, *Rev.Sci.Instruments*, **75**, p. 1675 (2004).
2. S.V. Golubev, I.V. Izotov, S.V. Razin, V.A. Skalyga, A.V. Vodopyanov, V.G. Zorin. "Multicharged Ion Generation in Plasma Created by Millimeter Waves and Confined in a CUSP Magnetic Trap."*Transactions of Fusion Science and Technology*, to be published.
3. V. Skalyga, V. Zorin. "Multicharged Ion Generation In Plasma Confined In a Cusp Magnetic Trap at Quasi-gasdynamic Regime."*These proceedings*.
4. N. N. Semashko, A. N. Vladimirov, V. V. Kuznetsov, V. M. Kulygin, A. A. Panasenkov, *Injectory bystryh atomov vodoroda,* Moscow: Energoizdat, 1981, p. 35.

Plasma Potential Measurements for a "Volume"-Type ECR Ion Source

Y. Kawai, Y. Liu, G. D. Alton, H. Z. Bilheux, J. M. Cole

Oak Ridge National Laboratory, *Oak Ridge, Tennessee, 37831-6368, U.S.A.*

Abstract. A "volume"-type ECRIS, based on the flat central-field (flat-B) concept, has been developed at the Holifield Radioactive Ion Beam Facility, Oak Ridge National Laboratory. The superiority of the "volume-type" ECRIS concept over its conventional minimum-B source counterpart, in terms of higher charge-state and intensity within a particular charge-state, has clearly been demonstrated. As evidenced by comparing X-ray spectra for the two source geometries, improvements in charge-state and intensity within a particular charge-state are directly attributable to the increased physical "volume" of the ECR zone in the flat-B source that provides better coupling of RF power to the plasma, resulting in the acceleration of a much larger population of electrons to much higher energies. In this report, we provide plasma potential measurement data, made at various gas pressures, gas mixing ratios, and RF power, for the flat-B source. The plasma potential is found to increase with pressure and RF power, and to decrease with increase in gas mixing ratio whenever He gas is used as the mixing gas.

INTRODUCTION

Conventional minimum-B electron cyclotron resonance ion sources (ECRISs) utilize narrow bandwidth, single-frequency microwave radiation. Their ECR zones are thin annular, fluted elliptical surfaces that surround and intersect the axis of symmetry in the mirror regions of these sources. Due to the locations, thinness and consequently small volumes that these zones occupy in the minimum-B configuration, RF power absorption is not very efficient. The "volume"-type ECRIS attempts to overcome the limitations of conventional minimum-B sources by employing a much larger and axially located resonant plasma volume [1]. Since the size, shape and location of the resonant zone in an ECR ion source is determined by the plasma confining magnetic field distribution, the field, in principle, can be designed to optimally absorb microwave radiation. The magnetic field design for the source described in Ref. 2 attempts to accomplish this objective.

The demonstrated performance of the "volume"-type ECRIS, developed at the Holifield Radioactive Ion Beam Facility (HRIBF), Oak Ridge National Laboratory (ORNL), is unquestionably superior to the conventional minimum-B form of the source in terms of charge-state and intensity within a given charge state [2]. These affects are directly attributable to the much larger resonance zone in the flat-B geometry which provides better coupling of RF-power to the electrons, resulting in a higher population density of higher energy electrons and consequently, an enhancement of the ionization rates in the source.

In this report, we provide experimentally measured plasma potentials for the "volume" magnetic field geometry form of the source, measured at various pressures, RF-power and gas mixing ratios.

EXPRIMENTAL APPARATUS

The all-permanent-magnet source incorporates a flat-central field that is resonance with 6 GHz microwave radiation [1]. Radial confinement is effected with a $N = 12$ multi-cusp magnetic field; plasma confinement in the axial direction is effected

---

* Managed by UT-Battelle, LLC, for the U.S. Department of Energy under contract DE-AC05-00OR22725.

by symmetric mirror magnets that each have a mirror ratio slightly greater than 2 [3]. The *RF* power supply consists of an *RF* signal generator and a traveling wave tube amplifier (TWTA) that produces *RF* power that can be varied between 5.85-6.40 GHz. The *RF* injection system is equipped with a polarizer so that Right-Hand Circularly Polarized (RHCP) or Left-Hand Circularly Polarized (LHCP) waves can be injected into the plasma chamber. Most of the experiments presented here were carried out with RHCP waves. The source and ancillary experimental equipment utilized in the present studies were identical to those described in Ref. 2 with the exception that magnetic *M/q* analyses were performed with a higher resolution system. More detailed description of the source and ancillary experimental equipment can be found elsewhere [2].

PLASMA POTENTIAL MEASUREMENTS

Experimental conditions for *RF* power (*P*), frequency (*f*), pressure, and ion optics were first optimized for generation of Ar^{8+} using pure argon. All measurements, unless otherwise noted, were performed with RHCP waves. The pressure, p_{ein}, was measured at the einzel lens located down-stream of the extraction electrode. During measurements, the extraction electrode position was set ~ 20 mm from the extraction aperture. Optimum operational parameters were found to be $P \sim$ 23-25 W, f = 6.096 GHz, and p_{ein} =1.1x10^{-7} Torr at a source bias of 20 kV.

Plasma potentials can be determined by measuring the total energy of ions extracted from an ion source as a function of source bias assuming a uniform potential difference between the plasma and the chamber wall [4]. The magnetic field, *B*, required for bending an ion beam is proportional to the ion velocity, *v*, therefore, proportional to the square root of the acceleration potential;

$$B^2 = A(m_i/q)\Delta V_{ext} \quad (1)$$

where *A* is a constant; *q* is the ion charge state; and $\Delta V_{ext} = V_s + V_p$ where V_s and V_p are, respectively, source bias and plasma potentials. The electric-field penetration through the aperture and into the plasma volume will vary depending on the field strength, *E*, imposed during extraction. In the present case, the extraction field penetration depth into the plasma for a fixed gap system will vary with source bias. If particles are formed within the penetration depth they will be born at different equipotentials and consequently, their final energies will be less than for particles accelerated from within the plasma at fixed plasma potential. Thus, the curvature of the plasma meniscus will vary with electric-field strength which will affect the extraction optics and consequently beam transport properties. However, as long as the ionization process occurs beyond the penetration region (as most probably is the case for an ECR ion source), then the energy of the ions should not depend on the electric field strength $E = \Delta V_{ext}/d$. Since the charge-state distribution was found to remain constant for all measurements, regardless of the voltage used during extraction, we assume that particles are not born within field penetration regions.

Plasma potentials were obtained from least square fits to B^2 versus source bias potential data for charge-states Ar^{1+} to Ar^{5+} (Fig. 1). B^2 and source bias voltage ΔV_S are linearly correlated despite the large variation in source bias used during the measurements as well as wide variation in extracted ion beam intensities. As can be seen from Eq. (1), the horizontal axis intercept of the B^2 versus V_s curves corresponds to the value of the plasma potential. The intercept points for different charge-states were nearly the same. According to Ref. 5, ions with different charge-states can be extracted from different regions of the plasma. The plasma potential may vary from point to point; however, the potential variation under steady-state conditions must be small in the limited region where ions can be extracted. We did not observe sequential potential shifts for different charge states. The plasma potential was taken as the average of values obtained for different charge-states.

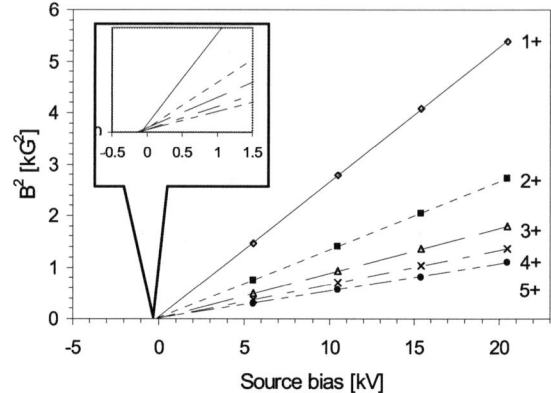

FIGURE 1. Measurement data for argon beams with charge-states 1+ to 5+. The least square fit intercepts of the B^2 versus V_S curves correspond to the value of the plasma potential. p_{ein}=1.1x10^{-7} Torr; P =23 W.

PLASMA POTENTIALS AND ION BEAM INTENSITIES

Figure 2 shows plasma potentials and ion beam intensities as functions of pressure measured at the einzel lens. The optimum pressure was $p_{ein}=1.1\times10^{-7}$ Torr. The plasma potential appears to increase with pressure. As noted, the beam intensities of the high-charge-state ions sharply drop beyond a pressure $p_{ein}=1.1\times10^{-7}$ Torr whereas the intensities of low-charge-state ions increase, as does the total current. When the pressure is increased without changing other parameters, the electron density can be increased in the region below $p_{ein}=1.1\times10^{-7}$ Torr (low-pressure regime). Thus, increases in plasma potential with pressure in this regime are attributable to increases in positive ion density.

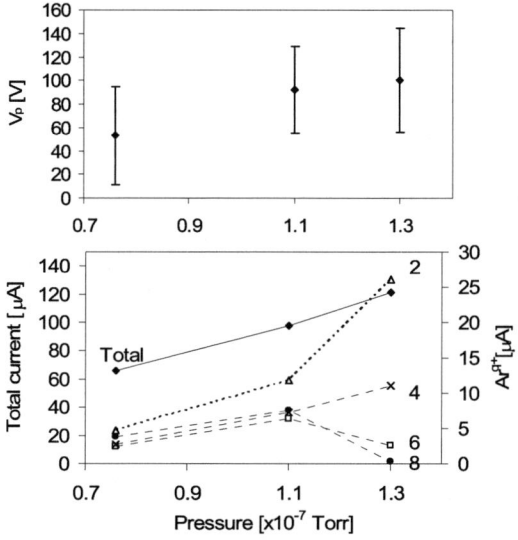

FIGURE 2. Plasma potential, total current, and Ar^{q+} intensity for an Ar plasma as a function of pressure, p_{ein}, at the einzel lens located next to the extraction electrode. RF frequency = 6.096 GHz; RF power = 23 W.

The dependencies of plasma potential and ion beam intensity on RF power are shown in Fig. 3. As noted, the plasma potential appears to increase while the intensities of highly-charged-ions rapidly decrease with RF power. According to these results, there is an optimum RF power above which the intensities drop and then typically reach a constant value. Beyond this value, total and lower charge-state ion beam intensities are observed to be almost independent of RF power. When the RF power is increased, both the electron energy and density increase, as does the plasma potential, V_p, as noted in Fig. 3. However, RF power itself can enhance the loss of electrons due to RF-induced scattering of electrons into the loss-cone in velocity space, as discussed, e.g., in Refs. [6-8]. This loss process is critical for hot electrons that are conjectured to create a negative potential-dip within the plasma that aids in confining positive ions. Thus, the loss of hot electrons results in a shallower potential-dip, and consequently, shorter confinement times for ions within the potential well. Therefore, an excess of RF power may result in a reduction in ion beam intensity.

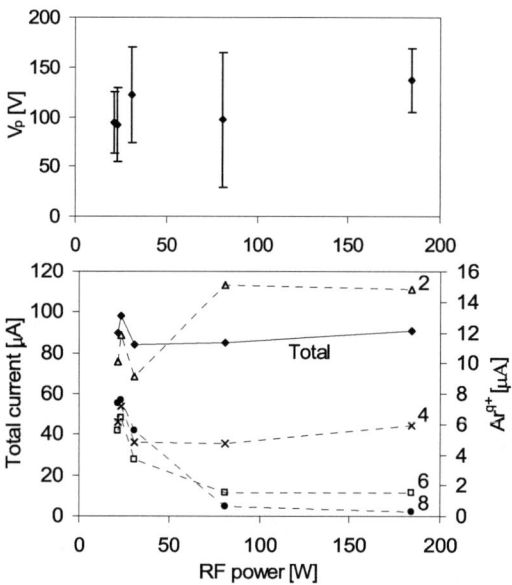

FIGURE 3. Plasma potential, total current, and intensity of Ar^{q+} for an Ar plasma as a function of RF power. RF frequency = 6.096 GHz; $p_{ein} = 1.1\times10^{-7}$ Torr.

We also investigated the gas mixing effect by adding known amounts of He into the discharge. The plasma potential decreases with increases in the He/Ar ratio as seen in Fig. 4. (These findings are consistent with former studies, e.g., [4,5].) While the intensities of high-charge-state ions were observed to increase with He/Ar ratio, the intensities of low-charge-state ions decreased. One of the reasons for this observation may be attributable to the so-called ion-cooling effect in which collisions between heavy and lighter ions results in heating the lighter collision partner while cooling the heavier partner. Arguments are often made that the above observations are attributable to increased confinement times of the heavy-ion population due to their lower energies. While collision cooling effects certainly take place, the most likely explanation of this effect is that collision induced momentum transfer from the heavier ions to lighter ions preferentially eject the lighter species from the plasma. Since quasi-neutrality conditions dictate that the electron/ion charge losses be balanced, the lighter particles serve to supply the positive charge losses required to maintain quasi-neutrality and are therefore sacrificial. Thus, this mechanism increases the confinement times of the heavier ions, resulting in higher charge-states and higher intensities within a particular charge-state while the intensities of low-charge-state ions decrease due to the fact that they become members of the high-charge-state population.

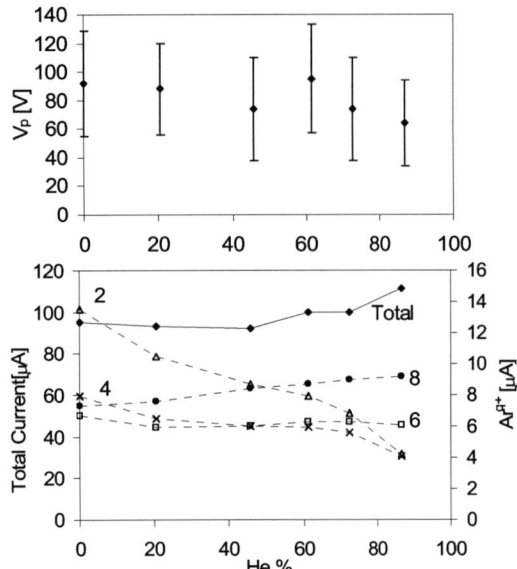

FIGURE 4. Plasma potential, total current, and intensity of Ar^{q+} as a function of He/Ar ratio. RF frequency = 6.096 GHz; $p_{ein} = 1 \times 10^{-7}$ Torr.

The average plasma potential was 48 ± 29 V under optimized conditions for Ar^{8+} with LHCP waves; $f = 6.19$ GHz, $P = 44$ W, $p_{ein} = 3.1 \times 10^{-8}$ Torr. As noted, when LHCP waves were utilized, optimum RF power and frequency were higher than those obtained with RHCP waves while optimum pressure and measured plasma potential were both lower. Although when independently optimized, the intensities for Ar^{8+} (~ 7 μA) were approximately the same for both RHCP and LHCP waves, their respective charge-state-distributions were quite different. The charge-state distribution was peaked at $q = 8+$ when LHCP waves were utilized and peaked at $q = 2+$ when RHCP waves were utilized, without gas mixing. LHCP waves cannot directly heat electrons on first pass through the plasma; however, they are reflected as RHCP waves on their return pass through the plasma and thus can heat electrons within the ECR zone. Thus adsorption of reflected LHCP waves occurs nearer the extraction aperture where they can be more easily extracted. As opposed to ions created by RHCP waves on entry into the injection side of the ECR zone, ions created nearer the aperture with reflected LHCP waves do not have to pass through a long plasma column during which charge-exchange interactions with neutrals can alter (reduce) the charge-state distribution. Due to this mechanism, the region where highly charged ions are produced was moved toward the extraction aperture making it possible to more efficiently extract these ions. Typically, RHCP and LHCP waves are distributed 50%/50%. Thus, in this scenario, ions are created from both sides of the ECR zone. Whenever, LHCP waves were used, more RF power was required to optimize Ar^{8+} intensities, probably due to losses due to wave scattering during reflection. The plasma density was also likely to be lower for LHCP waves than for RHCP waves because the pressure and plasma potential were both found to be lower. More detailed comparisons of results obtained with RHCP and LHCP waves can be found in Ref. 9.

CONCLUSIONS

Plasma potentials for the flat-field ECR ion source were measured as functions of pressure, RF power, and He/Ar ratio. Plasma potentials were observed to increase with pressure and RF power, and to decrease with He/Ar ratio. Overall, the measured plasma potentials for RHCP waves were higher than those for LHCP waves. Future plans call for measuring plasma potentials of the source with a sophisticated energy analysis system that will greatly improve the accuracy of these measurements.

REFERENCES

1. Alton, G.D. and Smithe, D.N., *Rev. Sci. Instrum.* **65**, 775-787 (1994)
2. Liu, Y., Alton, G.D., Bilheux, H., Cole, J.M., and Meyer, F.W., *Rev. Sci. Instrum.* (in press).
3. Liu, Y., Alton, G.D., Mills, G.D., Reed, C.A., and Haynes, D.L., *Rev. Sci. Instrum.* **69**, 1331-1315 (1998).
4. Xie, Z.Q. and Lyneis, C.M., *Rev. Sci. Instrum.* **65**, 2947-2952 (1994).
5. Nadzeyka, A., Meyer, D., and Wiesemann, K., *Proceeding of the 13th workshop on ECRIS* (1997).
6. Mauel, M.E., *Phys. Fluids* **27**, 2899-2911 (1984).
7. Hokin, S.A., Post, R.S., and Smatlak, D.L., *Phys. Fluids* B **1** 862-873 (1989).
8. Girard, A., Pernot, C., and Melin, G., *Phys. Rev. E,* **62** 1182-1189 (2000).
9. Bilheux, H.Z., Liu Y., Kawai, Y., Cole, J.M., Reed, C.A., and Alton, G.D., (these Proceedings)

Theoretical and Experimental Studies of the Extracted MCI Beam from an ECR Ion Source

L. T. Sun[1], Y. Cao[1], H. W. Zhao[1], X .H. Guo[1], Z. M. Zhang[1], Y. C. Feng[1], J. Y. Li[1], L. Ma[1], J. Li[1], H. Y. Zhao[1], W. He[1], X. X. Li[1], D. Hitz[2], A. Girard[2]

[1]Institute of Modern Physics (IMP), Chinese Academy of Sciences, Lanzhou, China
[2]CEA-Grenoble, Département de Recherche Fondamentale sur la Matière Condensée, Service des Basses Températures, 17 Rue des Martyrs, 38054 Grenoble Cedex 9 , France

Abstract. With the development of Electron Cyclotron Resonance Ion Source (ECRIS), very high performance ECRIS nowadays have been set up one by one around the world, such as the GTS in Grenoble, SERSE in Catania, LECR3 in Lanzhou and etc, which can produce very intense Multiply Charged Ion (MCI) beam. But till now, the study of the extracted MCI beam from an ECRIS remains open. In this article, we present a theoretical and experimental study of the extracted MCI beam. In the theoretical part, the influences of the extraction system on the extracted ion beam quality are mainly analyzed. The aspects that have influences on the extracted ion beam quality have been analyzed. With the instruction of the analysis, the PBGUNS code is used to simulate the influences of some important aspects concerning the extraction system. The influences of the extraction system geometry design, magnetic field, and the space charge effect will be detailedly presented in this article. In the experimental part, with an Electric-Sweep Scanner (ESS) emittance detection system, the influences on the extracted ion beam emittance of some typical parameters of ECRIS have been researched, such as the injected RF power, the RF frequency, the magnetic field and etc. The obtained results and the corresponding explanations are presented. Some of the results are well in accord with some empirical laws, but some other results seem to be disputed.

INTRODUCTION

As is known, the emittance of the ion beam extracted from ECRIS is influenced by many aspects, such as the feeding gas, RF power, RF frequency, magnetic field, the extraction system and etc. There are many good codes having been developed and used to simulate the extraction of MCI beam from ECRIS, but the ECR plasma is so caprice, it is really very difficult to make a good simulation. To get a precise simulation of the extracted ion beams, better study on the ECRIS working mechanism must be done. GTS source is a high performance ECRIS developed in CEA/Grenoble [1, 2]. To enhance the performance of the source, some modifications had to be done. The latest developed PBGUNS code [3] was adopted to simulate the ion beam extractions. With this useful tool, a global study has been made considering the plasma status, magnetic field distribution, CSD condition, electrode's geometry design, space charge neutralization condition and etc. In this paper, some of the interesting results obtained from the simulations are presented.

ECRISs are used in IMP to provide the HIRFL accelerators with intense MCI beams. The quality of the ion beam extracted from the ECRIS is always what we are concerning about, which is directly related to the running efficiency and beam quality of the accelerators. An ESS emittance scanner has been developed recently in IMP for better study of the ion beam quality of ECRIS [4]. LECR3 is a double-frequency heating high performance ECRIS used for atomic physics research [5]. The ESS system had been put on the beam line of LECR3. After ICIS'03, many measurements concerning the relationship between the main parameters of ECRIS and the extracted ion beam quality have been done, which might help us have a better understanding of the working module of ECRIS. According to the experiments, some of the results are of great interest and will be presented in this paper.

THEORETICAL SIMULATION STUDY

The purpose of this work is to find the influences of the extraction system on the extracted ion beam quality, so as to optimize the design of the extraction system. A triode extraction system is studied here, and the sketch plot of this system is shown in **Fig. 1**. There are many parts of the triode system related with the extracted ion beam quality and they are all included in my study. Here, we put special emphasis on **a**, **b** and **c** parts. For all the simulations, total current is 10emA, the charge state distribution (CSD) in the plasma is obtained from GTS source, the plasma electrode aperture diameter is set to 13mm and Ar^{8+} is the ion we investigate. Laboratory emittance is given.

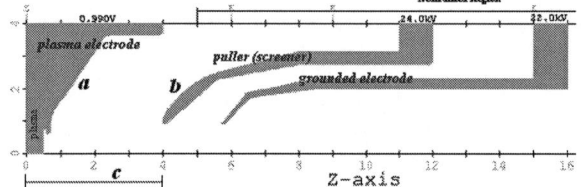

FIGURE 1. Sketch plot of the extraction system.

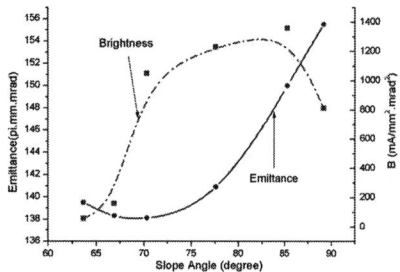

FIGURE 2. Influence of the slope angle.

a. Influence of the Slope of a part

In this study, the slope of *a* part (plasma electrode inner wall) varies from 62° to 90°, and then some rules are obviously shown in **Fig. 2**. The slope of the inner surface of plasma electrode influences the extracted ion beam quality by some degree. When the slope angle is around 70°, the emittance of the extracted ion beam gets the minimum. And with the increase of the slope angle, the brightness increases very fast till 75°. When the angle is larger than 75°, the brightness value almost saturates. Then to obtain an ion beam with low emittance and good brightness, it is better to design the slope angle of *a* part as 70°~75°.

b. Influence of Outer Surface of the Puller

To study the influence of *b* part, the angle is varied from 45° to 105°. With the variation of the line slope, the extracted ion beams quality changes. **Fig. 3** gives the variation curves. Easy to see, the smaller the slope angle of *b* part, the smaller the emittance value. From 45° to 105°, the emittance value changes a lot. But for the brightness, it does not alter in the same way as the emittance does. So, to design the puller, there will be a compromise of the brightness and the emittance. In fact, the loss of brightness can be compensated by modifying the other parts of the extraction system. The slope angle of the puller should be designed as small as possible if it does not influence the transmission of the extracted ion beam.

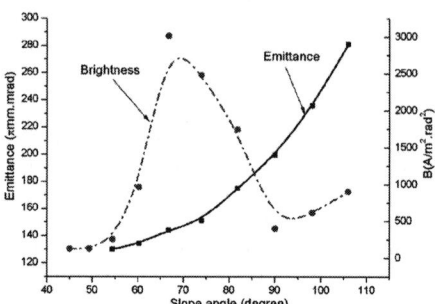

FIGURE 3. Influence of puller slope angle.

c. Influence of the Extraction Gap

In practice, we often vary the gap between the outlet electrode and the puller for a good optimization of ion beam extraction. In this study, the simulation of the influences of different gap value can give us something for reference. In **Fig. 4**, the gap influences on ion beam emittances of different charge state are given. Easy to see, for a certain charge state ion beam, the emittance decreases rapidly with the increase of the gap. But the emittance increases when the gap is larger than a critical value. The higher the charge state, the smaller the gap when the alteration occurs. It is obvious in **Fig. 4**, the higher the charge state, the larger the emittance, which does not consist with the empirical results. This is because the PUBGUNS code has not taken into account that the emission position of different charge state ions is different.

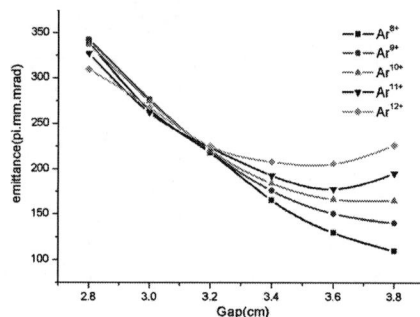

FIGURE 4. Influence of extraction gap.

RESULTS OF EXPERIMENTAL STUDY

All the experiments have been done on the beam line of LECR3. Two images (Image1 and Image2, Image 1 is put after the Galser lens, and Image2 is put near the Faraday cup) are used to observe the beam images of the mixed ion beams and the analyzed ion beams separately. The ESS system is installed after the Faraday cup. According to the experiments, many results concerning the relationship between the source parameters (such as the magnetic field, RF power, RF frequency and etc.), extraction system and the extracted ion beam quality has been investigated. In the following contents, some of them will be discussed.

Emittance Study under Different B Modes

In this research, we studied the emittance variation with the change of the RF power under H, M and L magnetic field modes. Here, H, M and L represent high (B_{inj}=1.58T, B_{min}=0.45T, B_{ext}=1.17T), moderate (1.53T, 0.41T, 1.05T) and low (1.45T, 0.36T, 0.93T) magnetic field configuration respectively. The feeding RF frequency is 14.5GHz, and the power varies from 100W to 700W for Ar^{8+} and from 400W to 800W for Ar^{12+}. No other parameters of the source have been modified during the experiments except for the necessary tuning. The results are given in **Fig. 5**.

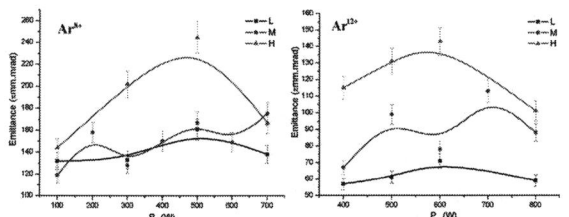

FIGURE 5. Influences of different B modes.

Easy to see, under the same B-mode, the beam emittance varies with the feeding RF power: the higher the RF power, the larger the ion beam emittance. This is consistent with G. Melin and A. G. Drentje's theory [6] on the calculation of ion temperature: the more RF power, the higher the ion temperature T_i. But when RF power is larger than a certain value, the variation tendency alters which indicates another regime of the plasma. We also notice the emittance variation when the same RF power is fed under different B modes: the higher magnetic field configuration induces larger emittance. And under different B-mode, the way that the emittance alters with the increase of RF power is different.

RF Frequency Influencces

Under the condition of 500W RF power feeding, (B_{inj}=1.6T, B_{min}=0.44T, B_{ext}=1.1T) for 14.5GHz running and (1.6T, 0.47T, 1.21T) for 18GHz running, we measured the influence of RF frequency on the ion beam emittance, as is shown in **Fig. 6**. The other parameters are kept unchanged except for the Glaser lens and analyzer.

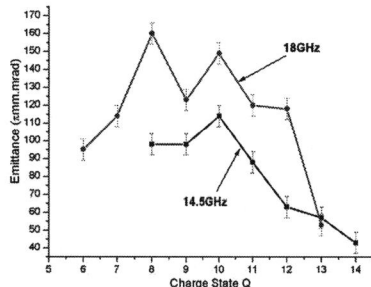

FIGURE 6. Influence of RF frequency.

Easy to see, higher frequency RF power feeding induces larger emittance for the same charge state ion beam. This might be the reason of the increment of the ion temperature T_i with the increase of RF frequency when the same RF power is fed, which can be deduced from G. Melin's formula to estimate T_i [6]. From **Fig.7** we can also conclude that the ion beam emittance decreases with the increase of charge state. This result is well consistent with D. Wutte's [7] and P. Sortais' [8] results.

Study under Different Extraction HV

The extraction high voltage will influence the plasma boundary at the extraction region and also the transmission of the ion beams being extracted. These aspects will probably influence the emittance of the extracted ion beam. In this research, the emittance variation of Ar^{8+} and Ar^{13+} is investigated with the increase of the HV from 10kV to 22kV. The extraction system is a triode system, and the biased voltage on the screen electrode is always -2kV. The other parameters of the source are kept constant during the experiments. The results are given in **Fig. 7**. Easy to see, the normalized emittance of Ar^{8+} almost saturates when HV is larger than 14kV. However, for the higher charge state ion Ar^{13+}, the normalized emittance increases slowly with the increment of the extraction HV. We have repeated the same experiments several times, and obtained the same results. It seems that the influences of extraction HV on the emittance of high charge state ions need further investigations.

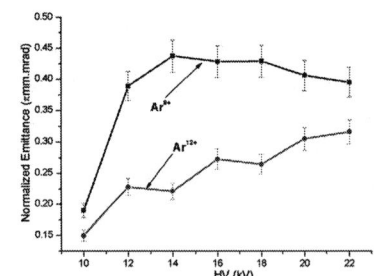

FIGURE 7. Influence of extraction HV.

The Influence of Plasma Electrode Aperture

In this study, we tested three plasma electrodes with different apertures: Ø10mm, Ø13mm and star shape. And the emittances of the electrodes with Ø10mm and Ø13mm aperture were measured. All these measurements have been done while feeding 18GHz RF power. When exchanging the plasma electrode, the vacuum had to be broken, so the working state might be different for different electrodes. To make the results more valid, the tunable parameters are all kept the same in the experiments.

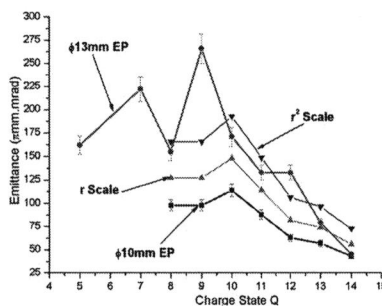

FIGURE 8. Plasma electrode aperture influence.

Fig. 8 indicates the evolution of the emittances of different charge state ion beams extracted from apertures of Ø10mm and Ø13mm. Easy to find, larger aperture plasma electrode induces larger ion beam emittance. The r (the aperture radius) scale and r^2 scale of the emittance value have been adopted to study the possible rules of the increment of the emittance with the increase of the aperture size. Easy to see, for medium charge state, r^2 scale is more suitable and while for higher charge state r scale seems to be better. Although larger aperture plasma electrode causes the increase of the emittance of extracted ion beam, intense ion beams are also expected which has been observed in our experiments. With Ø13mm or star shape (central aperture Ø10mm) aperture plasma electrode, we obtained 1.1emA Ar^{8+} beam at 18GHz.

CONCLUSION

A theoretical and experimental study has been done on the ion beam extraction from ECRIS. Some of the results are presented in this paper. The theoretical results are concerning the discussion of the geometrical design and adjustment to optimize MCI beam extraction, and trying to enhance the quality of the extracted ion beam. The experimental study is intending to reveal the relationship between the source working state and the extracted beam quality, and so as to find the working regime of ECRIS. In this paper, the influences of the feeding RF power, RF frequency, magnetic field, plasma electrode aperture, extraction HV and etc on the extracted ion beam have been discussed. These studies are just the preliminary investigations on the MCI beam extraction. Further global study is to be continued.

ACKNOWLEDGMENTS

The experiment is supported by National Natural Foundation for Distinguished Young Scientists under contract No.10225523 and National Natural Foundation for Young Scientists through the contract No.10305016.

REFERENCES

1. D. Hitz, D. Cormier, J. M. Mathennet et al, Proceedings of 15th Int. Workshop on ECRIS Ion Sources, Jyäskylä, Finland, pp. 53-55.
2. D. Hitz, A. Girard, K. Serebrenniko et al, *Production of Highly Charged Ion Beams with the Grenoble Test ECR Ion Source*, R. S. I., Vol. 75, No. 5, May, 2004.
3. J. E. Boers, *A Digital Computer Code for the Simulation of Electron and Ion Beams on A PC*, IEEE Cat. No. 93CH3334-0(1993) 213.
4. Y. Cao, H. W. Zhao, L. Ma et al, *Emittance Measurement at IMP LECR3*, R. S. I., Vol. 75, No.5, May 2004.
5. Z. M. Zhang, H. W. Zhao, X. Z. Zhang et al, Proceedings of 15th Int. Workshop on ECRIS Ion Sources, Jyäskylä, Finland, pp. 126-127.
6. G. Melin, A. G. Drentje, A. Girard, D. Hitz, Proceedings of 14th Int. Workshop on ECR Ion Sources, CERN, Geneva, 1999, pp.13-18.
7. D. Wutte, M. A. Leitner, C. M. Lyneis, *Emittance Measurements for High Charge State Ion Beams Extracted from the AECR-U Ion Source*, R. S. I., Vol. 75, No. 5, May 2004.
8. P. Sortais, L. Maunoury, A. C. C. Villari et al, Proceedings of 13th Int. Workshop on ECR Ion Sources, Texas, USA, 1997, pp.83-87.

An ESS system for ECRIS Emittance Research

Y.Cao L.T.Sun W.He L.Ma Z.M.Zhang H.Y.Zhao H.W.Zhao X.Z.Zhang
X.H.Guo B.H.Ma J.Li H.Wang J.Y.Li X.X.Li Y.C.Feng W.Lu

Institute of Modern Physics (IMP), Chinese Academy of Sciences
730000 Lanzhou, P.R.China

Abstract. An emittance scanner named Electric-Sweep Scanner had been designed and fabricated in IMP. And it has been set up on the LECR3 beam line for the ion beam quality study. With some development, the ESS system has become a relatively dependable and reliable emittance scanner. Its experiment error is about 10 percent. We have done a lot of experiments of emittance measurement on LECR3 ion source, and have researched the relations between ion beam emittance and the major parameters of ECR ion source. The reliability and accuracy test results are presented in this paper. And the performance analysis is also discussed.

INTRODUCTION

There are many characteristics to describe the performance of an ion source, such as beam current intensity, emittance, energy spread of the extracted ion beam and etc. Among them, the beam emittance is always the one that makes us suspicious. So it is especially important to measure and study the emittance of ion beams. Now, there are principally two kinds of methods to measure ion beam emittance. One of the measuring methods is to cut off the beam. With this method, the phase diagram of the emittance can be easily expressed. But its shortcoming is to cut off the beam, and the beam emittance can not be measured continuously. Another measuring method is to measure the beam emittance without cutting off the ion beam. Then it's possible to measure beam emittance continuously on line, but this method can not give the real phase diagram of the emittance.

A high performance Electron Cyclotron Resonance Ion Source (ECRIS) was built at IMP in 2002, named LECR3 (Lanzhou ECR No.3) [1]. The ion beam current extracted from LECR3 ion source is very strong, and the beam energy is low (extraction HV: 10~30kV). Under such a condition, we can assume the emittances at the horizontal and vertical direction do not couple with each other, and it is suitable to measure the beam emittance with the cut-off method. We have designed an emittance scanner named Electric-Sweep Scanner (ESS) [2], and have done a lot of experiments to study the beam emittance of LECR3 ion source.

MEASUREMENT SYSTEM

The system is mainly composed of four parts: scanner system, driving system, controlling system and data collection system. The sketch plot of ESS system is shown in **figure 1**.

Working principle: Spatial position of ion beam is confirmed by the front slit of the ESS; at each position, the voltage on the deflection plates scans automatically, so as to measure the divergence angle of the ion beams. And the beam current which can pass through the rear slit in ESS system can be measured at the same time. Accordingly with the data obtained we can get the ion beam emittance [2].

The horizontal and vertical emittances of the ion beam are separately measured, so two ESS systems are set up in a vacuum chamber at both the horizontal and the vertical directions. The layout of the experiment line is shown in **figure 2**.

RESULTS OF THE EXPERIMENT AND THE DISCUSSION

We had measured the emittance of the ion beams extracted from LECR3 ion source by the ESS system in 2003, and achieved some good results [3]. For better study of the ion beam emittance of ECRIS and examining the reliability and performance of the scanner, we had done a lot of emittance measurements. We had done repetition measurement experiments of the horizontal emittance for 86% beam of Ar^{8+} and Ar^{12+} ion beam, and the results are given in **figure 3**.

From **figure 3**, we can see the repetition error is about ±10 percent. On the other side, the system error of the scanner is very small, and it is about ±1.8 percent [2], which can be neglected, comparing with the repetition error. So, we may think the error of the scanner is less than ±10 percent.

Figure 4 indicates the distributions of the emittance of different charge state argon ion beam measured at different time. Two measurement results both demonstrate the beam emittances of the ECR ion source are smaller for higher charge state ions. It does not agree with traditional theory analysis and the results of some program simulations, but it agrees well with experiential analysis on ECR ion sources and the measurement results of other laboratories [4][5]. It seems that our experiments were successful, the scanner is reliable, and the results truly reveal the physical phenomena.

It is very important to study the relationship between the ion beam emittances and the parameters of ECR ion source, such as the mirror magnetic field, bias voltage, feeding gas, microwave power and RF frequency, the extraction system and etc. We had done a lot of corresponding research experiments [6][7]. Because of the lackness of enough experiment time and the limited condition, some experiments have not given good results. However, most of the experiments are successful, and the experimental results are of very good value.

SUMMARY

We have developed an ESS system to measure the ion beam emittance of ECRIS, and it has been successfully used to measure the ion beam emittance of LECR3. Those experiment results proved that the scanner is reliable, and the measuring results are significative. But with these experiments, we find that there are still many aspects of the ESS system to be improved. We will modify the structure of the scanner, better the controlling system, enhance the performance of anti-jamming, and minimize the experimental error. Through these modifications and improvement, more precise and reliable emittance measurement can be done with the future ESS system. With such a good ESS system, we can do better study on the emittance of ECRIS.

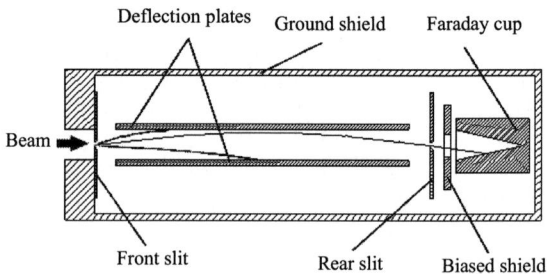

FIGURE 1. The sketch plot of the ESS system.

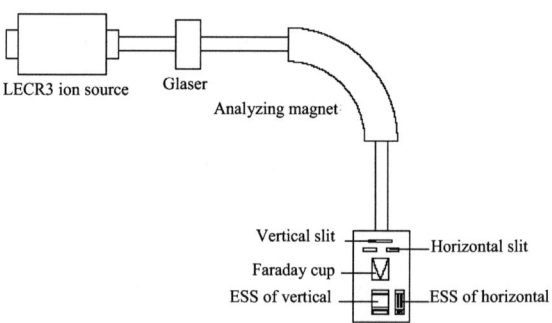

FIGURE 2. The layout of the experiment line.

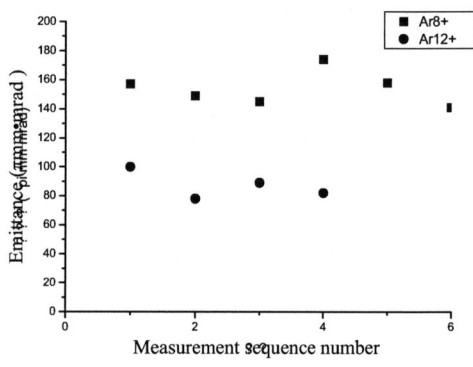

FIGURE 3. Repetition test of the ESS system.

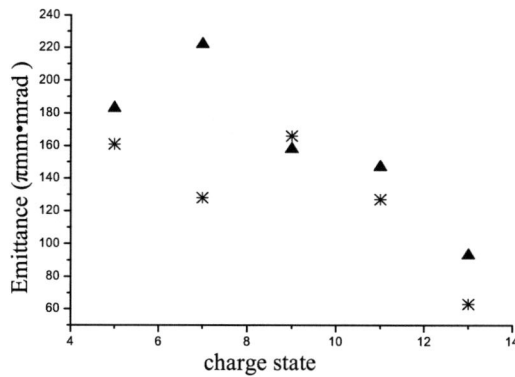

FIGURE 4. Beam emittances of different charge state argon ions

ACKNOWLEDGMENTS

This work was supported by National Natural Foundation for Distinguished Young Scientists under contract No.10225523 and National Natural Foundation for Young Scientists under contract No.10305016.

REFERENCES

1. L.T.Sun, Z.M.Zhang, H.W.Zhao, et al, The design of the double-RF heating LECR3 ion source and its preliminary results, *Nuclear Techniques,* Shang Hai: Science Press, 2002, pp. 674-677.
2. Y.Cao, L.Ma, H.W.Zhao, Emittance Measurement of Highly Charged Ion Beams Extracted from ECR Ion Source Using Electric Sweep Scanner, *High Energy Physics and Nuclear Physics,* 2004, pp.885-888.
3. Y.Cao, H.W.Zhao, L.Ma, et al, Emittance Measurement at IMP LECR3, *Rev.Sci.Instrum.,* edited by Albert T. Macrander, Argonne: 2004, p.1443-1445.
4. D.Wutte, M.A.Leitner, C.M.Lyneis, *Physica Scripta,* 2001, pp.247-249.
5. P.Sortais, L.Maunoury, A.C.C.Villari et al, the Proceeding of 13th Int.Workshop on ECR Ion Source, Texas: 1997, pp. 83.
6. L.T.Sun, Doctor Paper, Lanzhou: 2004, pp.94-104.
7. L.T.Sun et al, proceeding of 16th International Conference on ECR Ion Sources.

The Development Of New Space Charge Compensation Methods For Multi-Components Ion Beam Extracted From ECR Ion Source at IMP

L.Ma, H.W.Zhao, Y.Cao, H.Y.Zhao, T.M.Song, W.He, Z.M.Zhang

Institute of Modern Physics (IMP), CAS, Lanzhou, 730000 China

Abstract. Two new space charge compensation methods developed in IMP are discussed in this paper. There are negative high voltage electrode method (NHVEM) and electronegative charge gas method (EGM). Some valuable experimental data have been achieved, especially using electronegative gas method in O^{6+} and O^{7+} dramatic and stable increasing of ion current was observed.

INTRODUCTION

Limitation of beam intensity and performance by the space charge effect is a complex problem in the transport of ion beam especially for low energy and highly charged heavy ion beam extracted from ECR ion source. Nowadays the use of heavy ion beam becomes more and more popular especially for heavy particle physics and atomic physics. In recent years, ECRIS has got great development in IMP, and as a focused point of the development of our lab we have devoted to developing some effective methods to compensate the space charge of ion beam in order to increase the current intensity and the transport efficiency of the ion beams extracted from ECR ion source [1].

1. NEGATIVE HIGH VOLTAGE ELECTRODE METHOD FOR SCN DEVELOPED AT IMP

It is well known that some kinds of methods have been developed for space charge compensation (SCN) and some good results have been achieved in recent years mainly including electrons ejection (at both vertical and parallel directions) and residual gas ionization method (for high energy and low charge state ion beams). But all the mentioned methods have to be equipped with some kind of electron guns or need high ion energy. For low energy ion beams we have to need some equipments to generate electrons and thus the whole facility will become complicated. In order to avoid this defect and also combine the characters of ECR ion source we developed the negative high voltage anode method.

1.1. Analysis of NHVEM to compensate the space charge of ion beam extracted from ECR ion source [2, 3, 4]

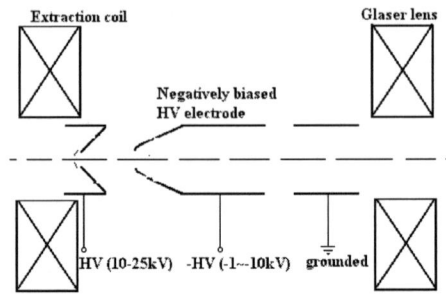

FIGURE 1. A sketch plot of the NHVEM system.

The principle is that using a negative high voltage electrode and a magnetic mirror to form a space charge lens. From numerical calculation and theoretic analysis we found this method not only avoids some defects of Gabor space charge lens [1] but also succeeds in capturing a large quantity of electrons and then a powerful space charge lens is created. The whole structure looks like a two-electrode extraction system and the only difference between them is that a negative high voltage is applied on the first electrode. The negative electrode and magnetic mirror composed the electron-capture system, and the electrons are mainly secondary electrons (coming from the tube and the ionization of residual gas). In principle the electronic field of plasma electrode (positive high voltage) is shielded by the first electrode (see **figure 1**), and at the same time the energy of secondary electrons is very low (10-10^{2}ev), so by the effect of electronic field of the negative high voltage electrode the electrons can't move to the left and can only move to the right and come into the area of magnetic mirror. In the magnetic mirror field, because of the existence of magnetic grads the direction of the magnetic force on the electrons is to the left, so the electrons will be bounced to the left. In one cycle of turning the average force applied on the electrons can be expressed as follow:

$$\overline{F_{11}} = -\mu \nabla_{11} B \qquad (1)$$

here B is the intensity of magnetic field, and μ is the magnetic moment of electron expressed as:

$$\mu = \frac{1}{2} m V_\perp^2 / B \qquad (2)$$

$V_\perp$ is the electron velocity component perpendicular to the magnetic field. On other hand, $V_{11} B = \frac{\partial B}{\partial S}$ and S is the line element along the $\vec{B}$. In the following we will do a brief analysis about the strength of this force. Assuming the magnetic field has a circular distribution which means the magnetic isodynam is circular then:

$$B_z = \frac{\sqrt{Z(2r-Z)}}{r} \qquad (3)$$

$$\theta = \arccos \frac{r-Z}{r} \qquad (4)$$

$$\frac{\partial B_z}{\partial Z} = \frac{r-Z}{r\sqrt{Z(2r-Z)}} \qquad (5)$$

So the force exerted to the electrons by the magnetic field can be expressed as the equation (6):

$$\overline{F_z} = -\mu_r \frac{r-Z}{\sqrt{Z(2r-Z)}}. \qquad (6)$$

The change of F_z along z can be presented as shown in **figure 2**. In fact the magnetic isodynam is not ideally circular and the edge effect of magnetic field must be considered but the tendency of force change along z is the same. The electrons are exposed in the combination of an electronic field and a magnetic field, so the force exerted on the electrons includes two parts and the force should be expressed as:

$$F = eE_z - \mu_r \frac{r-Z}{\sqrt{Z(2r-Z)}} \qquad (7)$$

From the discussion, it is not difficult to see that in this structure electrons are possibly confined by the negative electronic potential and magnetic field surrounding them, on the other hand the secondary electrons generated in the internal part of magnetic mirror field and the electrons energy is low so they might be captured and thus to increase the density of electrons in the ion beam extraction region.

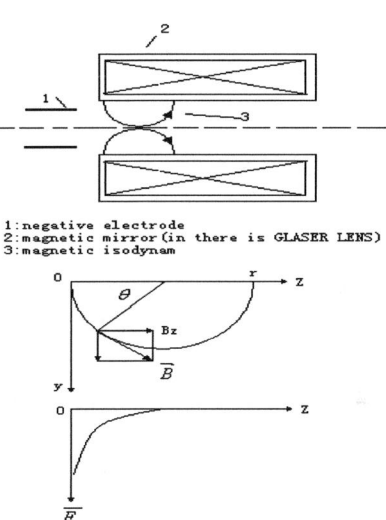

FIGURE 2. Schematic plot of space charge lens and the distribution of the magnetic strength F_z along z.

1.2. The Typical Experimental Data Using NHVEM and Analysis

A series of experiments using NHVEM to compensate the space charge of oxygen and argon ion beams extracted from ECR ion source have been done at IMP on the beam line of LECR3 [1] and some valuable results have been achieved. Some typical experimental data are given in **figure 3, 4** and **5**.

From the experimental data we can easily find that when applying the high voltage on the negative electrode the beam current intensity can be increased dramatically by about 20% or more. When the negative voltage is raised to a special value the ion beam becomes unstable. On the other hand, different species of ion beam need different negative high voltage value in order to get the maximum beam intensity.

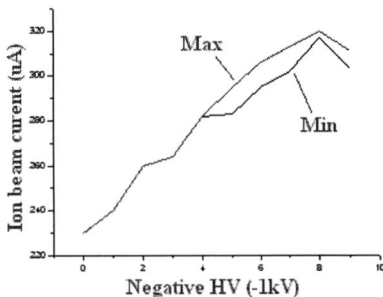

FIGURE 3. Relationship between ion beam intensity and HV when optimized for O^{6+}.

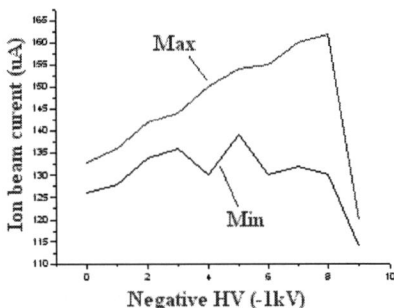

FIGURE 4. Relationship between ion beam intensity and HV when optimized for Ar^{8+}.

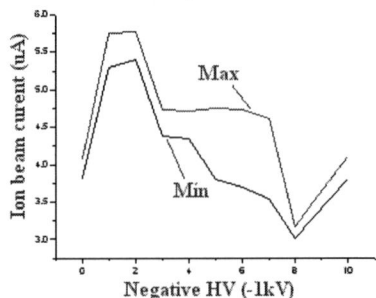

FIGURE 5. Relationship between ion beam intensity and HV when optimized for Ar^{16+}.

2. ELECTRONEGATIVE GAS METHOD FOR SCN DEVELOPED AT IMP [5]

At the enlightenment of the work had been done by V. Dudnikov using electronegative gas to compensate the space charge of low charge state ion beam in ion implant industry, we employed electronegative gas to help compensate the space charge of multi-component ion beams extracted from ECR ion source and have gotten some valuable results.

2.1. Consideration and Motivation

For positive ion beam neutralization, free electrons from residual gas ionization are commonly used, such as the secondary electrons generated by the ionization of residual gas or the electrons emitted by an extra electron gun and so on. But the disadvantages of these SCN methods are obvious such as the short lifetime of electrons in ion beams and the limited production ratio of electrons. On the other hand, it is difficult to use free electrons to compensate space charge of the ion beams in magnetic field because the electron plasma in magnetic field could be unstable and this instability will result in the dramatic loss of ion beam. Especially for the highly charged and low energy ion beams the compound cross section between the electrons and highly charged ions are very large, so using traditional free electron method to compensate the space charge of this kind of beam seems improper. Negative ions have been considered. Electronegative gas behaves as a source of negative ions because it has large affinity with electron and easily absorbs electrons to become negatively charged and hard to be ionized. Negative ions in ion beam are formed by the collision of electrons with molecules and by secondary emission initiated by bombardment of tube surface by ion beam.

2.2. The Design of Experiment

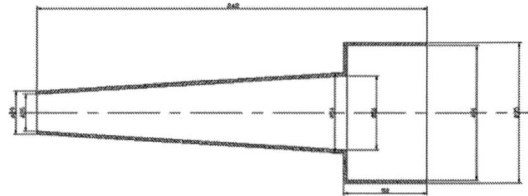

FIGURE 6. Sketch plot of the vacuum difference tube.

In this experiment we mainly considered two points in the process of experiment design. Firstly, try our best to avoid the leak of electronegative gas into the ion source so as to avoid the influences onto the state of ion source. Secondly, generate a large quantity of electrons as much as possible. For the first consideration we have designed a vacuum difference structure tube as shown in **figure 6** installed on the transport line after the extraction system. For the second consideration we use aluminum as the material of the electrode and the vacuum difference tube.

2.3. The Experimental Data and Analysis

In order to test the validity of this method we have done a series of experiments using this method to compensate the space charge of different kinds of beams especially for the highly charged ion beams. In our experiment we tested SF_6 and BF_3 as inpoured electronegative gas separatedly and found when inpouring SF_6 into the tube the beam current has a dramatic increase for O^{7+} and Ar^{11+}. The experiment results are given in **figure 7**. In **figure 8**, three CSD spectrums obtained at different transmission space vacuum conditions are also given, which might help understand the working mechanism of EGM.

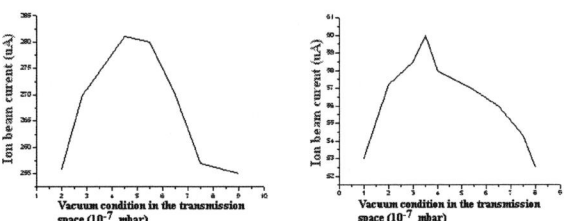

FIGURE 7. Relationship between O^{7+} (the left plot) and Ar^{11+} (the right plot) ion beam intensity and the relative SF_6 dose inpoured into the transmission space.

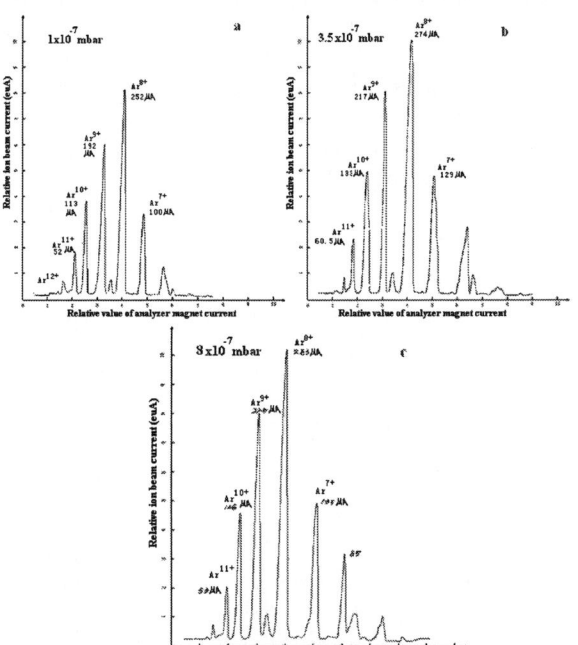

FIGURE 8. Oxygen ion beam spectrums when optimized for O^{7+} ion beam production at different relative SF_6 dose inpoured into the transmission space (a- 1×10^{-7} mbar, b- 3.5×10^{-7} mbar, c- 8×10^{-7} mbar).

From the experimental data it is not difficult to see that considerable increase of beam current occurs for O^{7+} and Ar^{11+} when inpouring SF_6 in the beam transport space and the amplitude is no less than 15%. Then we can see that an overcompensation phenomenon occurs when the quantity of inpoured electronegative gas reaches a special value and this value for highly charged ions is less than the value for low charge state ions. The reason is that high charge state ions have larger compound cross section than that of low charge state ions.

3. CONCLUSION AND PROSPECTS

From the above discussions and the experimental data presented, the methods we developed include NHVEM and EGM, and they may be the efficient ways to compensate the space charge of ion beams especially the ion beams extracted from ECRIS. But lots of shortcomings still exist in the two methods. Taking NHVEM as an example, when the negative high voltage reached a special value the beam current becomes unstable though the beam current increased dramatically. On the other hand, only a few ion species have been tested and more ion beam species should be tested to verify the effectiveness of these two methods and during the experiments, the extracted ion beam intensity is not strong enough. So, some further research and test of the two methods is to be done at IMP and some new results on these two methods are expected.

4. ACKNOWLEDGMENTS

The experiment is supported by National Natural Foundation for Distinguished Young Scientists under contract No.10225523 and National Natural Foundation for Young Scientists through the contract No.10305016.

5. REFERENCES

1. Thesis for master degree, L.Ma 2004.
2. Nature, 160, 89(1947), Gabor, D.
3. IEEE Trans., NS-26,3115(1979), Lefevre, H.W.
4. Atomic Energy Science and Technology Vol.24.No.2 Mar 1990 Y.B.Du et al.
5. Review of Scientific Instruments Vol.73.No.2 Feb 2002.

Recent Development of IMP LECR3 Ion Source

Z. M. Zhang, H. W. Zhao, J. Y. Li, L. T. Sun, Y. C. Feng, H. Wang, B. H. Ma, X. Z. Zhang, X. H. Guo, X. X. Li, Y. Cao, H. Y. Zhao

Institute of Modern Physics (IMP), Chinese Academy of Science, Lanzhou, 730000, China

Abstract. 18GHz microwave has been fed to the LECR3 ion source to produce intense highly charged ion beams although this ion source was designed for 14.5GHz. Then 1.1 emA Ar^{8+} and 325 eμA Ar^{11+} were obtained at 18GHz. During the source running for atomic physics experiment, some higher charge state ion beams such as Ar^{17+} and Ar^{18+} were detected and have been validated by atomic physics method. Furthermore, a few special gases, e.g. SiH_4 and SF_6, were tested on LECR3 ion source to produce required ion beams to satisfy the requirements of atomic physics experiments.

INTRODUCTION

LECR3 (Lanzhou Electron Cyclotron Resonance ion source no.3), a versatile ECR ion source, was built in 2001.[1,2] This ion source was designed with high B mode to produce highly charged ion beams to satisfy the need of atomic physics research. We have made great effort to enhance the performance of LECR3 ion source. During the tuning many pivotal techniques, such as biased disk and gas mixing, were used to improve the production efficiency of highly charged ion beams. In order to increase the intensity of medium charge state ion beams, 18 GHz microwave power was fed into the ion source. The production of metallic ion beams is always one of the most important points of a high performance ECR ion source. So we made great efforts to produce metallic ion beams. This ion source is almost running all the time for atomic physics experiments only with short interval for checking. In 2003 it has been running for more than 6000 hrs.

PROCUTION OF INTENSE ION BEAMS WITH 18GHZ MICROWAVE

Formerly the LECR3 ion source was running at 14.5GHz and some good results have been obtained such as 240eμA of Ar^{11+} and 160eμA of Xe^{20+}. Although the magnetic field of this ion source was designed and calculated to optimize 14.5GHz frequency running, we have succeeded in trying feeding 18GHz frequency microwave power into the plasma chamber. After a long time tuning, some surprising results were gotten when we elevated the mirror field as much as possible by increasing the output current of power supply. 1100eμA Ar^{8+} and 325 Ar^{11+} have been delivered successfully at 18GHz, which are much higher than those produced at 14.5GHz microwave. Figure 1 shows the spectrum when optimizing the production of 1100 eμA Ar^{8+}.

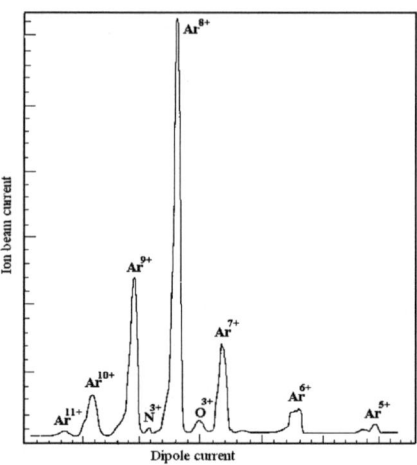

FIGURE 1. Ar spectrum when optimizing Ar^{8+} production.

EFFORT ON VERY HIGH CHARGE STATE ION BEAM PRODUCTION

During the tuning of argon ions for the atomic physics experiments, the source baking was divided into two steps. At first, the source was baked without extraction high voltage and then it was baked with extraction high voltage. After a long time (about 2 days) tuning without mixing gas feeding, C, N and O ion peaks almost disappeared in the argon ion beam spectrum. Under this condition, the source was principally optimized for the production of highly charged ion beams such as Ar^{17+} and Ar^{18+}. By matching the magnetic field and gas mixing technique and so on, 500enA Ar^{17+} was detected on the faraday cup and confirmed by the atomic physics experiments. Ar^{17+} and Ar^{18+} beams have been finally provided for the atomic physics experiments.

Figure 2 shows the Ar ion charge state distribution spectrum when optimizing Ar^{17+} production without mixing gas. But sometimes gas mixing effect is very efficient to produce the highly charged ions. In addition, the highly charged ion beams of xenon were also tuned. $^{129}Xe^{33+}$ was measured on the faraday cup and has been provided for experiments successfully.

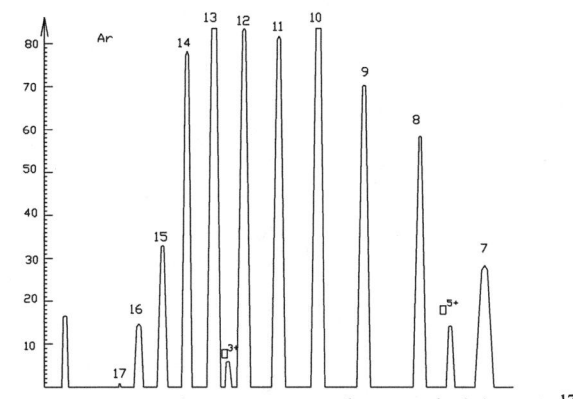

FIGURE 2. Ar spectrum when optimizing Ar^{17+} production.

PRODUCTION OF THE METALLIC ION BEAMS

We have tried different methods to produce metallic ion beams with the LECR3 ion source to satisfy the requirements of atomic physics research, and have obtained many good results in the last several years. Till now, more than ten species of metallic ion beams of different metallic elements have been successfully delivered to the experimental terminal. We have designed three types of ovens working from several hundred degrees to 1300 degrees [3]. Furthermore, the MIVOC method and plasma sputtering method have also been tested on LECR3. Considering the cyclotron operation, we think the oven method is more effective for the production of metallic ions which require a temperature less than 1500 degrees. As a selective method, MIVOC is very good for producing some kinds of metallic ions [4]. For example, with this method, we obtained 210eμA Fe^{11+} and 74eμA Ni^{12+} last year. The results of some metallic ion beams produced with oven method are shown in **table 1,** and the results of some metallic ion beams produced with MIVOC method are shown in **table** 2.

TABLE 1. The metallic ion beam current with oven method (unit: eμA)

Ions	10+	11+	12+	13+	14+	15+	16+	27+
Cu	20	31		39	34	10		
Zn				50		30		
Ni	29		26.5		40	27	10	
Fe	74	30	25	20		19		
Pb								

TABLE 2. The Metallic ion beam current with MiVOC method (unit: eμA)

Ion	11+	12+	13+	16+
Fe	210	175	141	25
Ni		74	57	17

SOURCE TUNING WITH SPECIAL ELEMENTS

Because some experiments need the injection of the ions of some special elements such as Si, Cl, S, F and etc, we have tested the production of these ion beams on LECR3. Most of these elements exist with the compound state in nature. Some of them are suitable to be used for ion beam production on ECRIS. So we choose SiH_4, CCl_4 and SF_6 as the working compounds fed into ion source to produce those special ion beams. Liquid SiH_4 is the material often adopted by ECRIS to produce Si ion beams, for its strong volatility under room temperature. The liquid sample is carefully put inside a special container, and the outgoing SiH_4 gas can be tuned by a vacuum valve. According to the preliminary experiments, $5e\mu A Si^{10+}$, $13e\mu A Si^{9+}$, $45e\mu A Si^{8+}$, $110e\mu A Si^{7+}$ have been obtained. In addition, CCl_4 and SF_6 were also fed into the ion source to produce the ion beams of Cl, F and S. Finally $19e\mu A Cl^{12+}$ and $70e\mu A F^{7+}$ have been successfully delivered for atomic physics experiments.

CONCLUSION

After three years' operation and running, LECR3 ion source has been testified with good results for the medium charge states like Ar^{8+} and results for very high charge states like Ar^{17+} or Pb^{40+}. Furthermore a lots of metallic ions beams were obtained by the oven method and MIVOVC method.

Because of the high performance of LECR3 source, it will be used to provide intense highly charged ion beams for HIRFL accelerator research.

ACKNOWLEDGMENTS

The work was supported by National Natural Science Foundation for Young Scientists under contract No.10305016, Knowledge Innovation Program of Chinese Academy Sciences under contract No. KJCX1-09 and National Natural Foundation for Distinguished Young Scientists under contract No. 10225523.

REFERENCES

1. Z.M. Zhang, H.W. Zhao, et. al. *Rev. Sci. Instrum.* **73**, 580-581, (2002).
2. Z. M. Zhang, H. W. Zhao, et al. *Proceedings of the 15th International Workshop on ECR ion Sources,* Finland, 12-14 June, 126-127 (2002).
3. H.W.Zhao, X.Z.Zhang, Z.W.Liu, et al, *Proceedings of the 14th International Conference on ECR Ion Source,* CERN, Geneva, Switzerland, 216-218 (1999).
4. H. Koivisto, J. Arje, M. Nurmia, *Nucl. Instr. and Meth.* B94, 291-296 (1994).

Experiments on Beam Extraction from the CAPRICE ECRIS [1]

K. Tinschert, P. Spädtke, J. Bossler, R. Iannucci, R. Lang

Gesellschaft für Schwerionenforschung (GSI), Planckstraße 1, D-64291 Darmstadt, Germany

Abstract. Classical compact ECRIS of small volume operating at 14 to 18 GHz provide an ion current density of several mA/cm^2 at the extraction aperture (e.g. CAPRICE at 14.5 GHz). This is considerably less than typical values for high current ion sources which are in the order of up to several 100 mA/cm^2 proton equivalent for ion beams of lowly charged ions.

However, the present development of powerful ECRIS with large volumes and higher plasma densities working at 28 GHz (or even higher frequencies) will lead to higher current densities which may approach the values of high current sources. Therefore a careful beam formation within the extraction which provides space charge compensation downstream behind the extraction system will be essential. The GSI CAPRICE ECRIS was equipped with a movable accel-decel extraction system in order to investigate the influence of electric field gradient and space charge compensation as well as further effects of ion extraction. First results will be presented and will be compared with computer simulations.

---

[1] Paper included in "Use of simulations based on experimental data",
P. Spädtke, K. Tinschert, R. Lang, R. Iannucci; these proceedings

Beam Simulation Studies of the LEBT for RIA Driver Linac

Q. Zhao, X. Wu, V. Andreev, A. Balabin, M. Doleans, D. Gorelov, T. L. Grimm,
W. Hartung, D. Leitner[#], C. M. Lyneis[#], F. Marti, S. O. Schriber, R. C. York

National Superconducting Cyclotron Laboratory, Michigan State University, East Lansing, MI 48824, USA
[#]*Lawrence Berkeley National Laboratory, Berkeley, CA 94704, USA*

Abstract. The low energy beam transport (LEBT) system in the front-end of the Rare Isotope Accelerator (RIA) uses a 70 kV platform to pre-accelerate the ion beam from a 30 kV Electron Cyclotron Resonance (ECR) ion source, followed by an achromatic charge selection system. The selected beam is then pre-bunched and matched into the entrance of a Radio Frequency Quadrupole (RFQ) with a multi-harmonic buncher. To meet the beam power requirements for heavy ions, high current (several mA), multi-species beams will be extracted from the ECR. Therefore, it is crucial to control space charge effects in order to obtain the low emittance beam required for RIA. The PARMELA code is used to perform the LEBT simulations for the multi-species beams with 3D space charge calculations. The results of the beam dynamics simulations are presented, and the key issues of emittance growth in the LEBT and its possible compensation are discussed.

INTRODUCTION

To meet the final beam power requirement of 400 kW for various ions, from protons to uranium, high intensity and low emittance beams are required from the Electron Cyclotron Resonance (ECR) ion source, which is still very challenging, especially for heavy ions [1]. Research and development directed towards improved ECR performance is currently conducted at Lawrence Berkeley National Laboratory's (LBNL) VENUS test facility [2]. To circumvent ion source limitations on beam current for heavier ions, a two-charge-state injection mode was proposed [3,4]. Meanwhile, the simultaneous acceleration of multiple charge-state beams has been adopted in the design of RIA driver linac in order to increase accelerator efficiency [5,6]. To minimize beam loss in the superconducting driver linac, the RIA front-end is required to provide small transverse and longitudinal emittances for the driver linac. For a multi-component, high current (total several mA) beam extracted from the ECR, space-charge effects will have significant impact on transverse beam emittance. 6D particle tracking with space-charge effects is therefore required to describe the evolution of beam in the low energy beam transport (LEBT). The DIMAD code [7] was used for the optical design of the LEBT, including the correction of higher order aberrations. The PARMELA code [8], which includes 3D space-charge calculations, was used to simultaneously track the multi-species particles from the ECR to the entrance of the Radio Frequency Quadrupole (RFQ). Transport of the heaviest ion beams is the most challenging, thus the LEBT design effort has primarily focused on uranium beams.

LEBT LAYOUT

The front-end of the RIA driver linac is comprised of ECR ion sources, LEBT lines, an RFQ, and a medium energy beam transport system. The dc ion beam from the ECR is first analyzed by an achromatic charge-to-mass selection system. The beam of the selected charge-to-mass ration is then pre-bunched by a multi-harmonic buncher and matched into the RFQ for further acceleration. A schematic layout of the LEBT is shown in Figure 1. The ECR and the 1st achromat (up to position P) are located horizontally on ground level for easy access, and a 2nd vertical achromat, not shown Figure 1, transports the beam from ground level down to the RIA driver linac tunnel (after position P). The design can accommodate four ion sources and beam transport systems on ground level. Since the performance of ion source is critical for the driver linac, multiple ion sources will ensure reliability through redundancy, and allow off-line beam development, along with the use of sources that are optimized for specific beam (e.g., metallic or gas).

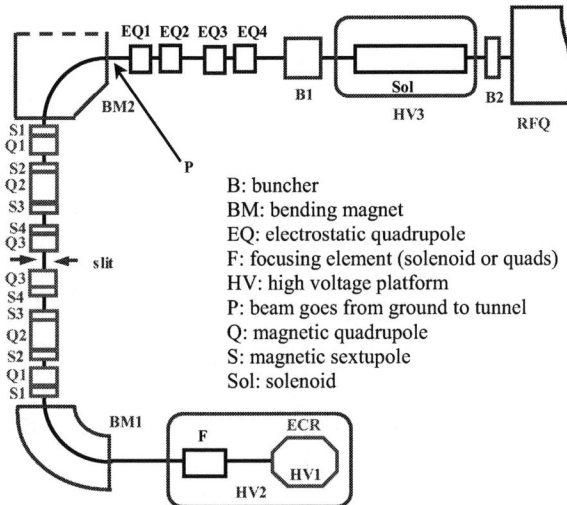

FIGURE 1. Schematic layout of the LEBT (top view). The ECR and 1st achromat (up to position P) are located horizontally on ground level; a 2nd achromat (not shown, but included in the simulation) is used to vertically transport the beam down to the linac tunnel (after position P).

To limit the impact of space-charge effects and to ease beam transport, a higher ECR extraction voltage is preferred. Based on experience with VENUS, a nominal extraction voltage of 30 kV is chosen, and the ECR is placed on a 70 kV high voltage platform for further acceleration. Due to space-charge effects, a high current beam diverges quickly after extraction from the ECR. Therefore, proper focusing near the ECR exit is needed to control the beam envelope and limit the emittance growth. In the design, a focusing element is used on the 70 kV platform to match the beam from the source to the downstream achromatic system.

The beam extracted from an ECR ion source is a multi-component ion beam which must be filtered for the required ion and charge state(s). For two-charge-state beam transport, the charge-to-mass selection system must be achromatic to minimize emittance growth. A symmetric achromatic system is proposed with two 90° dipoles, six quadrupoles, and eight sextupoles, as shown in Figure 1. The 90° analyzing magnet has a 63.7 cm radius of curvature to produce a high dispersion. To achieve a high charge-to-mass resolution, a small horizontal (dispersion plane) beam waist was achieved at the mid-plane ("slit" position in Figure 1) while maintaining the beam envelopes within a reasonable range. Four pairs of sextupoles were used to correct the 2nd order chromatic aberrations in both horizontal and vertical planes.

The selected beam is then pre-bunched and matched into the RFQ. Four electrostatic quadrupoles are used to convert the asymmetric beam from the upstream analysis section to an axisymmetric beam at the multi-harmonic buncher (MHB). They provide a large tuning range to accommodate beams with different Twiss parameters.

After charge selection from achromatic system, the beam current is low enough so that the dc beam can be efficiently pre-bunched by an external MHB before injection into the RFQ. The 80.5 MHz RFQ can produce a smaller longitudinal output emittance with a pre-bunched beam. The buncher can be a single gridded gap applied by a saw-tooth waveform or resonant cavities operating at the fundamental and multiple harmonics. In this design, the MHB will operate at a fundamental frequency of 40.25 MHz in the two-charge-state mode or 80.5 MHz for the single-charge-state mode. In the two-charge-state case, after the MHB, the two charge-state beams are bunched as well as longitudinally separated from each other due to the velocity difference. The separation is equal to one RF period at the entrance of the RFQ, so that each charge-state bunch can be accelerated in every other bucket in the RFQ [4]. Therefore, the required bunching distance is different for different ions. However, this distance cannot be easily changed after installation, so a bipolar high-voltage platform is used in the drift space to adjust the drift time duration and thus keep bunching distance the same [4]. For two-charge-state-beam operation for ions heavier than xenon, the distance was set for $_{197}Au^{23+}$ and $_{197}Au^{24+}$ ions so the maximum platform voltage is within ±50kV. In addition, a 2nd buncher in front of the RFQ operating at 40.25 MHz is required for two-charge-state beam injection. This buncher will provide each of the two-charge-state beams with the same energy at the RFQ entrance. To simplify the system design and future operation, both bunchers are located off the high voltage platform in this design. For ions that can reach the final beam power with single charge-state injection, neither the high voltage platform nor the 2nd buncher is required.

BEAM DYNAMICS

To perform beam simulations for LEBT, it is essential to have reasonable initial beam parameters (e.g. particle distribution) from the ion source. However, there is no experimental data available for a uranium beam comparable to the RIA specification. As part of the RIA collaboration, a bismuth beam was systematically tested at VENUS. A normalized rms emittance of 0.08 π·mm·mrad was measured for the 3 pμA $_{209}Bi^{27+}$ beam with total extracted current of ~1.2 mA for an ECR platform voltage of 20 kV [9]. Figure 2 shows the measured horizontal phase space of the bismuth beam after the analysis magnet.

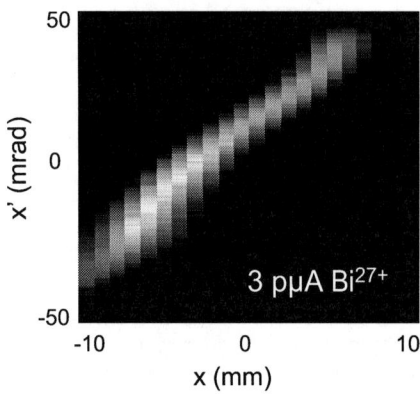

FIGURE 2. Measured horizontal phase space of 3 puA bismuth beam at VENUS [9].

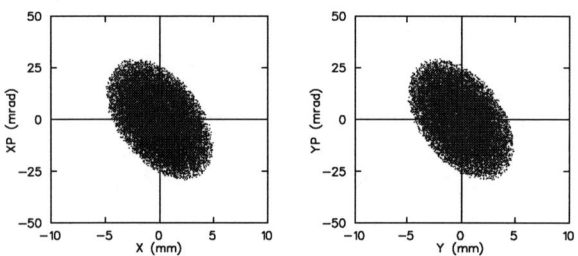

FIGURE 3. Calculated initial phase spaces for the bismuth beam at VENUS.

Beam simulations with PARMELA were performed using the experimental settings and measured data to obtain the initial beam parameters at the VENUS ion source. An initial normalized rms emittance of ~0.05 π·mm·mrad was obtained from the beam simulations for the 3puA $_{209}Bi^{27+}$ beam, as shown in Figure 3. The bismuth data was then scaled for uranium. We assume uranium will have a charge-state distribution similar to bismuth. The current was scaled up, as shown in Figure 4, so that both $_{238}U^{28+}$ and $_{238}U^{29+}$ reached 5 puA, producing a total uranium current of 1.32 mA. In addition, a total oxygen current of 1.5 mA from the support gas was assumed. In the simulation, we also assumed all ion species had the same Twiss parameters at the exit of ECR with a beam waist radius of 5.7 mm, a typical radius of plasma extraction aperture. The normalized uranium beam emittance is assumed to be similar to that of bismuth, and an intrinsic energy spread of 0.1% is included for all charge-state particles. A total of ~1.5 million macro-particles were tracked from the ECR in the PARMELA simulations. These particles were initially distributed in a 4D water-bag transverse hyperspace with a uniform distribution in phase and energy spread.

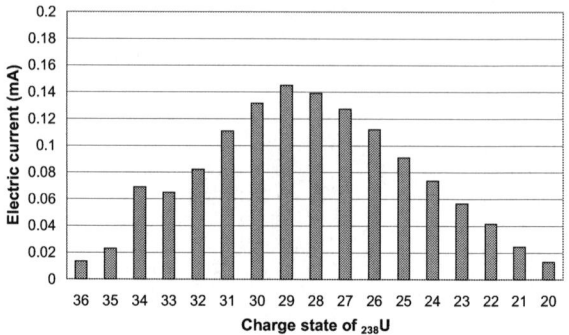

FIGURE 4. Charge-state distribution of uranium scaled from the experimental data for bismuth. The 1.5 mA of oxygen, not shown, was included in the beam simulation.

An electrostatic triplet, a magnetic triplet and a solenoid were investigated as possible focusing elements at the exit of ECR. Figure 5 shows the beam spot of the two-charge- state uranium beam at the entrance of the first analysis magnet with different solenoid focusing strengths. A hollow beam was formed with the solenoid at 0.345 T, and the rms emittance increased by nearly a fact of 2. The hollow became smaller and then disappeared as the solenoid field increased; the rms emittance growth also decreased. A similar hollow beam was produced with magnetic triplet focusing, but not with the electrostatic triplet. The hollow beam formation and emittance growth are the result of space-charge effects from the higher charge-to-mass ions. These beams have a shorter focal length with magnetic focusing and are focused into a very small size around the focal point where the beams of interest ($_{238}U^{28+}$ and $_{238}U^{29+}$) are also small. Most of the particles see the strong space-charge field, which deteriorates the beam quality. The separation between the focal lengths of the different charge-state ions increases with the focusing magnetic field, therefore, the beam hollow formation and beam emittance growth were reduced with stronger magnetic focusing. Minimum hollow formation and beam emittance growth were obtained with a solenoid field of ~0.45 T, which was used in the LEBT beam simulations. Since the focusing strength of an electrostatic quadrupole does not depend on the particle's charge state, no hollow beam developed. However, emittance growth was still observed due to the focusing variation (change of velocity) for particles close to the pole tips. Further studies are still required to compare solenoid and electrostatic triplet focusing at the exit of ECR.

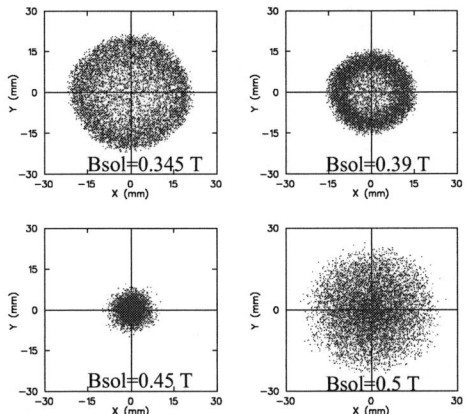

FIGURE 5. Beam spots of the two charge-state uranium beam ($_{238}U^{28+}$ and $_{238}U^{29+}$) at the entrance to the first analysis magnet with different solenoid focusing strengths.

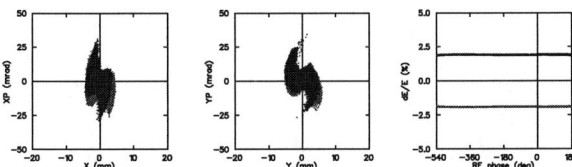

FIGURE 6. Horizontal (left), vertical (middle), and longitudinal (right) phase spaces of the two-charge-state uranium beam at the exit of the 2nd achromatic system.

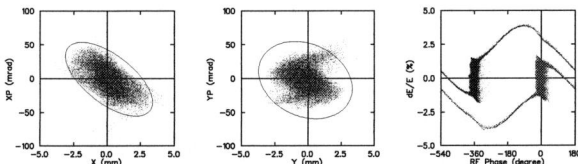

FIGURE 7. Horizontal (left), vertical (middle), and longitudinal (right) phase spaces of the two-charge-state uranium beam at the entrance to RFQ.

The phase spaces of the two-charge-state uranium beam at the end of the achromatic transport are shown in Figure 6. The transverse emittance growth in the achromatic section is small. Figure 7 shows the phase spaces of the $_{238}U^{28+}$ and $_{238}U^{29+}$ beams at the entrance of the RFQ. More than 80% of the particles are bunched within 60° RF phase by the MHB. The normalized 99.5% transverse emittance (effective emittance including 99.5% of the total particles) is ~1 π·mm·mrad for the two charge-state beam. The normalized rms emittance is about 10 times smaller than the 99.5% emittance. The cores of the two charge-state beams are overlapped well, but the tail particles are slightly shifted with respect to each other, causing an increased total effective emittance. Similar results were obtained for single charged $_{136}Xe^{17+}$ and $_3He^+$ beam from the charge selection to the RFQ entrance with the required current.

SUMMARY

Beam simulation results show the LEBT design for the RIA driver linac provides adequate beam matching and bunching for both single- and two-charge-state beam operations. Small transverse and longitudinal emittances can be achieved at the entrance of RFQ for acceleration. Further beam simulations in the LEBT will include space-charge neutralization effects and experimentally-based initial beam phase spaces and parameters for uranium beam, possibly from VENUS or other ion sources. Other particle tracking codes will also be used in future beam simulations.

ACKNOWLEDGMENTS

We thank T. Wangler at LANL, N. Kazarinov at JINR, P. Zavodszky, and J. Stetson at NSCL for helpful discussions. We thank L. Young and J. Billen at LANL for maintaining the PARMELA code. This work was supported by the State of Michigan and Michigan State University.

REFERENCES

1. Drentje, A. G., Rev. Sci. Instrum. 74, 2631 (2003).
2. Leitner, D., Abbott, S.R., Dwinell, R.D., et al., Proc. 20th PAC, Portland, OR, May 2003, p. 86; Leitner, D., Lyneis, C.M., Abbott, S.R., et al., RIA R&D Workshop, Bethesda, MD, Aug. 2003.
3. Deitinghoff, H., Proc. 16th PAC, Dallas, TX, May 1995, p. 1158.
4. Ostroumov, P. N., et al., Proc. 20th Linac Conf., Monterey, CA, Aug. 2000, SLAC-R-561, p. 202.
5. Ostroumov, P. N., et al., Proc. 20th Linac conf., Monterey, CA, Aug. 2000, SLAC-R-561, p. 1018.
6. Wu, X., et al., Proc. 22th Linac Conf., Luebeck, Germany, Aug. 2004.
7. Servranckx, R., et al., "User's Guide to the Program DIMAD", SLAC Report 285, UC-28, May 1985.
8. Young, L., et al., "PARMELA", LA-UR-96-1835, Aug. 2004.
9. Leitner, D., RIA Facility Workshop, East Lansing, MI, Mar. 2004. http://meetings.nscl.msu.edu/ria2004/talks/dl//wedpm/Leitner.ppt

Efficient Plasma Ion Source Modeling With Adaptive Mesh Refinement (Abstract)

J.S. Kim[1], J.L. Vay[2], A. Friedman[3], D.P. Grote[3]

[1] *FAR-TECH, Inc., San Diego, CA 92121, USA*
[2] *LBNL, Berkeley, USA*
[3] *LLNL, Livermore, USA*

Abstract. Ion beam drivers for high energy density physics and inertial fusion energy research require high brightness beams, so there is little margin of error allowed for aberration at the emitter. Thus, accurate plasma ion source computer modeling is required to model the plasma sheath region and time-dependent effects correctly.

A computer plasma source simulation module that can be used with a powerful heavy ion fusion code, WARP, or as a standalone code, is being developed. In order to treat the plasma sheath region accurately and efficiently, the module will have the capability of handling multiple spatial scale problems by using Adaptive Mesh Refinement (AMR). We will report on our progress on the project.

Acknowledgement: The authors thank Dr. J.W. Kwan for his contribution to the project.

Work at FAR-TECH, Inc. is performed under the SBIR grant DE-FG02-04ER83953 and work at LLNL and LBNL is performed under the auspices of the U.S. Department of Energy by the University of California, Lawrence Livermore and Lawrence Berkeley National Laboratories under Contract Nos. W-7405-Eng-48 and DE-AC03-76SF00098.

Development of the 3D Parallel Particle-In-Cell Code IMPACT to Simulate the Ion Beam Transport System of VENUS (Abstract)

J. Qiang, D. Leitner, D. S. Todd, R. D. Ryne

LBNL, Berkeley, USA

Abstract. The superconducting ECR ion source VENUS serves as the prototype injector ion source for the Rare Isotope Accelerator (RIA) driver linac. The RIA driver linac requires a great variety of high charge state ion beams with up to an order of magnitude higher intensity than currently achievable with conventional ECR ion sources. In order to design the beam line optics of the low energy beam line for the RIA front end for the wide parameter range required for the RIA driver accelerator, reliable simulations of the ion beam extraction from the ECR ion source through the ion mass analyzing system are essential. The RIA low energy beam transport line must be able to transport intense beams (up to 10 mA) of light and heavy ions at 30 keV.

For this purpose, LBNL is developing the parallel 3D particle-in-cell code IMPACT to simulate the ion beam transport from the ECR extraction aperture through the analyzing section of the low energy transport system. IMPACT, a parallel, particle-in-cell code, is currently used to model the superconducting RF linac section of RIA and is being modified in order to simulate DC beams from the ECR ion source extraction. By using the high performance of parallel supercomputing we will be able to account consistently for the changing space charge in the extraction region and the analyzing section.

A progress report and early results in the modeling of the VENUS source will be presented.

A Microwave Driven Ion Source for Continuous-Flow AMS (Abstract)

J. Wills[1],
R.J. Schneider[2], K.F. von Reden[2], J.M. Hayes[2], M.L. Roberts[2], A. Benthien[2]

[1] *Atomic Energy of Canada Ltd. Chalk River Laboratories, Chalk River, Ontario, Canada*
[2] *Woods Hole Oceanographic Institution, Woods Hole, MA, USA*

Abstract. A microwave-driven, gas-fed ion source originally developed as a high-current positive ion injector for a Tandem accelerator at Chalk River has been the subject of a three-year development program at the Woods Hole Oceanographic Institution NOSAMS facility. Off-line tests have demonstrated positive carbon currents of 1 mA and negative carbon currents of 80 µA from CO_2 gas feed. This source and a magnesium charge-exchange canal were coupled to the recombinator of the NOSAMS Tandetron for on-line tests, with the source fed with reference gasses and a combustion device.

The promising results obtained have prompted the redesign of the microwave source for use as an on-line, continuous-flow injector for a new AMS facility under construction at NOSAMS. The new design is optimized for best transmission of the extracted positive-ion beam through the charge-exchange canal and for reliable operation at 40 kV extraction voltage. Other goals of the re-design include improved lifetime of the microwave window and the elimination of dead volumes in the plasma generator that increase sample hold-up time.

This talk will include a summary of results obtained to date at NOSAMS with the Chalk River source and a detailed description of the new design.

ECRIS'04 Registered Attendees

Gerald Alton
Oak Ridge National Laboratory
1 Bethel Valley Road, P. O. Box 2008
Oak Ridge, TN 37831-6368
US
Phone: (865) 574-4751
Email: gda@ornl.gov

Friedhelm Ames
TRIUMF
4004 Wesbrook Mall
Vanvouver, BC V6T 2A3
CANADA
Phone: (604) 222-7581
Email: ames@triumf.ca

Vladislav Asseev
Argonne National Laboratory
9700 South Cass Av.
Argonne, IL 60439-4803
US
Phone: (630) 252-6120
Email: aseev@phy.anl.gov

Christophe Barué
GANIL / Caen
Bd Henri Becquerel
Caen, 14076
FRANCE
Phone: 2 31 45 44 57
Email: barue@ganil.fr

Hans Beijers
Kernfysisch Versneller Instituut
Zernikelaan 25
Groningen, 9747 AA
NETHERLANDS
Phone: +31-50-3633600
Email: beijers@kvi.nl

Hassina Bilheux
Oak Ridge National Laboratory
Bethel Valley Road
Oak Ridge, TN 37831-6368
US
Phone: (865) 574-4112
Email: bilheuxhn@ornl.gov

Sandor Biri
ATOMKI
Bem ter 18/C
Debrecen, H-4026
HUNGARY
Phone: + (365) 241-7266
Email: biri@atomki.hu

Pierre Bricault
TRIUMF
4004 Wesbrook Mall
Vanvouver, BC V6T 2A3
CANADA
Phone: (604) 222-7417
Email: bricault@triumf.ca

Luigi Celona
INFN-LNS
Via S. Sofia 64
Catania, 95123
ITALY
Phone: +39095542261
Email: celona@lns.infn.it

Edwin Chacon-Golcher
Los Alamos National Lab
MS H838
Los Alamos, CA 87545
US
Phone: 505-667-4008
Email: edcg@lanl.gov

Dallas Cole
NSCL
Mich. State Univ.
East Lansing, MI 48824
US
Phone: (517) 333-6342
Email: cole@nscl.msu.edu

Wayne Cornelius
Scientific Solutions
PO Box 500207
San Diego, CA 92150
US
Phone: (619) 840-6411
Email: wcornelius@ssolutions.cc

Greg Derrig
Texas A&M University
Cyclotron Institute
College Station, TX 77843-3366
US
Phone: (979) 845-1411
Email: derrig@comp.tamu.edu

Marvin Eberhardt
Eimac Division / CPI, Inc.
301 Industrial Road
San Carlos, CA 94070
US
Phone: (650) 594-4124
Email: marvin.eberhardt@cpii.com

Victor Erukhimov
Institute of Applied Physics, RAS
Ulyanova str., 46
Nizhny Novgorod, 603950
RUSSIA
Phone: +7(910)7978414
Email: eruhimov@appl.sci-nnov.ru

Steven Evans
Communcations & Power Industries
150 Sohier Rd.
Beverly, MA 01915
US
Phone: (978) 922-6000
Email: steve.evans@bmd.cpii.com

Alex Friedman
LLNL & LBNL
1 Cyclotron Road MS 47R0112
Berkeley, CA 94720
US
Phone: (510) 486-5592
Email: afriedman@lbl.gov

Michelle Galloway
Lawrence Berkeley National Lab
1 Cyclotron Road
Berkeley, CA 94720
US
Phone: (510) 486-7570
Email: mlgalloway@lbl.gov

Santo Gammino
INFN
Via S. Sofia 62
Catania, 95123
ITALY
Phone: + 39 095 542270
Email: gammino@lns.infn.it

Richard Gough
Lawrence Berkeley National Lab
1 Cyclotron Road
Berkeley, CA 94720
US
Phone: (510) 486-4573
Email: ragough@lbl.gov

David Grote
Lawrence Berkeley National Lab
1 Cyclotron Road MS47-0112
Berkeley, CA 94720
US
Phone: (510) 495-2961
Email: dpgrote@lbl.gov

Sami Hahto
Lawrence Berkeley National Lab
1 Cyclotron Road, MS 5R0121
Berkeley, CA 94720
US
Phone: (510) 486-6958
Email: skhahto@lbl.gov

Yoshihide Higurashi
RIKEN
Hirosawa 2-1
Wako, 351-0198
JAPAN
Phone: 8 (148) 467-4880
Email: higurasi@riken.jp

Charles Hill
CERN
PO box 23
Geneva, 1211
SWITZERLAND
Phone: + 41 22 767 3659
Email: charles.hill@cern.ch

Denis Hitz
CEA-Grenoble
DRFMC-SBT
Grenoble, 38054
FRANCE
Phone: 3 (343) 878-4476
Email: dhitz@cea.fr

Howard Jory
CPI, Inc.
811 Hansen Way
Palo Alto, CA 94305
US
Phone: (650) 846-3295
Email: howard.jory@cpii.com

Yoko Kawai
Oak Ridge National Laboratory
PO Box 2008 MS6371
Oak Ridge, TN 37831
US
Phone: (865) 576-9058
Email: kawaiy@ornl.gov

Roderich Keller
Lawrence Berkeley National Lab
1 Cyclotron Rd. MS 71R0259
Berkeley, CA 94720
US
Phone: (510) 486-5223
Email: R_keller@lbl.gov

Masanori Kidera
RIKEN
Hirosawa 2-1
Saitama, 351-0198
JAPAN
Phone: +81-48-467-4880
Email: kidera@kindex.riken.go.jp

Jin-Soo Kim
FAR-TECH, Inc.
10350 Science Center Drive
San Diego, CA 92121
US
Phone: (858) 455-6655
Email: kimjs@far-tech.com

Hannu Koivisto
University of Jyväskylä
Survontie 9
Jyväskylä, 40500
FINLAND
Phone: +358 14 2602371
Email: Hannu.Koivisto@phys.jyu.fi

Daniela Leitner
Lawrence Berkeley National Lab
1 Cyclotron Rd. MS 88R0192
Berkeley, CA 94720
US
Phone: (510) 486-7814
Email: dleitner@lbl.gov

Matthaeus Leitner
Lawrence Berkeley National Lab
1 Cyclotron Road Bldg 47R0112
Berkeley, CA 94720-8201
US
Phone: (510) 486-4090
Email: MLeitner@lbl.gov

Renan Leroy
GANIL
Bd H. Becquerel
Caen, 14076
FRANCE
Phone: 3 (323) 145-4505
Email: leroy@ganil.fr

Yuan Liu
Oak Ridge National Lab
Bldg. 6000, MS-6368
Oak Ridge, TN 37831-6368
US
Phone: (865) 574-4761
Email: liuy@ornl.gov

Claude Lyneis
Lawrence Berkeley National Lab
MS 88R0192
Berkeley, CA 94720
US
Phone: (510) 486-7815
Email: cmlyneis@lbl.gov

Guillaume Machicoane
NSCL
Michigan State University
East Lansing, MI 48824
US
Phone: (517) 333-6419
Email: machicoa@nscl.msu.edu

Michael Marks
CPI, Inc.
150 Sohier Rd.
Beverly, MA 01915-5595
US
Phone: (978) 922-6000
Email: mike.marks@bmd.cpii.com

Donald May
Texas A&M University
Cyclotron Institute
College Station, TX 77843
US
Phone: (979) 845-1411
Email: may@comp.tamu.edu

Margaret McMahan Norris
Lawrence Berkeley National Lab
MS88R0192
Berkeley, CA 94618
US
Phone: (510) 486-5980
Email: p_mcmahan@lbl.gov

Jim Morel
Lawrence Berkeley National Lab
1 Cyclotron Rd
Berkeley, CA 94720
US
Phone: (510) 486-7995
Email: jrmorel@lbl.gov

Masayuki Muramatsu
National Inst. of Radiological Sci.
4-9-1 Anagawa
Chiba, 263-8555
JAPAN
Phone: +81-43-206-4031
Email: m_mura@nirs.go.jp

Takahide Nakagawa
RIKEN
Hirosawa 2-1
Wako, 351-0198
JAPAN
Phone: 8 (148) 467-4880
Email: nakagawa@riken.jp

Hiroyuki Ogawa
Accelerator Engineering Corp.
4-9-1 Anagwa, Inage-ku, Chiba 263-
Chiba, 263-8555
JAPAN
Phone: +81-43-251-6440
Email: ogawa@aec-beam.co.jp

Richard Pardo
Argonne National Laboratory
9700 S. Cass Avenue
Argonne, IL 60637
US
Phone: (630) 252-4029
Email: pardo@phy.anl.gov

Doug Parent
Communications & Power
11301 W. Olympic Blvd, #121
Los Angeles, CA 90064
US
Phone: (310) 268-1226
Email: doug.parent@cpii-us.com

Ji Qiang
Lawrence Berkeley National Lab
1 Cyclotron Rd.
Berkeley, CA 94720
US
Phone: (510) 495-2608
Email: jqiang@lbl.gov

Brian Richter
GMW Associates
955 Industrial Road
San Carlos, CA 94070
US
Phone: (650) 802-8292
Email: brian@gmw.com

Leon Schachter
Inst. for Physics & Nuclear Eng.
Str. Atomistilor 407, Magurele, CP.
Bucharest, RO 077125
ROMANIA
Phone: 4 (021) 404-2331
Email: lsch@tandem.nipne.ro

Alfred Schlachter
Lawrence Berkeley National Lab
1 Cyclotron Road
Berkeley, CA 94720
US
Phone: (510) 486-4892
Email: fsschlachter@lbl.gov

Robert Schneider
Woods Hole Oceanographic Inst.
WHOI / Mail Stop 8
Woods Hole, MA 02543
US
Phone: (508) 289-2756
Email: rschneider@whoi.edu

Robert Scott
Argonne National Laboratory
9700 S Cass
Argonne, IL 60439
US
Phone: (630) 252-4115
Email: scott@phy.anl.gov

Vadim Skalyga
IAP RAS
46 Ulyanov St., Nizhny Novgorod,
Nizhny Novgorod, 603950
RUSSIA
Phone: 7 (831) 216-4835
Email: sva1@appl.sci-nnov.ru

Pascal Sortais
CNRS
LPSC
Grenoble cedex, 38026
FRANCE
Phone: 3 (347) 628-4188
Email: sortais@lpsc.in2p3.fr

Peter Spädtke
GSI Darmstadt
Planckstr. 1
Darmstadt, HI 64291
GERMANY
Phone: +49 6159 71 2320
Email: p.spaedtke@gsi.de

Michael Stalder
University of Kiel
Leibnizstrasse 11
Kiel, 24118
GERMANY
Phone: 00494318803020
Email: stalder@physik.uni-kiel.de

Pekka Suominen
University of Jyvaskyla
Survontie 9
Jyvaskyla, 40500
FINLAND
Phone: +358 50 305 2292
Email:

Olli Tarvainen
University of Jyväskylä
Department of Physics
Jyväskylä, FIN-40014
FINLAND
Phone: +358503096814
Email: olli.tarvainen@phys.jyu.fi

Teck Hing Teo
ISOFLEX USA
P.O. Box 29475
San Francisco, CA 94129
US
Phone: (415) 440-4433
Email: iusa@isoflex.com

Thomas Thuillier
CNRS
LPSC
Grenoble cedex , 38026
FRANCE
Phone: 3 (347) 628-4022
Email: thuillier@lpsc.in2p3.fr

Dr. Klaus Tinschert
GSI Darmstadt
Planckstr. 1
Darmstadt, HI 64221
GERMANY
Phone: +49 6159 71 2321
Email: k.tinschert@gsi.de

Damon Todd
Lawrence Berkeley National Lab
1 Cyclotron Rd. MSR0192
Berkeley, CA 94720
US
Phone: (510) 486-5707
Email: dstodd@lbl.gov

Martino Trassinelli
Laboratoire Kastler Brossel
case 74, 4 place Jussieu
Paris, F-75005
FRANCE
Phone: +33 1 44274396
Email:

Todd Treado
CPI
150 Sohier Road
Beverly, MA 01915
US
Phone: (978) 279-0458
Email: todd.treado@bmd.cpii.com

Jean-Luc Vay
Lawrence Berkeley National Lab
1 Cyclotron Road
Berkeley, CA 94804-8201
US
Phone: (510) 486-4934
Email: JLVay@lbl.gov

Nikolai Vinogradov
Argonne National Laboratory
9700 S. Cass Ave.
Argonne, IL 60439
US
Phone: (630) 252-5643
Email: vinogradov@phy.anl.gov

Alexander Vodopyanov
Institute of Applied Physics RAS
46, Ul'yanova str.
Nizhny Novgorod, 603950
RUSSIA
Phone: +7 8312 164835
Email: avod@appl.sci-nnov.ru

Karl Von Reden
Woods Hole Oceanographic Inst.
McLean Lab 245
Woods Hole, MA 02543
US
Phone: (508) 289-3384
Email: kvonreden@whoi.edu

Richard Vondrasek
Argonne National Laboratory
9700 So. Cass Ave
Argonne, IL 60439
US
Phone: (630) 252-5972
Email: vondrasek@phy.anl.gov

Peter Zavodszky
Michigan State University
1 Cyclotron Laboratory
East Lansing, MI 48824
US
Phone: (517) 333-6460
Email: zavodszky@nscl.msu.edu

Hongwei Zhao
Institute of Modern Physics, CAS
Nanchang Road No.509
Lanzhou, 730000
CHINA
Phone: +86 931 4969210
Email: zhaohw@impcas.ac.cn

Qiang Zhao
Michigan State University
1 Cyclotron
East Lansing, MI 48824-1321
US
Phone: (517) 324-8136
Email: zhao@nscl.msu.edu

Vladimir Zorin
Institute of Applied Physics
46 Ulyanov St.
Nizhny Novgorod, 603950
RUSSIA
Phone: +7 8312 164835
Email: zorin@appl.sci-nnov.ru

Author Index

A

Abbott, S. R., 3
Abegglen, F. P., 196
Aguilar, A., 88
Aihara, T., 23, 71
Alton, G. D., 103, 191, 195, 223
Alvarez, I., 88
Ames, F., 147
Andò, L., 15
Andreev, V., 242
Andresen, C., 161
Arend, B., 131
Ärje, J., 27
Aseev, V. N., 215

B

Baartman, R., 147
Baker, S., 75
Balabin, A., 242
Barue, C., 137
Beck, P., 207
Beijers, J. P. M., 35, 175
Benthien, A., 248
Bhandari, R. K., 202
Bieth, C., 19
Bilheux, H., 103
Bilheux, H. Z., 191, 195, 223
Biri, S., 67, 81
Bossler, J., 241
Boucard, S., 81
Brandenburg, S., 35, 175
Bricault, P., 143, 147
Bruch, R., 179

C

Canet, C., 137
Cao, Y., 10, 227, 231, 234, 238
Celona, L., 15, 99
Chamings, J., 161
Chartier, J., 123
Ciavola, G., 15, 99
Cisneros, C., 88

Cisto, B., 207
Coco, V., 161
Cole, D., 131
Cole, J. M., 187, 195, 223
Collins, D., 3, 207
Consoli, F., 99
Covita, D. S., 81
Curdy, J. C., 41

D

DeKamp, J., 131
Delaunay, M., 123
Derrig, G. J., 196
Dobrescu, S., 211
Doleans, M., 242
dos Santos, J. M. F., 81
Drentje, A. G., 183
Dubois, M., 137
Dupuis, M., 137
Durantel, F., 137
Dwinell, R. D., 3, 207

E

Edgell, D. H., 31
Emmons, E. D., 88
Enomoto, S., 85
Erukhimov, V. L., 92
Evans, S., 207

F

Farabolini, W., 137
Feng, Y. C., 227, 231, 238
Flambard, J.-L., 137
Formanoy, I., 35, 175
Friedman, A., 55, 246
Frondelius, P., 27, 171
Fujimaki, M., 85

G

Galatà, A., 99
Gammino, S., 15, 99
Gaubert, G., 137
Geller, R., 219
Gharaibeh, M. F., 88
Gibouin, S., 137
Girard, A., 123, 157, 227
Golubev, S. V., 116
Gorelov, D., 242
Gotta, D., 81
Grimm, T. L., 242
Grote, D. P., 55, 246
Guillaume, D., 157
Guillemet, L., 123, 127, 157
Guo, X. H., 10, 227, 231, 238

H

Haber, I., 55
Hahto, S. K., 179
Hartung, W., 242
Hayes, J. M., 248
He, W., 10, 168, 227, 231, 234
Hecht, A. A., 75
Higurashi, Y., 23, 71
Hill, C. E., 127, 161
Hinojosa, G., 88
Hirtl, A., 81
Hirunuma, R., 85
Hitz, D., 15, 123, 127, 157, 227
Holstein, D., 207
Huet-Equilbec, C., 137
Huguet, Y., 137

I

Iannucci, R., 47, 241
Igarashi, K., 85
Ikezawa, E., 85
Indelicato, P., 81
Izotov, I., 219

J

Jardin, P., 137
Jayamanna, K., 143, 147
Ji, Q., 179

Jory, H., 207

K

Kalvas, T., 27
Kamigaito, O., 85
Kanjilal, D., 19
Kantas, S., 19
Karácsony, J., 67
Kase, M., 23, 71, 85
Kawai, Y., 103, 223
Keller, R., 108
Kenéz, L., 67
Kern, M. R. L., 215
Kidera, M., 23, 71, 85
Kilcoyne, A. L. D., 88
Kim, J. S., 246
Kitagawa, A., 67, 183
Koivisto, H., 27, 31, 61, 171, 175
Kondagari, S., 179
Krása, J., 15
Kremers, H. R., 35, 175
Küchler, D., 127, 161
Kumar, P., 19

L

Lachaize, A., 41
Lammentausta, E., 27
Lamy, T., 41, 147, 219
Lang, R., 47, 241
Lappalainen, P., 27
Láska, L., 15
Le Bigot, E.-O., 81
Lecesne, N., 137
Leherissier, P., 137
Leitner, D., 3, 151, 206, 207, 242, 247
Leitner, M., 3, 108
Lemagnen, F., 137
Leoni, B., 81
Leroy, R., 127, 137
Leung, K.-N., 179
Li, J., 227, 231
Li, J. Y., 227, 231, 238
Li, X. X., 227, 231, 238
Liu, Y., 103, 187, 191, 195, 223
Lombardi, A., 161
Lu, W., 231

Lyneis, C. M., 3, 206, 207, 242

M

Ma, B. H., 165, 231, 238
Ma, B. W., 168
Ma, L., 10, 227, 231, 234
Ma, X. W., 168
Machicoane, G., 131
Mallik, C., 202
Manciagli, S., 15
Mandal, A., 19
Mansfeld, D. A., 116
Marks, M., 207
Marti, F., 131, 242
Mathonnet, J. M., 123
May, D. P., 196
McDonald, M., 147
McLaughlin, B. M., 88
McMahan, M. A., 151
Merabet, H., 179
Meyer, F. M., 187, 191
Meyer, F. W., 123
Mezzasalma, A. M., 15
Miller, P., 131
Moehs, D. P., 108
Moore, E. F., 75
Moskalik, J., 131
Mulder, J., 35, 175
Müller, A., 88
Muramatsu, M., 67, 183

N

Nakagawa, T., 23, 71, 85
Nieminen, V., 171
Nikolaev, A. G., 116

O

Ogawa, Hirotsugu, 183
Ogawa, Hiroyuki, 183
Oks, E. M., 116
Olivo, M., 143, 147
Ostroumov, P. N., 215
Ottarson, J., 131

P

Pacquet, J. Y., 127, 137
Pálinkás, J., 67
Pardo, R., 31
Pardo, R. C., 75, 215
Parys, P., 15
Peeler, H., 196
Pellemoine, F., 137
Pfeifer, M., 15
Phaneuf, R. A., 88
Picciotto, A., 15
Poncet, J. M., 157
Ponton, A., 41
Powell, J., 151
Presti, M., 15

Q

Qiang, J., 247

R

Radics, B., 67
Rantilla, K., 175
Rao, U. K., 19
Rashid, M. H., 202
Razin, S., 219
Razin, S. V., 116
Reijonen, J., 179
Rizzo, R., 207
Roberts, M. L., 248
Rodrigues, G. O., 19
Ropponen, T., 27, 61
Roy, A., 19
Ryne, R. D., 247

S

Safvan, C. P., 19
Saint Laurent, M. G., 137
Sakamoto, Y., 183
Sargsyan, E., 161
Sato, S., 183
Sato, Y., 183
Savard, G., 75
Savkin, K. P., 116

Schachter, L., 211
Schippers, S., 88
Schlachter, A. S., 88
Schmor, P., 143, 147
Schneider, D., 179
Schneider, R. J., 248
Schriber, S. O., 242
Scott, R., 31, 215
Scott, R. H., 61
Scrivens, R., 127, 161
Scully, S. W. J., 88
Semenov, V. E., 92
Seyfert, P., 157
Shirkov, G. D., 15
Sidorov, A., 219
Sijbring, J., 35, 175
Silver, C., 151
Simons, L. M., 81
Skalyga, V., 112, 219
Sole, P., 41
Song, M. T., 10
Song, T. M., 234
Sortais, P., 19, 41, 219
Spädtke, P., 47, 241
Stalder, M., 199
Steigies, C. T., 199
Stiebing, K. E., 211
Stingelin, L., 81
Stockli, M. P., 108
Sun, L. T., 10, 157, 165, 168, 227, 231, 238
Suominen, P., 27, 31, 61, 171

T

Takács, E., 67
Takahashi, K., 85
Tarvainen, O., 27, 31, 61, 171
Thuillier, T., 41, 219
Tinschert, K., 47, 241
Todd, D. S., 247
Torrisi, L., 15
Trassinelli, M., 81
Tuske, O., 137

V

Valek, A., 67
Vay, J. L., 246
Vay, J.-L., 55

Veloso, J. F. C. A., 81
Vieux-Rochaz, J. L., 41
Villari, A. C. C., 137
Vincent, J., 131
Vinogradov, N. E., 215
Vodopyanov, A. V., 116
Vondrasek, R., 31
Vondrasek, R. C., 61, 215
von Reden, K. F., 248

W

Wang, H., 165, 168, 231, 238
Wasser, A., 81
Wei, B. W., 10
Welton, R. F., 108
Wills, J., 248
Wimmer-Schweingruber, R. F., 199
Wolowski, J., 15
Woryna, E., 15
Wu, X., 242

Y

Yamada, S., 183
Yano, Y., 23, 71, 85
York, R. C., 242
Yoshida, Y., 183
Yuan, D. H. L., 143, 147
Yuan, P., 10

Z

Zavodszky, P. A., 131
Zeller, A., 131
Zhan, W. L., 10
Zhang, X. Z., 10, 165, 231, 238
Zhang, Z. M., 10, 165, 168, 227, 231, 234, 238
Zhao, H. W., 10, 165, 168, 227, 231, 234, 238
Zhao, H. Y., 165, 227, 231, 234, 238
Zhao, Q., 242
Zmeskal, J., 81
Zorin, V., 112, 219